33rd International Symposium on Space Terahertz Technology (ISSTT 2024)

Charlottesville, Virginia, USA
8 - 11 April 2025

ISBN: 979-8-3313-1993-9

Printed from e-media with permission by:

Curran Associates, Inc.
57 Morehouse Lane
Red Hook, NY 12571

Some format issues inherent in the e-media version may also appear in this print version.

International Symposium on Space Terahertz Technology
c/o NRAO Headquarters
520 Edgemont Road
Charlottesville, VA 22903-2475
USA

Phone: (434) 296-0254
Fax: (434) 296-0278

mbishop@nrao.edu

Additional copies of this publication are available from:

Curran Associates, Inc.
57 Morehouse Lane
Red Hook, NY 12571 USA
Phone: 845-758-0400
Fax: 845-758-2633
Email: curran@proceedings.com
Web: www.proceedings.com

The 33rd International Symposium on Space Terahertz Technology

ISSTT 2024

Charlottesville, Virginia, USA

The 33rd International Symposium on Space Terahertz Technology was held in Charlottesville, Virginia USA from April 7th to April 11th, 2024, featuring excellent presentations on millimeter, submillimeter-wave and Terahertz technologies and applications in astrophysics, planetary science, Earth science and remote sensing. A special session on metamaterials was also held. Students submitted abstracts for the Student Poster Competition in which winners earned a ten-minute oral presentation as well as a cash prize. New this year was two "Speed Geeking" sessions to help facilitate focus on additional posters in a more inclusive, interactive format.

In addition, all in-person conference participants were treated to a welcome reception, a fabulous symposium dinner at a local venue, ample networking opportunities, a Solar Eclipse party, and two incredible excursions on the final day to visit nearby area sites.

Scientific Organizing Committee Members

- Patricio Mena, Chair (National Radio Astronomy Observatory, USA)
- Andrey Baryshev (Kapteyn Astronomical Institute, Netherlands)
- Victor Belintisky (Chalmers University of Technology, Sweden)
- Michael Cyberey (University of Virginia, USA)
- Brian Ellison (Science and Technology Facilities Council, UK)
- Jian-Rong Gao (Delft University of Technology, Netherlands)
- Gregory Goltsman (Moscow State Pedagogical University, Russia)
- Christopher Groppi (Arizona State University, USA)
- Jeffrey Hesler (Virginia Diodes, USA)
- Netty Honingh (University of Cologne, Germany)
- Heinz-Wilhelm Hübers (German Aerospace Center, Germany)
- Boris Karasik (NASA Jet Propulsion Laboratory, USA)
- Andrey Khudchenko (Aerospace Center, Russian Academy of Sciences, Russia)
- Alain Maestrini (NASA Jet Propulsion Laboratory, USA)
- Hiroshi Matsuo (National Astronomical Observatory of Japan)
- Imran Mehdi (NASA Jet Propulsion Laboratory, USA)
- Christophe Risacher (Institute of Millimetric Radio Astronomy, France)
- Kameljeet Saini (National Radio Astronomy Observatory, USA)

- Shengcai Shi (Purple Mountain Observatory, Chinese Academy of Sciences, China)
- José Siles (NASA Jet Propulsion Laboratory, USA)
- Jan Stake (Chalmers University of Technology, Sweden)
- Edward Tong (Center for Astrophysics, Harvard & Smithsonian, USA)
- Yoshinori Uzawa (National Astronomical Observatory of Japan)
- Min-Jye Wang (Academia Sinica Institute of Astronomy and Astrophysics, Taiwan)
- Ghassan Yassin (University of Oxford, UK)

Local Organizing Committee Members

- Bert Hawkins, Chair (National Radio Astronomy Observatory, USA)
- Sheryl Donley (National Radio Astronomy Observatory, USA)
- Jeffrey Hesler (Virginia Diodes, Inc., USA)
- Patricio Mena (National Radio Astronomy Observatory, USA)
- Karen Prairie (National Radio Astronomy Observatory, USA)
- Kamaljeet Saini (National Radio Astronomy Observatory, USA)

Our Sponsors

ISSTT2024 Official Photo

Program Overview

Monday, April 8, 2024

SESSION 1

The Integration of LNA Based Receivers for Millimeter and Sub-millimeter Wavelength Radio Astronomy

William McGenn[1*], Hui Wang[2], Elle Franks[1], Long Jiang[3], Byron Alderman[2], Brian Ellison[1], Gary A. Fuller[3], Peter G. Huggard[2], Danielle George[1]

Abstract— **Due to improvements in pHEMT technology it is now possible to realize Low Noise Amplifiers (LNAs) at millimeter and sub-millimeter wavelengths. This permits the design and manufacture of cryogenic, LNA-based, sideband separating heterodyne receivers for radioastronomy at frequencies above 100 GHz. There are, however, a variety of design challenges to overcome to enable multi-pixel array receiver topologies. We present the lessons learnt from the recently completed, sixteen-pixel, W-band (70 - 116 GHz) CARUSO receiver and show how we are further miniaturizing these components as we work towards a future integrated receiver technology.**

Keywords— **Millimeter-wave, submillimeter-wave, radio astronomy, Atacama Large Millimeter submillimeter Array, Low Noise Amplifier, Subharmonic Image Rejection Mixer, Integration.**

I. Introduction

RECENT developments in sub-100 nm gate length Indium Phosphide (InP) pseudomorphic High Electron Mobility Transistor (pHEMT) technology has pushed the boundaries of Low Noise Amplifiers (LNAs) to operating frequencies approaching the terahertz region [1][2][3]. These developments permit the design and fabrication of LNA based heterodyne receivers for the millimeter and sub-millimeter wavelength frequency bands.

Typically, millimeter and sub-millimeter radio telescope receivers are constructed from individual waveguide components that are packaged in separate mechanical housings, as is the case for the European Southern Observatory's (ESO's) recent state-of-the-art receiver for the Atacama Large Millimetre/sub-millimeter Array (ALMA) Band 2 frequency range (67 – 116 GHz) [4]. This allows each of the receiver components to be developed by specialist groups and suppliers, and also allows the best performing / best matched components to be selected for each receiver. However, this discrete component approach does not result in the most space efficient receiver format. For instance, each of the sub-components needs to contain all of the associated circuity within its housing and to have predefined mechanical interfaces. Whilst this discrete approach is viable at lower frequencies, for prototyping, or for a single receiver 'pixel', as operating frequencies increase and complex multi-pixel receiver topologies, such as focal plane arrays (FPA) and phased array feeds (PAF), are required for enhanced observation speed, a more space efficient and integrated approach to receiver design is needed.

The Millimetre-wave Technology Group and the Technology Department, both at the STFC Rutherford Appleton Laboratory, and the University of Manchester Advanced Radio Instrumentation Group (ARIG) have recently completed the sixteen-pixel, 70 – 116 GHz, Cryogenic Array Receiver for Users of the Sardinia Observatory (CARUSO) [5][6].

The sixteen receiving elements of CARUSO are located within the focal plane of the Sardinian 64 m diameter Gregorian telescope. To meet necessary spatial sampling requirements, individual pixel feedhorns that form the array are positioned in close proximity to each other. The randomly polarized signal entering a feedhorn is then divided by an orthomode transducer (OMT) into two linear components of polarization. Each polarization signal is then amplified by two cascaded millimeter-wave LNAs [1] and frequency down-converted by a Subharmonic Image Rejection Mixer (SHIRM) [5][7]. A single receiver dual polarization pixel chain therefore comprises a feedhorn, OMT, four LNAs and two SHIRMs with four intermediate frequency outputs corresponding to upper and lower sidebands for each polarization component. Furthermore, local oscillator power is delivered to each SHIRM via coaxial cables and all 16 pixels are enclosed within a cryogenic system that allows low temperature operation to approximately < 20 K.

Due to a highly demanding development timescale, using individual and previously proven, millimeter-wave component technology greatly lessened the project risk. However, close pixel spacing and telescope quasi-optical interface requirements imposed a volume constraint on the receiver architecture. For instance, each pixel was required to fit within the footprint of the feedhorn with a diameter of 25 mm and a feedhorn center-to-center separation of 31 mm. Moreover, all feedhorn entrance apertures needed to be positioned in close proximity to the cryostat's millimeter-wave signal input window. Complying with these dimensional constraints therefore required bespoke packaging design and miniaturization, but with compactness limited by the need to adopt a standard waveguide flange (WR10) interface format for the majority of the discrete components. Overall integration of

[1]Advanced Radio Instrumentation Group, Department of Electrical and Electronic Engineering, University of Manchester, Manchester, M13 9PL, UK; [2]STFC Rutherford Appleton Laboratory, Didcot, Ox11 0QX, UK; [3]Advanced Radio Instrumentation Group, Jodrell Bank Centre for Astrophysics, Department of Physics and Astronomy, University of Manchester, Manchester, M13 9PL, UK.
[*]William McGenn (email: william.mcgenn@manchester.ac.uk).

NOTES:

the receiver pixel chain was therefore limited by the use of separate components. Nevertheless, the approach established a first step towards a future more closely integrated approach and based on the success of the CARUSO receiver, we are now exploring methods of realizing new fully integrated millimeter-wave receivers.

Our new development is working Towards ALMA System-on-chip European Receivers (TASER) and is supported by ESO and additionally through the Horizon Europe RADIOBLOCKS project. A key objective is the creation of a highly integrated hybrid receiver configuration that unifies LNA and SHIRM devices within a single integrated waveguide block. This work will utilize the proven LNA and SHIRM technologies developed for CARUSO, with a view to future translation of the developed technologies to higher frequencies. The performance of the CARUSO LNA and SHIRM components will be used in this study as a baseline for assessing and quantifying the performance of the integrated LNA+SHIRM devices. In this paper we will present and analyze several alternative routes towards achieving this goal that can be used as a basis for future highly integrated receiver components.

II. LNA and SHIRM Technologies

The LNAs used for this work are based on those developed in the ARIG for ESO's ALMA Band 2 project [1][4]. The LNAs each incorporate 2-stage MMICs fabricated using the Northrup Grumman 35 nm gate length InP HEMT process [3]. The gate and drain of both transistor stages on the MMICs can be independently biased. The input and output ports of the miniaturized LNA package use a modified version of the standard UG-387/m waveguide flange with two of the screw holes removed. The LNAs use WR10 waveguide and custom waveguide-to-microstrip transitions are used to couple waveguide signals to and from the MMIC.

The SHIRM [5][7] comprises a pair of InGaAs Schottky barrier diode subharmonic mixers, bespoke signal splitting and recombination circuitry, and necessary signal, LO, and IF connections. All components and structures that form a SHIRM unit are integrated within a single mechanical package, providing a very compact solution. The RF signal is divided with a 90-degree phase difference between the two mixers, while the LO power is coupled with a 90-degree phase difference to the subharmonic mixers. The IF outputs are combined in an IF quadrature hybrid, at whose output ports the downconverter upper and lower sideband signals appear separately.

III. LNA+SHIRM Integration

The intention is to develop an integrated LNA+SHIRM using the well proven LNA and SHIRM technologies that have been previous demonstrated during the CARUSO project. This LNA+SHIRM will serve as both a building block for highly integrated receiver technologies in the 67 – 116 GHz frequency band, as well as a steppingstone to translate this integrated technology to higher frequencies in the future.

One potential use for such integrated receiver technologies would be in future upgrades of the receivers in the ALMA observatory, in particular there is desire to upgrade and combine the Band 4 and 5 frequency ranges (125 – 211 GHz) into a single receiver. Recently the ALMA wideband sensitivity upgrade [8] has been published and future receiver upgrades would be expected to be compatible with these specifications. In particular there is a requirement for 16 GHz of IF bandwidth for each sideband of the receiver, typically this would cover a frequency band of 2 – 18 GHz or 4 – 20 GHz. The CAURSO SHIRM currently has an IF bandwidth of 4 – 12 GHz, which is limited by the on-chip IF hybrid coupler, and as such an upgrade of this component would be necessary to make the integrated LNA+SHIRM compatible with the ALMA WSU specifications.

The RF and LO hybrids in the SHIRM are currently implemented using waveguides and are wavelength dependent structures, so the physical size of the coupler is significantly larger for the LO than for the RF. Switching the LO waveguide hybrid to an on-chip hybrid delivers a major benefit because it will greatly reduce the overall length of the integrated LNA+SHIRM and will therefore be a key focus for our projects. Inserting a 40 – 60 GHz LO hybrid is in some ways more straightforward than the RF equivalent, as frequencies and fractional bandwidth are lower and loss requirements less demanding. However, when considering SHIRMs for higher frequencies a higher frequency LO will be required, consequentially reducing the size of a waveguide hybrid and the miniaturization benefit of using an on-chip coupler. As the choice of LO hybrid implementation is independent of choice for the RF hybrid, all three of the integration options studied later in this paper can utilize a waveguide structure or an on-chip component for the LO hybrid.

The choice of waveguide or on-chip implementation for the RF hybrid is complicated by the higher frequency and also by the integration of the LNA with the SHIRM. We will present in this paper on the three integration options that have been studied for the RF, section of the LNA+SHIRM, these are:

1. Waveguide RF Hybrid: Maintain the LNA followed by the RF waveguide hybrid coupler: Fig. 1a.
2. On-chip RF Hybrid: Maintain the LNA and follow it with an on-chip RF hybrid coupler in place of the waveguide structure: Fig. 1b.
3. Balanced Approach: Insert the LNA MMICs inside the SHIRM structure in the RF paths of the Schottky diode mixers: Fig. 1c.

NOTES:

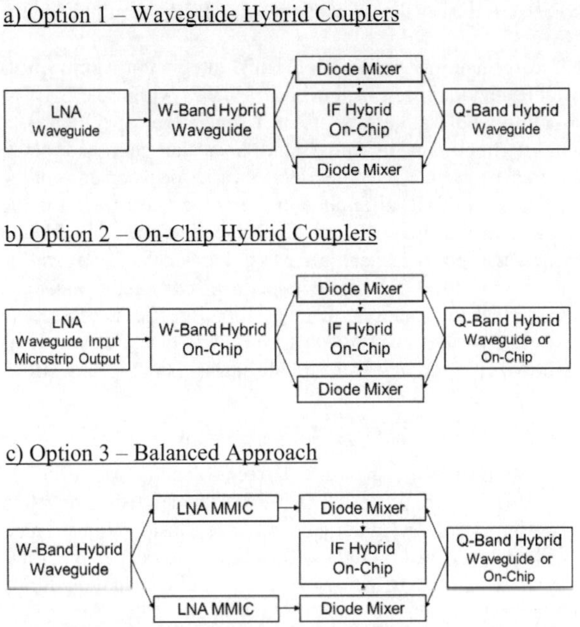

Fig. 1. Block diagrams of the three integration options for an LNA+SHIRM that are being studied.

Each of these options will be described in more detail through the rest of this paper:

Option 1 – RF Waveguide Hybrid Coupler

The first of the integration options will make minimal changes to the existing component designs. All of the components in the LNA and SHIRM offer very high performance and have been extensively modelled in the HFSS 3D EM simulator as well as manufactured in larger numbers as part of the CARUSO project. This includes the RF waveguide hybrid structure which offers excellent low loss performance with good amplitude and phase balance.

The integration of the LNA and SHIRM into a single block allows for the removal of the intervening waveguide flange interface, the space required in the block for two sets of screw and dowel pin holes, and the short length (11.4 mm) of waveguide between the output of the LNA and the input of the hybrid coupler that had been necessary to accommodate these. The overall length of this integrated LNA+SHIRM design is 72.8 mm. The CAD model of this LNA+SHIRM integration utilizing waveguide hybrids for both the RF and LO is shown in Fig. 2.

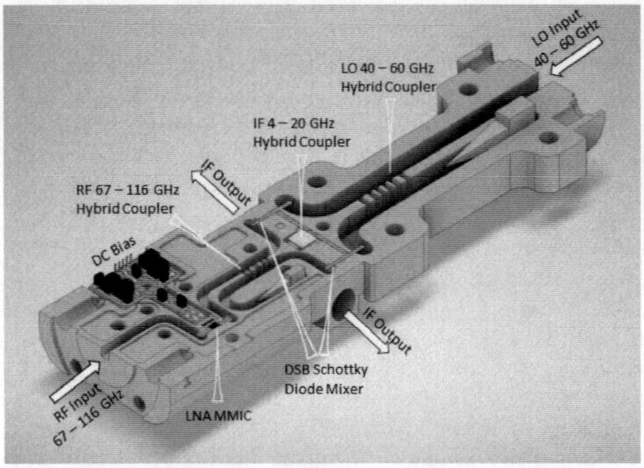

Fig. 2. CAD model of the LNA+SHIRM integration using the first integration option described in this paper, with the RF and LO hybrid couplers implanted with waveguide structures. The overall length of the LNA+SHIRM is 72.8mm.

Option 2 – RF On-Chip Hybrid Coupler

The RF waveguide hybrid structure discussed above has a fixed size and shape that limits further miniaturization. Using a waveguide structure also requires a transition from microstrip to waveguide between the output of the LNA and the input of the RF hybrid and then another opposite direction transition at the outputs of the hybrid. Each transition introduces unwanted losses and reflections. The use of a waveguide structure also limits future possibilities for further on-chip integration. It is therefore desirable to investigate other integration techniques that could offer potential further miniaturization and integrations of the LNA+SHIRM.

The second integration option will investigate the potential for using an on-chip RF hybrid coupler. The performance of the on-chip hybrid will be critical to the success of this integration, especially over the wide bandwidth involved.

Option 3 – Balanced Amplifier Approach

The third integration option will take a more holistic approach to the integration of the LNA and SHIRM components. With individually packaged LNA and SHIRM components there is no choice but to have the LNA before the entire SHIRM structure. An integrated design approach allows for the adjustment of the topology of the entire LNA+SHIRM structure to reach the ultimate goals. A pair of LNA MMICs can be inserted after the RF hybrid and in-front of the Schottky diodes within the SHIRM structure. The benefits of this approach will be the introduction of a balanced amplifier structure into the design, potentially offering improvements in the reflection coefficient of the RF input of the LNA+SHIRM. This location of the LNA and mixer also opens up potential for future on-chip integration of these components, that may be

NOTES:

attractive when translating this LNA+SHIRM technology to higher frequencies.

Although, this study of LNA+SHIRM integration is currently only considering a single LNA gain stage in the design, for a full receiver it is likely that two stages of LNAs will be needed to provide sufficient gain to minimize the noise performance of the receiver. Typically, an isolator would be required between LNA gain stages to prevent gain ripples. However, with the improvements in input reflection coefficient of the LNA+SHIRM offered by the balanced amplifier structure, the isolator may no longer be necessary. The inclusion of an external RF LNA in front of the LNA+SHIRM would also reduce the effect of the RF hybrid losses on the overall system noise, allowing the hybrid to be implemented in either a waveguide structure or on-chip component. This choice of hybrid implementation is essentially the same consideration that is being studied for integration options 1 and 2 described earlier in the paper; the waveguide structure would offer the best performance, but the on-chip component would offer better miniaturization and further integration options. This consideration will be dependent on the RF frequency of the LNA+SHIRM: as the frequency increases the size of the waveguide hybrid will reduce and the difficulty of implementing an on-chip hybrid with acceptable performance will increase.

The main disadvantage of this approach is that twice the number of LNA MMICs are required. At a system level this will have both technical and cost implications, especially for focal plane array systems where multiple individual receiver pixels (and therefore multiple LNA+SHIRM components) are being used. It is typical that all of the transistor stages in the LNAs are biased independently in order to have the most control over the gain and noise performance. However, in receivers with large numbers of LNAs and receiver pixels, this strategy will lead to a very large number of bias cables that need to be assembled in the cryostat and to pass through its vacuum case, increasing the complexity of the system. The increase in the number of MMICs will also result in a corresponding increase in the power, and therefore heat dissipation from the LNAs in the cryostat, something that will be exacerbated as the number of pixels is increased for larger FPA or PAF receivers. Another drawback is that a second MMIC bias circuit must be contained within the LNA+SHIRM, which will require careful design to keep the size of the integrated LNA+SHIRM block within the acceptable envelope for FPA and PAF receivers. However, for the most part, these challenges are well understood or can be quantified during the design of the components and receiver pixel(s). For example, the cost of the required MMICs and heat dissipation can be factored into the design of a receiver, and steps can be taken to bias multiple LNA stages from a single set of bias lines, reducing the number of cables required in the cryostat and also the size of the bias PCBs required for the MMICs.

The remaining unknown to this integration approach is whether the inclusion of the LNA MMICs within the SHIRM structure would adversely affect its operation. Specifically, it will be important to understand how the gain and phase difference introduced by the MMICs varies between units of the same design. If the difference between the two MMICs is too great, then this may disrupt the balance of the SHIRM and reduce the performance, affecting the sideband separation rejection ratio and causing an imbalance between the upper and lower sideband outputs. It may therefore be necessary to perform on-wafer pre-screening phase and gain measurements of the MMICs and to select suitable pairs for the LNA+SHIRM.

Summary of RF Hybrid Integration Options

A summary of the three integration options for an LNA+SHIRM are presented in Table 1, which lists a selection of the important design and performance considerations. These considerations are highlighted for each option depending on if they are a positive, negative, or need more investigation during our current projects.

The first integration, using an RF waveguide hybrid structure, requires the least changes to the pre-existing LNA and SHIRM designs, but also offers the least benefits of integrating the components (due to the waveguide coupler) and also the least potential for future developments towards on-chip integration.

The second integration option, replacing the RF waveguide hybrid coupler with an on-chip component, offers more potential benefits from the integration of LNA and SHIRM, and also offers a clear path towards future integration of multiple components onto the same chip. However, the performance of the RF on-chip hybrid coupler needs to be investigated in order to better understand the potential performance trade off that would need to be made in order to allow for this integration.

The third integration option, adopting a balanced approach and placing an LNA MMIC with each of the diode mixers within the SHIRM structure, offers some attractive upsides such as the potential for on-chip integration of LNA MMIC and mixer, and introducing the balanced amplifier topology to the LNA+SHIRM that should provide very good input reflection coefficient on the RF port. The implementation of the RF hybrid coupler in this option will take into account the findings of the first two integration options of this study. A waveguide hybrid will offer the best performance but will be limit the miniaturization potential, however this will be less of a factor at higher frequencies as the size of the hybrid is reduced. An on-chip hybrid will reduce the size of the LNA+SHIRM integration and offer the possibility for future on-chip

NOTES:

integration, however the performance trade off to enable this may become prohibitive at higher frequencies.

TABLE I. SUMMARY OF THE INTEGRATION OPTIONS

	#1	#2	#3
RF Hybrid	Waveguide	On-Chip^	Waveguide or On-Chip^
LNA MMIC	One MMIC	One MMIC	Two MMICs*
Bias PCB	No Issue	No Issue	Space Constraints&
MMIC Phase	No Issue	No Issue	Phase Difference#
Integration	Least Integration	More Integration	More Integration
Future On-chip Integration	No	Yes	Yes

Positive | Needs Investigation | Negative

^ Performance of on-chip RF hybrids will need to be investigated
* Two MMICs would lead to twice the power dissipation, twice the number of bias PCBs, connectors and cables
& Careful design will be needed to fit the bias PCB for the second MMIC in the layout
# Concern if possible phase and gain differences between MMICs could affect SHIRM performance.

IV. CONCLUSIONS

This paper has presented the outcomes of our initial study on the integration of the LNA and SHIRM components of a cryogenic millimeter-wave receiver for radioastronomy. Currently the RF and LO hybrid couplers of the SHIRM are implemented as waveguide structures, however changing these to on-chip hybrid components will offer miniaturization and integration benefits but the performance trade offs need to be assessed. Three different options for the integration of the LNA MMIC with the RF section of the SHIRM have been presented, each of which offer different benefits and have been assessed with a view towards future highly integrated millimeter and sub-millimeter wave receiver instrumentation. More work will now be carried out to assess the performance tradeoff between using a waveguide or on-chip hybrid components at the RF and LO sections of the SHIRM, and whether the inclusion of the LNA MMICs within the SHIRM topology will affect the down conversion performance.

ACKNOWLEDGEMENTS

The authors of this paper would like to thank Gie Han Tan, Carlos de Breuck, and Tony Mroczkowski from ESO for useful discussions as part of the TASER project. They would also like to acknowledge the contributions of Nart Daghestani, David Cuadrado-Calle and Claudio Jarufe to the design and development of the LNAs and the components for CARUSO.

This work has been carried out as part of the TASER: *Towards ALMA System-on-a-chip European Receivers* project, which is funded by the ESO European ALMA Development Fund; and also as a part of the RADIOBLOCKS project, which has received funding from the European Union's Horizon Europe research and innovation programme under grant agreement No 101093934.

REFERENCES

[1] D. Cuadrado-Calle *et al.*, *"Broadband MMIC LNAs for ALMA Band 2+3 With Noise Temperature Below 28 K,"* IEEE Trans. on Microwave Theory and Techniques 65, pp. 1589-1597, 2017, doi: 10.1109/TMTT.2016.2639018.

[2] D. White *et al. "125 - 211 GHz low noise MMIC amplifier design for radio astronomy"*. Exp Astron 48, 137–143 (2019), https://doi.org/10.1007/s10686-019-09641-z.

[3] X. Mei *et al.*, "*First demonstration of amplification at 1 THz using 25-nm InP high electron mobility transistor process*," IEEE Electron Device Letters, vol. 36, no. 4, pp. 327–329, 2015.

[4] P. Yagoubov *et al. "Wideband 67-116 GHz receiver development for ALMA Band 2."* Astronomy & Astrophysics, 634:A46, 2020. doi: 10.1051/0004-6361/201936777

[5] N. Dagestani *et al.*, "*Extended-W-band Single Pixel for Cryogenic Array Receiver for Users of the Sardinia Observatory (CARUSO) Astronomical Instrument"*, 4th URSI AT-RASC, Gran Canaria, 19-24 May 2024

[6] N. Dagestani *et al.*, "*Multipixel Cryogenic Extended-W-band Astronomical Receiver for the Sardinia Radio Telescope*", 4th URSI AT-RASC, Gran Canaria, 19-24 May 2024

[7] H. Wang *et al.*, *"Pre-prototype ALMA Band 2+3 Down-Converter & Local Oscillator System"*, proceedings of the 28th International Symposium on Space Terahertz Technology, Cologne, Germany, March 13-15, 2017.

[8] J. Carpenter *et al.* "*The ALMA2030 wideband sensitivity upgrade.*" ALMA Memo 621, arXiv preprint arXiv:2211.00195 (2022).

NOTES:

wSMA Receiver Cartridges

Lingzhen Zeng[1*] on behalf of the wSMA team

Abstract—**The wideband Submillimeter Array (wSMA) is an ongoing upgrade project for the SMA to replace the aging original receiver cryostats and receiver cartridges with all new cryostats and wideband 230 and 345 GHz receivers. In this report, we will describe the design, fabrication, assembly, and the measurements of the wSMA receiver cartridges. The receiver elements are installed on a floating 4 K stage, which maintains alignment of the receiver in the cold.**

Keywords— **Coherent detectors, SIS, submillimeter wave technology, radio astronomy, receivers.**

I. Introduction

The Submillimeter Array (SMA) is an array of eight antennas operating at millimeter and submillimeter frequencies on Mauna Kea, Hawaii. First commissioned in 2003, the SMA has been operating for over 20 years. At present, the frequency coverage of the SMA is from 180 to 420 GHz. The wideband Submillimeter Array (wSMA) is an ongoing upgrade project to replace the SMA cryostats and receiver cartridges with all new cryostats and wideband 230 and 345 GHz receiver cartridges.

Each wSMA cryostat consists of one 230 GHz (low band) and one 345 GHz (high band) receiver cartridge. As shown in Figure 1, both low band and high band cartridges share identical receiver optics. The low band receiver has a Local Oscillator (LO) covering 210 – 270 GHz and the high band receiver has a LO range from 280 GHz to 360 GHz. The output Intermediate Frequency (IF) bandwidth is 4 – 16 GHz in the current operation. It will be expanded to 4 – 20 GHz in future upgrades.

The wSMA receiver cartridge bears two major mechanical features. One is a pair of Automatic Thermal Links (ATL), through which the 50 K plate and a fixed 4 K plate are thermally connected to the cryostat in operation. The other feature is the introduction of a separate floating-4K stage, which carries the receiver optics, frontend assembly, and mixers. It is designed to be able to be mechanically registered to the top 50 K plate of the cryostat so that the receiver beams may achieve the required optical alignment accuracy. There are thermal links connecting the floating-4K and fixed-4K plates.

Both low band and high band receivers are dual polarization receivers. The frontend assemblies consist of a profiled-corrugated horn, an Orthomode Transducer (OMT), two couplers based on silicon chip technology and a 90 deg waveguide twist for compact integration.

We will present the design, fabrication, and measurements of each component of the wSMA receiver cartridges, along with the measurement results assessing the overall receiver performance.

Fig. 1. Photo of a wSMA receiver cartridge. The upper section comprises the receiver optics, coupling the signals from the cryostat optics to the frontend assembly located inside the base of the receiver optics. The 4K plate is divided into a floating-4K and a fixed-4K stage for optical alignment purposes. The image also shows the 50K plate, the lower 300K plate, and the G10 shells providing support between the stages.

[1]All authors are with: Center for Astrophysics | Harvard & Smithsonian, Cambridge, MA, USA.

[*]Corresponding author (email: lingzhen.zeng@cfa.harvard.edu).

NOTES:

Phoenix: A Far-IR Mission for Cosmology

A. Kogut[1*], J. Chluba[2], D. Fixsen[1], E. Wollack[1] for the Phoenix team

***Abstract*—Spectral distortions in the cosmic microwave background open a new window to the structure, content, and evolution of the universe. Detecting cosmological signals at the few part-per-billion level requires background-limited sensitivity with careful control of instrumental signatures. PHOENIX is a mission concept to map the CMB and diffuse astrophysical foregrounds with nK sensitivity at microwave through far-IR wavelengths. We describe the scientific goals of the mission and provide a summary of the predicted instrument performance.**

***Keywords*—Cosmic microwave background, Fourier transform spectrometer, instrumentation**

I. Introduction

Observations of cosmological radiation backgrounds have played a key role in our understanding of the universe. Measurements of the CMB blackbody frequency spectrum support the thermal hot big bang model while ruling out alternatives such as steady state models. Measurements of CMB anisotropies in temperature and polarization have provided insight into the contents of the universe and their evolution from primordial density perturbations to matter clustering, reionization, and the growth of large scale structure, consistent with a single 6-parameter cosmological model. Despite its success, this model is manifestly incomplete. It requires both dark energy and dark matter, neither of which exist within the Standard Model of particle physics. The observed flat geometry and nearly scale-invariant distribution of density perturbations hint at an origin in a period of exponential expansion called inflation, but direct evidence for inflation is missing.

II. Phoenix Mission

Precise measurements of the CMB introduce new opportunities to study the early universe and its evolution. Phoenix is a mission concept to measure the frequency spectrum and polarization of the CMB and astrophysical foregrounds. It will use 3 polarizing Fourier transform spectrometers (Fig 1) to map the full sky in Stokes I, Q, and U parameters to nK precision at 1.6 deg angular resolution from frequencies 20 GHz to 6 THz. With sensitivity over 1000x better than the seminal FIRAS measurements (Fig 2), Phoenix will detect the distortion from electron pressure and temperature in groups and clusters of galaxies while searching for evidence of new physics from the

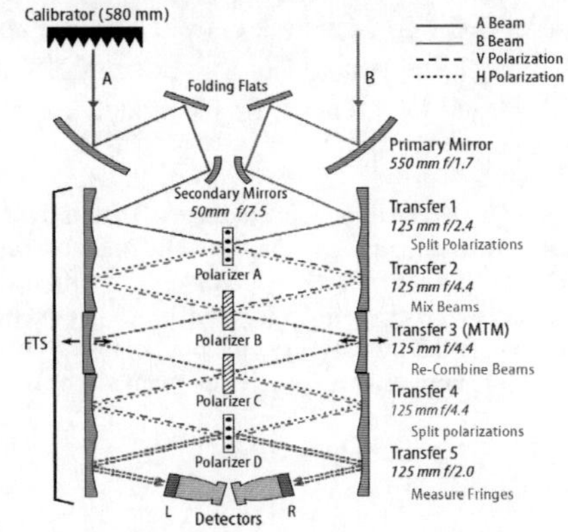

Fig 1: Schematic of the Phoenix Fourier transform spectrometer.

decay of primordial density perturbations or dark matter.

The instrument design provides multiple levels of null operation, signal modulation, and signal differences, with only few-percent systematic error suppression required at each level. Detailed time-ordered simulations evaluate the projected instrument performance. We describe the Phoenix mission and discuss the measures used to optimize performance.

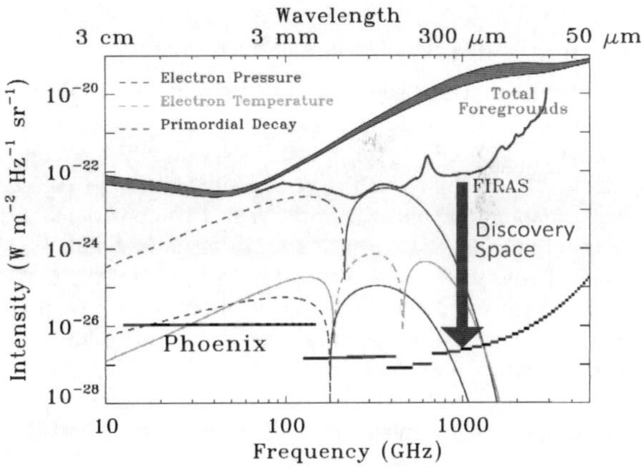

Fig 2: Phoenix will measure distortions from the CMB blackbody at sensitivity 1000x better than FIRAS, opening an enormous discovery space for cosmology and new physics while measuring the diffuse spectral signals from the low-redshift universe and our galaxy to unprecedented precision.

[1]NASA Goddard Space Flight Center, Greenbelt, 20715 USA; [2]Jodrell Bank Centre for Astrophysics, Department of Physics and Astronomy, The University of Manchester, Manchester M13 9PL, UK. [*]Corresponding author (email: alan.j.kogut@nasa.gov).

NOTES:

Heterodyne Spectrometer Instrument (HSI) for Far-IR Spectroscopy Space Telescope (FIRSST)

Martina C. Wiedner[1*], Andrey Baryshev[2], Paul Grimes[3], Vicotor Belitsky[4], Vincent Desmaris[4], Yan Delorme[1], Juan-Daniel Gallego[5], Cristina García Miró[5], Paul Hartogh[6], Netty Honingh[7], Bernd Klein[8], Jean-Michel Krieg[1], Gary Melnick[3], Sebastiano Ligori[9], Boon-Kok Tan[10], Bertrand Thomas[11], Volker Tolls[9], Jeanne Treuttel[1], Jérôme Valentin[1]; David Clements[12], Ray Blundell[3], Asantha Cooray[13], Meredith McGregor[14]

***Abstract*—The Heterdyne Spectrometer Instrument (HSI) is one of two instruments on the Far-IR Spectroscopy Space Telescope (FIRSST) proposal to NASA. It would be the first heterodyne array receiver in space and has 3 frequency bands, each containing two 5-pixel arrays, one per polarizations. HSI uses high TRL (>6) components and is an innovative, but low-risk instrument.**

***Keywords*—THz, heterodyne array.**

I. INTRODUCTION

The Far-IR Spectroscopy Space Telescope (FIRSST) is proposal submitted to the NASA Astrophysics Probe Explorer call in Nov 2023. FIRSST will determine how planets form in disks, will explain how water arrives on planets, and how galaxy grow. FIRSST has a 1.8m cryogenically cooled mirror and carries two instruments: the Direct Detection Spectrometer Instrument (DDSI) employing MKIDs behind a grating or a VIPA, as well as the Heterodyne Spectrometer Instrument (HSI) described here.

II. THE HETERODYNE SPECTROMETER INSTRUMENT

The Heterodyne Spectrometer Instrument (HSI) is designed around the trail of water science case, which requires wide RF bandwidth to tune to many water lines, but only moderate IF bandwidth (4GHz) at high spectral resolution (up to 10^7), as well as some mapping capacity. With an international team from 6 European countries and the US we designed the HSI to have 3 frequency bands covering 500 to 790 GHz, 882 to 1250 GHz and 1500 to 2000 GHz. Each band has two 5-pixel arrays, one for each linear polarization. The optics directs the astronomical signal to all of the arrays simultaneously using dichroic filters and grids. HSI uses double sideband state-of-the-art superconducting mixers, SIS mixers for band 1, HEB for band 3; for band 2 currently both options SIS and HEB are pursued.

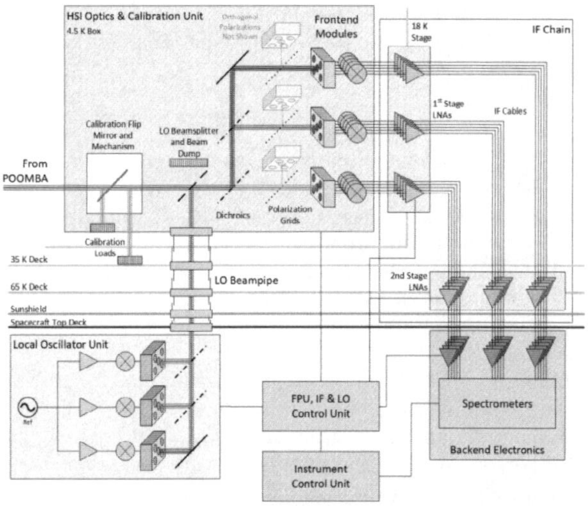

Fig. 1. Schematics of the Heterodyne Spectroscopy Instrument (HIS) on the Far-IR Spectroscopy Space Telescope (FIRSST) proposal. The light from the astronomical object enters the instrument on the left and is then superimposed with the artificial monochromatic signals from the three Local Oscillators. The Local Oscillator Unit is located in the warm space bus and consist of amplifier multiplier chains. Dichroics split the light into the three different bands. (Grids split the signal further into the two linear polarizations, so that 3 bands in 2 polarizations can be observed simultaneously (the components for the other polarization is not drawn, but there is a hint in light grey). In each band the signals (sky and local oscillator) are mixed and the beat signal is further amplified and then detected with spectrometers. The Instrument Control Unit steers the receiver and transmits the data to the central computer of the satellite.

[1]Observatoire de Paris, Paris, 75014, France, [2]Kapteyn Astronomical Institute University of Groningen, Groningen, Netherlands; [3]Center for Astrophysics Harvard & Smithsonian, Cambridge, *MA 02138*, USA; [4]Group for Advanced Receiver Development (GARD), *41296, Gothenburg, Sweden*; 5 Yebes Observatory, 19141 Yebes, Guadalajara, Spain; 6Max Planck Insitut for Solar System Research, Goettingen, Germany; 7Universitaet zu Koeln Cologne, Germany; 8Max Planck Institut fuer Radioastronomie, Bonn, Germany; 9NAF-OA Torino, Italy; 10Oxford University, Oxford, UK; 11 RPG, Meckenheim, Germany; 12Imperial College, London, UK; 13University of California, Irvine, CA 92697, USA; 14Johns Hopkins University, Baltimore, Maryland 21218, USA *Corresponding author (email: martina.wiedner@obspm.fr).

NOTES:

The biggest challenges for the design of heterodyne arrays in space are the limited availability of cryogenic cooling and electrical power. The cryogenic heat load is largely determined by the first amplifiers. For HSI we intend to use < 2mW InP amplifiers that will be heat sunk to the 18K stage, but are physically close to the 4.5K mixers. The subsequent IF will have further amplification around 70K and 300K. The electrical power consumption is mostly due to the spectrometer backends and the local oscillators. We selected Chirp transform spectrometers and autocorrelation spectrometers (ACS), very similar to those of SWI/JUICE, as backends, each requiring below 10W. HSI has amplifier-multiplier Local Oscillators, also building on the SWI/JUICE heritage. We attempt to limit the power consumption to below 2W per pixel. The Local Oscillators is located in the space bus at ambient temperature and their signal is directed to the cold optics assembly via several baffles and filters in order to limit thermal radiation and stray light to the neighboring instrument. HSI is managed by an ICU, that controls the cold optics, mixers, LO, IF, and spectrometer subsystems. It also processes and transfers data and controls the calibration sequence. As the mission is very quick with a launch in 2032, HSI needs components with TRL > 6.

HSI has at least two observing modes: one mapping mode, where 10 pixels of one band will be observing simultaneously and scanned across the sky, and a multifrequency pointed observing mode, where all three bands will observe in both polarizations simultaneously, but only with 2 pixels (i.e. 3 x 2 x 2 pixels in total). During the pointed observations HSI will carry out a double pixel switch, so that one pixel is always on the astronomical object while the other is off-source.

In conclusion, it is now possible to design small heterodyne array receivers with high TRL components for future space missions.

HSI PARAMETERS

PARAMETER		BAND		
		BAND 1	BAND 2	BAND 3
Wavelength (μm)		380 - 600	240 - 340	150 - 200
Frequency (GHz)		790 - 500	1250 - 882	2000 - 1500
Resolving power (λ/Δλ)*		10^6 to 10^7		
Beam size		52″ - 83″	33″ - 47″	21″ - 28″
Instantaneous FoV		300″×200″	150″×100″	150″×100″
Spectral channels*		1024 or 10,000		
Array size		5 pixels × 2 polarizations		
Aperture efficiency		80%		
Mixer Type		SIS	HEB	HEB
Receiver noise temperature (DSB)	CBE	60K	300K	400K
	MEV	72K	400K	500K
	Sci. Reqt.	80K	430K	525K
IF bandwidth		4GHz		
Optical bench temperature		4.7K with ±0.1K stability (not critical)		
LNA temperature (1st stage)		18K with ±0.1K stability during Allan time		
Mixer temperature		4.7K with ±10mK stability during Allan time		
RMS WFE budget (nm)	Requirement	<7500		
	Allocated	3000		
	Margin	250%		

Fig. 2. Parameters of the Heterodyne Spectroscopy Instrument.

NOTES:

A Compact Terahertz Instrument for Continuity Microwave Limb Sounding of Atmosphere

Subash Khanal[1*], Adrian Tang[1], Jacob Kooi[1], Nathaniel Livesey[1], and Goutam Chattopadhyay[1]

***Abstract*— Spaceborne observations have been pivotal in enhancing our understanding of Earth's atmosphere. Atmospheric limb sounding provides excellent vertical resolution and a robust signal-to-noise ratio for measuring trace gases. Here, we provide a brief overview of the Aura MLS instrument currently in space and then introduce the next generation "Continuity-MLS" in development now. This new instrument achieves nearly all Aura MLS measurements but with a significantly smaller form factor and better sensitivity, leveraging on the recent advances in CMOS, microwave and terahertz technologies.**

***Keywords*—Microwave limb sounding, Terahertz, Schottky, CMOS.**

I. Introduction

SINCE its launch in 2004, the AURA Microwave Limb Sounder (MLS) has provided a wealth of highly reliable and stable atmospheric observations with excellent vertical resolution [1]. Even after almost two decades of being in space, the measurement data from MLS instruments remain crucial for understanding the composition of the stratosphere and upper troposphere and their impacts on Earth's surface climate. Acknowledging the importance of this data, NASA's Earth Science Technology Office (ESTO) is actively supporting technology development initiatives for the next generation of MLS-like instruments. These efforts are directed towards crafting CubeSat/SmallSat-class instruments with the capability to investigate issues associated with the stability of the stratospheric ozone layer, climate change, and the impact of stratospheric ozone on air quality.

The Continuity-MLS (C-MLS) initiative seeks to uphold the MLS instrument's legacy while achieving better sensitivity and significant reductions in mass, power, volume, and cost. This is reflected in its name, Continuity-MLS, which achieves almost all the measurements performed by Aura MLS but with a much more compact form factor, weighing about 60kg and requiring around 80W of power. In contrast, the larger and more power-intensive Aura MLS weighs 500kg and demands 500W. These significant improvements have been possible due to the advancements and utilization of recent innovations in the areas of CMOS, microwave, and terahertz technology, including CMOS System on Chip (SoC) digital spectrometers, and CMOS W-band synthesizers, compact intermediate-frequency (IF) processors, state-of-the-art Schottky diode mixers and multipliers, as well as low-loss waveguide-based diplexers.

Aura MLS covered measurements in seven spectral regions from 118 GHz to 2.5 THz. The Continuity-MLS instrument focuses specifically on the 340 GHz and 640 GHz spectral regions due to its relevance for key molecules like H_2O, CO and HCl. This region allows measurements down to ~10 km altitude with low atmospheric absorption.

The C-MLS instrument comprises an antenna conducting limb scans over a tangent altitude range of 0 to 100 km approximately every 26 seconds. Calibration is performed through a switching mirror that provides perspectives of a spectrally flat calibration target and space. These views are spectrally divided by a polarizing beam-splitter into two receiver subsystems operating at 340 GHz and 640 GHz. The 340 GHz subsystem employs a sideband-separating architecture, utilizing a low-loss waveguide-based diplexer as a passive high/low pass filter for the incoming RF signal. The filtered RF bands undergo mixing with the LO signal through sub-harmonically pumped Schottky diode mixers, enabling down-conversion of observations to an IF of 0–18 GHz. A compact and low-power IF processor refines the IF signal and directs it to dedicated digital spectrometers. The spectral refinement is significantly improved through the use of spectrometers employing all-digital signal processing. Each digital spectrometer has a bandwidth of 3 GHz, comprising 4096 channels, and consumes 1.5 W. The 340 GHz receiver channel lacks spectral lines for hydrogen chloride (HCl), an essential molecule for understanding ozone destruction in the stratosphere and as a marker for stratospheric air transport. To address this, a second Schottky-based heterodyne receiver operating at 640 GHz is incorporated into C-MLS.

This presentation will provide an overview of the Continuity-MLS (C-MLS) instrument architecture, emphasizing key enabling technologies as these technologies are instrumental in maintaining the continuity of AURA MLS observations while significantly reducing the form factor and overall cost of the instrument.

References

[1] J. W. Waters *et al.*, "The Earth observing system microwave limb sounder (EOS MLS) on the aura Satellite," in *IEEE Transactions on Geoscience and Remote Sensing*, vol. 44, no. 5, pp. 1075-1092, May 2006, doi: 10.1109/TGRS.2006.873771.

[1]Jet Propulsion Laboratory, California Institute of Technology, Pasadena, CA 91109, USA
*Corresponding author (email: Subash.khanal@jpl.nasa.gov

U-WISHeS and V-WiSHeS: Terahertz Heterodyne Flight Spectrometers under Development Targeting the Uranus and Venus Systems

Carrie M. Anderson[1*], Negar Ehsan[1], Gordon L. Bjoraker[1], Melissa S. Ugelow[1], Chase E. Kielbasa[1], Tilak Hewagama[1], Gordon Chin[1], and Timothy A. Livengood[2]

***Abstract*— New technology developments overcome critical limitations of heritage flight heterodyne spectrometers that sacrificed broadband spectral coverage for high spectral resolution at a fixed frequency and narrow frequency band. We are developing two flight instruments optimized for Uranus' stratosphere and Venus' middle atmosphere – the Uranus Wideband far-Infrared Spectrometer at Heterodyne reSolution and the Venus Wideband Submillimeter Heterodyne Spectrometer – that will revolutionize the present-day planetary flight spectrometers in that they achieve both high spectral resolution and broadband spectral coverage at terahertz frequencies in a low size, weight, and power configuration.**

I. SCIENCE RATIONALE

REMOTE sensing spectroscopy at high resolving power ($R \sim 10^6$), is a powerful tool to investigate physical and chemical processes within environments throughout our solar system. Spectroscopy at heterodyne resolutions at long wavelengths in the submillimeter (sub-mm) and far-infrared (far-IR), enables retrieving temperature, dynamics, and chemical composition in atmospheres, including trace species that are more difficult to observe at shorter IR wavelengths, as well as cold surface temperatures of shadowed regions and small bodies in the outer solar system. The terahertz (THz) spectral region is optimal for remote sensing of thermal emission from cold objects, like those in the Uranus System, as well as from relatively cool regions, like Venus' atmosphere above its cloud tops, in which Venus' far-IR properties affect its climate system.

We present two flight instruments under development: the Venus Wideband Submillimeter Heterodyne Spectrometer (V-WiSHeS) and the Uranus Wideband far-Infrared Spectrometer at Heterodyne reSolution (U-WISHeS). The V-WiSHeS maturation effort is under NASA's Maturation of Instruments for Solar System Exploration (MatISSE) program and will achieve TRL 6 in September 2024. V-WiSHeS enables high sensitivity and high resolving power measurements of Venus' middle atmosphere to address key questions related to Venus' chemistry, dynamics, and thermal structure, and provides powerful diagnostics of trace gases that affect terrestrial planet climate systems. Similarly, U-WISHeS will measure the composition and temperatures in Uranus' cold stratosphere, as well as meridional circulation and the decay of zonal winds with height. High spectral resolution is essential to measure stratospheric zonal winds by Doppler shift, complementing cloud-tracking that probes deeper atmospheric levels, and contributes to constraining vertical temperature-pressure profiles. U-WISHeS will map surface temperature spatial variations on Uranus' moons by continuum emission and derive thermal inertia from day- and night-side temperature contrast.

II. TECHNOLOGY ADVANCEMENT

V-WiSHeS spans 64 GHz between 529 and 600 GHz, with 500 kHz spectral sampling ($R>10^6$), enabling measurements of H_2O, $H_2{}^{17}O$, $H_2{}^{18}O$, HDO, $CO^{18}O$, $CO^{17}O$, $^{13}CO^{18}O$, $^{13}CO^{17}O$, CO, ^{13}CO, $C^{18}O$, $O^{18}O$, O_3, $OO^{17}O$, $OO^{18}O$, H_2S, $H_2{}^{34}S$, $H_2{}^{33}S$, H_2SO_4, ClO, ^{37}ClO, H_2O_2, SO, SO_2, $^{34}SO_2$, OCS, $O^{13}CS$, $OC^{34}S$, $OC^{33}S$, NO, NO_2, and PH_3 in Venus' middle atmosphere. The high spectral resolution and broadband coverage directly results from advancements enabled in part by incorporating Application-Specific Integrated Circuit (ASIC) spectrometers into the receiver's back-end, critically needed for high-speed computation. We achieve wide spectral coverage with a time-multiplexed frequency-switching scheme using four 4-GHz ASIC spectrometers to achieve a 16-GHz instantaneous bandwidth, switched four times in frequency for a total coverage of 64-GHz.

U-WISHeS leverages the V-WiSHeS technology. It is a low TRL instrument spanning 300 GHz between 1.05 and 1.35 THz, with 2 MHz spectral resolution ($R>5 \times 10^5$), enabling measurements of CH_4, CO, ^{13}CO, $C^{18}O$, H_2O, HDO, NH_3, H_2S, HCN, $H^{13}CN$, and $HC^{15}N$. Transitions of PH_3, HF, HCl, and HI reside in the U-WISHeS bandpass, opening a window to discovery science. We will summarize both the V-WiSHeS and U-WISHeS instrument architectures and their science.

Acknowledgments—We extend our thanks to NASA's MatISSE program, which is providing funding for the V-WiSHeS maturation effort.

[1]NASA Goddard Space Flight Center, Greenbelt, 20771, USA; [2]University of Maryland, College Park, 20742, USA. [*]Corresponding author (email: carrie.m.anderson@nasa.gov).

NOTES:

SESSION 2

Metasurface-based terahertz quantum-cascade lasers operating beyond 5 THz

A. D. Kim[1*], M. Shahili[1], S. Addamane[2], C. A. Curwen[3], J. H. Kawamura[3], and B. S. Williams[1]

***Abstract*—A terahertz quantum-cascade (QC) metasurface VECSEL that operates above 5 THz is demonstrated. The amplifying metasurface is loaded with a QC active material that was separately shown to lase in continuous-wave mode up to 5.71 THz at 45 K. The challenge of the VECSEL includes overcoming the net loss due to proximity to the *Reststrahlen* band, and the reduction in metasurface amplification at higher frequency designs. A metasurface in a tunable external cavity was observed to lase in single-mode operation from 5.40-5.71 THz in pulsed mode operation at 5 K heatsink temperature.**

***Keywords*— local oscillator, metasurface, quantum cascade laser, terahertz, VECSEL.**

I. Introduction

One of the key components to a successful heterodyne spectroscopic system is a stable local oscillator (LO) with sufficient power to pump the mixer. In the THz regime, Schottky diode multiplier chains have been the leading LO source for decades. But since their power output decays with frequency, they have not been featured as LOs above 3 THz. Quantum-cascade lasers (QCLs) are an alternative LO candidate at these high frequencies; they have been used in the GREAT spectrometer and GUSTO/STO-2 to observe [OI] lines at 4.74 THz. However, beyond 5 THz, there is a technological gap in LO candidates, despite the regime being populated by astrophysically interesting lines such as [NIII] (5.23 THz), [SI] (5.32 THz), [FeI] (5.52 THz), [OIII] (5.79 THz), and [FeIII] (5.8 THz) [1]. However, to design QCLs above 5 THz, it becomes important to address the increased losses and reduced gain due to proximity to the *Reststrahlen* band of GaAs. This band is a consequence of strong optical-phonon resonances in the 8-9 THz range. The first demonstration of a THz QCL operating in continuous-wave (cw) above 5 THz was in 2022, with a maximum operating temperature of 15 K at 5.26 THz [2]. Since then, a metal-metal (MM) ridge waveguide operating up to 5.71 THz with a maximum operating temperature of 68 K was demonstrated [3]. This demonstration motivates the work to realize a metasurface-based external-cavity laser at these frequencies — a necessary progression to realize a technology more suitable for a local oscillator, as the novel architecture comes with the benefits of single-mode operation, scalable power output, high beam-quality, and frequency tunability [4].

II. Results

Initial results of the MM ridge waveguide were obtained by careful design of a GaAs/AlGaAs quantum-cascade active region, as well as the waveguide itself to minimize losses. Fig. 1 shows its power-current-voltage characteristics operating in cw at various temperatures. The device showed a maximum output power of 1 mW at 45 K, which was the lowest temperature able to be achieved using a Stirling cooler.

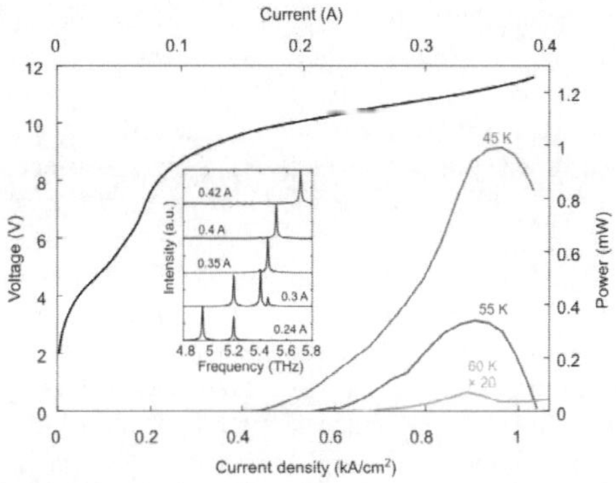

Fig. 1. Power-current-voltage data for a 0.5 mm x 75 μm metal-metal waveguide operating in continuous-wave mode at various heatsink temperatures. The inset shows the measured spectrum from an FTIR at various bias points, with modes spanning from 4.95 to 5.71 THz.

Additionally, the device demonstrated broadband gain with various lasing modes observed spanning from 4.76 – 6.03 THz in pulsed mode operation, and 4.95 – 5.71 THz in cw.

Following these MM waveguide results, metasurface-based QC-VECSELs were designed and fabricated using the same active region. The QC-VECSEL is enabled by a metasurface composed of a subwavelength array of elongated microstrip antennas loaded with the QC gain material. Fig. 2a shows an SEM image of such a metasurface. The dimensions of the antenna are chosen to operate at the TM_{01} cutoff frequency ($w = 6.5\ \mu m \cong \lambda_0/2n$), allowing for coupling to surface-incident radiation. In principle, scaling the metasurface to

[1]Department of Electrical and Computer Engineering, University of California, Los Angeles, California 90095, USA; [2]Sandia National Laboratories, Center of Integrated Nanotechnologies, Albuquerque, New Mexico 87185, USA; [3]Jet Propulsion Laboratory, California Institute of Technology, Pasadena, California 91109, USA; [*]Corresponding author (email: dkim95@ucla.edu).

NOTES:

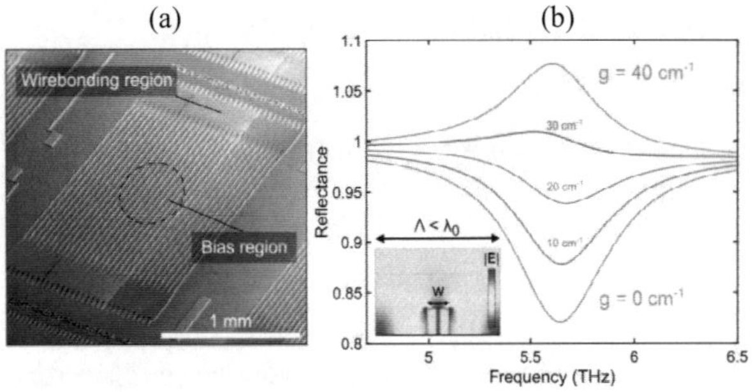

Fig. 2. (a) SEM image of a representative metasurface fabricated for operation above 5 THz. The blue circle corresponds to the region of the metasurface that is selectively biased to encourage lasing in the fundamental mode. (b) An FEM simulation of the metasurface reflectance as uniform gain is applied to the semiconductor. The inset shows the electric field profile of the corresponding eigenmode of a single unit cell of the metasurface with subwavelength periodicity. For the design in this work, w = 6.5 μm and h = 7 μm.

operate at higher frequencies is a straightforward matter of scaling the metasurface dimensions (e.g. width, period, height) by the wavelength. However, this is not necessarily true [5]. At higher frequencies, metallic loss makes it undesirable to scale down the metasurface active region thickness; however, if it is not scaled down, the quality factor of the metasurface drops significantly, which in turn reduces the effective gain interaction length. This results in a weaker metasurface amplification per reflection. An FEM simulation of the reflectance for varying levels of applied gain is shown in Fig. 2b. The results show a transparency gain of 28 cm^{-1}.

Initial results of the QC-VECSEL were successful in pulsed mode operation. Fig. 3a demonstrates broadband single-mode tuning from 5.40 – 5.72 THz as the output coupler is mechanically stepped outward via an intracryostat piezoelectric stepper motor, as illustrated by the VECSEL configuration in the inset. To achieve this, the cavity length was reduced to less than 500 μm to extend the free-spectral-range beyond the threshold gain-bandwidth. Fig. 3b shows a corresponding power-current-voltage plot at a particular lasing frequency of 5.69 THz operating at 5 K. The device lased up to a maximum heatsink temperature of 50 K. A high reflectance output coupler was needed (~99%), however this had the consequence of reducing the slope-efficiency due to the lower mirror-loss and the loss from the metals of the output coupler.

The initial results are promising and prove that QC-VECSELs can be made to operate beyond 5 THz. Additionally, these results inform subsequent designs to optimize the efficiency, power output, and achieve cw operation for LO candidates.

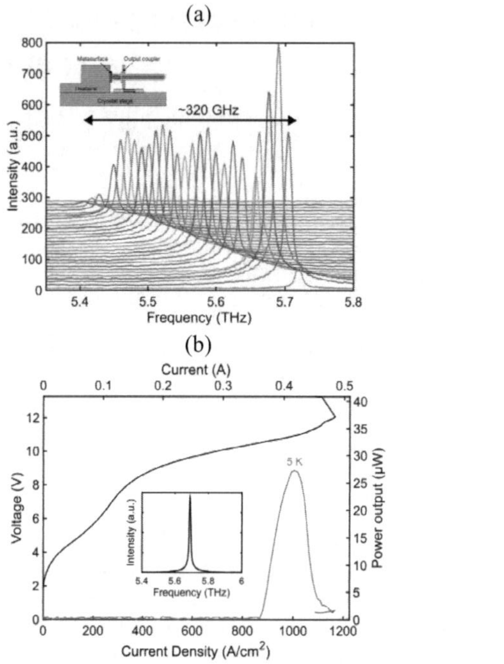

Fig. 3. (a) Stacked spectra of the QC-VECSEL output collected via FTIR that demonstrates single-mode tuning over a 320 GHz bandwidth. The spectra were collected in pulsed mode at 5 K. The inset shows a diagram of the VECSEL configuration, where the output coupler is mounted on a piezoelectric stepper motor to step the cavity length intra-cryostat (b) Power-current-voltage plot of the QC-VECSEL operating at 5.69 THz in pulsed mode.

REFERENCES

[1] E. Peeters et al., "ISO spectroscopy of compact HII regions in the Galaxy - I. The catalogue," *A&A*, vol. 381, no. 2, Art. no. 2, Jan. 2002.

[2] W. Li et al., "Continuous-wave single-mode quantum cascade laser at 5.1 THz based on graded sampled grating design," *Photon. Res., PRJ*, vol. 10, no. 12, pp. 2686–2692, Dec. 2022.

[3] M. Shahili, S. Addamane, C. A. Curwen, J. H. Kawamura, and B. S. Williams, "GaAs-based quantum cascade laser emitting above 5 THz," *The International Infrared and Terahertz Quantum Workshop 2023 (ITQW)*, 2023.

[4] C. A. Curwen, J. L. Reno, and B. S. Williams, "Broadband continuous single-mode tuning of a short-cavity quantum-cascade VECSEL," *Nat. Photonics*, vol. 13, no. 12, Art. no. 12, Dec. 2019.

[5] A. D. Kim, C. A. Curwen, Y. Wu, J. L. Reno, S. J. Addamane, and B. S. Williams, "Wavelength Scaling of Widely-Tunable Terahertz Quantum-Cascade Metasurface Lasers," *IEEE Journal of Microwaves*, vol. 3, no. 1, pp. 305–318, Jan. 2023.

NOTES:

Superconducting Glide-Symmetric Bifilar Transmission Lines for Tunable Stop-Band and Filtering Applications

Jorge Cárdenas[1*], F. Patricio Mena[2] and Francisco Pizarro[1]

Abstract—**This work shows a glide-symmetric corrugated transmission line, based on previous studies, but this time implemented using superconductive characteristics. The scope is to investigate if stop-bands are still present when the transmission line has superconductive features. The simulations demonstrate that stop-bands are indeed observed, opening the possibility of using them in superconducting circuits, particularly in traveling-wave kinetic-inductance parametric amplifiers.**

Keywords— **Glide-symmetry, metasurface, superconductor.**

I. INTRODUCTION

SUPERCONDUCTING circuits are essential in state-of-the-art fields such as quantum computing and radioastronomy [1]. In both cases, noise plays a major role thus the need of superconducting devices capable of operating up to quantum-level noise constraints. In order to come up with superconducting devices, for example parametric amplifiers and resonators, microstrip transmission lines are frequently used hence the importance of exploring their topologies and properties.

When symmetry considerations are added to non-superconducting transmission lines, new properties can emerge, for instance low-dispersive transmission lines and tunable stop-bands. Furthermore, non-superconducting symmetric strip lines have been previously studied as periodic structures [2]. Longitudinal and translational symmetries were considered to design stub-loaded microstrips to produce group delay changes and, consequently, achieve a tunable band gap filter [3]. In this work, we present a glide-symmetric line, based on [4], that includes a simulated superconducting condition. We demonstrate that the tunable stop-band property achieved for the non-superconducting case, can be obtained while imposing the superconducting condition to the strip line.

II. GLIDE SYMMETRY WITH SUPERCONDUCTING CONDITION

Glide symmetry enables the control of dispersion characteristics of the transmission line. Though this condition can be achieved by different means, we have considered a geometry in which the symmetry property is realized from added corrugations to the main line in a double-sided bifilar line. To study the effect of symmetry breaking in the band structure, simulations were performed using Ansys HFSS commercial software. A unit cell with a period p=400 μm was defined, as presented in Fig. 1. This cell also represents the Irreducible Brillouin Zone containing the whole information of the wave vector $\vec{k}$. The corrugations are set such that symmetry can be broken by either altering the corrugation heights and the widths at each side of the bifilar line or altering the separation between corrugations [2]. The glide-symmetric state is the one in which the corrugations are separated half period apart, while a longitudinal symmetry is that in which there is no separation in between.

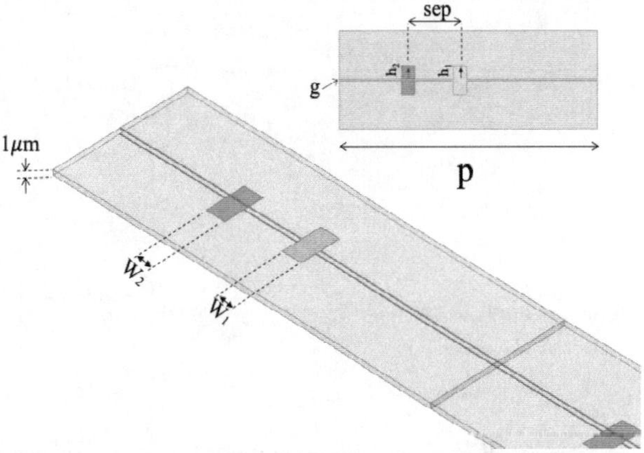

Fig. 1. Glide-symmetric bifilar line with a SOI substrate with 1 μm height. It is used along with a line width of 1 μm. The separation parameter, is measured between the front and rear corrugations as fractions of the period p.

Here we present the propagation analysis of a periodic glide symmetric bifilar with a superconducting condition on the surface impedance of the conductor [4],

$$Z_s = 2\,\pi\, j\, f\, \mu_0\, \lambda \coth{(t/\lambda)}, \qquad (1)$$

The penetration depth was set to $\lambda = 314$ nm [5] and the conductor thickness to $t = 60$ nm. The lengths of the

[1]Escuela de Ingeniería Eléctrica, Pontificia Universidad Católica de Valparaíso, Valparaíso, 2340000, Chile; [2]NRAO, Charlottesville, 22901, USA. [*]Corresponding author (email: jorge.cardenas.l@mail.pucv.cl).

NOTES:

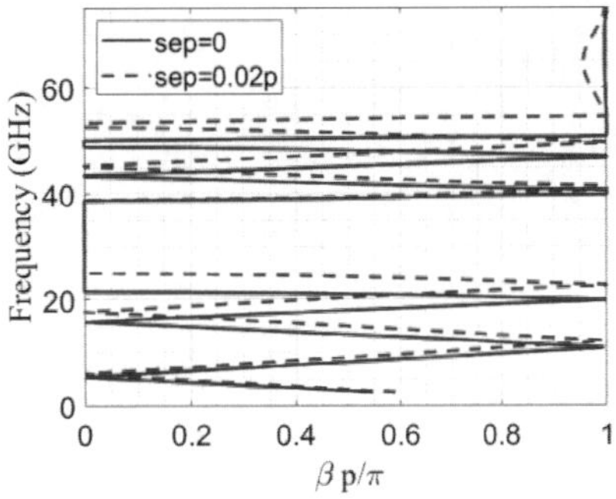

a)

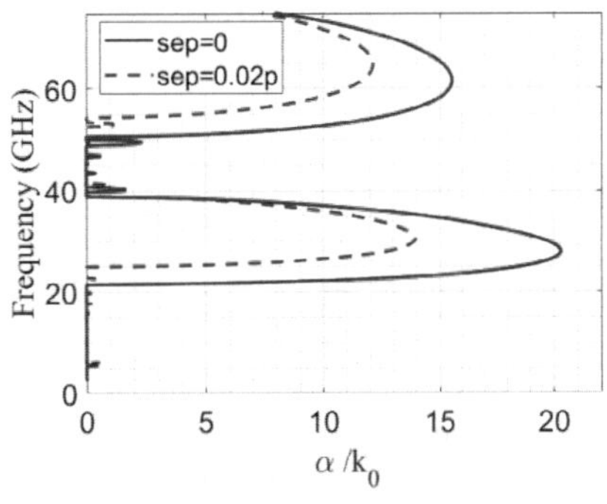

b)

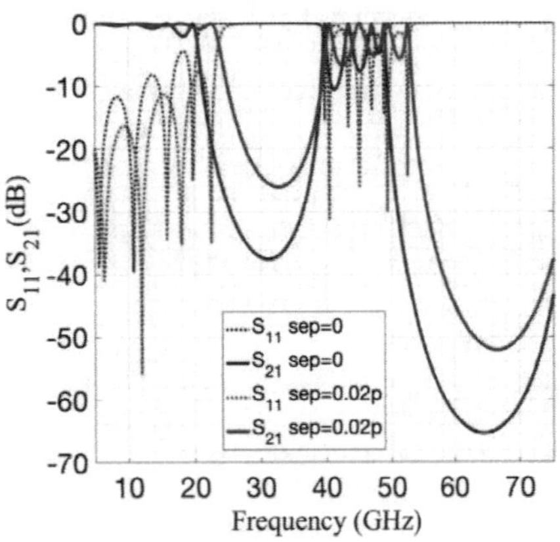

c)

Fig. 2. Simulated results of the proposed structure under superconductor surface impedance condition. (a) normalized β and (b) normalized α obtained for two separations between corrugations. (c) S-parameters obtained for the same separation values.

corrugations were set to $h_1 = h_2 = 20\ \mu$m and their widths to $w_1 = w_2 = 20\ \mu$m, as well.

III. Results

First, we have obtained the dispersion diagram of the glide symmetric structure under the superconductive condition using the Multimodal Transfer Matrix method [6]. We can see the presence of stop-bands, which is demonstrated on the propagation and attenuation constants shown in Fig. 2.a. and Fig. 2.b. Moreover, varying the separation between corrugations yields a tunable stop-band. The transmission and reflection coefficients S_{11} and S_{21} are presented in Fig. 2.c. The tunable stop-band was confirmed with a full-wave simulation. By modifying the separation to zero hence breaking the glide symmetry, the transmission is modified thus moving the stop-band.

IV. Conclusions

This article presented a tunable superconductive glide-symmetric bifilar line. The tunable stop-band property was achieved under superconducting conditions. The tuning was realized by varying the separation between corrugations, though further tuning can be achieved by modifying other parameters of the unit cell. A superconducting device with such capability provides benefits for low-noise applications under cryogenic conditions, for instance radioastronomy, parametric amplification, and quantum computing. Future work will show the validation of the structure in higher frequency bands and explore properties of other non-superconducting microstrip designs under superconducting conditions.

References

[1] Razmkhah, S., Bozbey, A., & Febvre, P. "Superconductor modulation circuits for Qubit control at microwave frequencies," 2023. Applied Superconductivity Conference (ASC 2022).

[2] P. Padilla, et al., "Glide-symmetric printed corrugated transmission lines with controlable stopband," 2019. 13th European Conference on Antenna|s and Propagation (EuCAP), Krakow, Poland, 2019, pp. 1-4.

[3] A. Hessel, Ming Hui Chen, R. C. M. Li and A. A. Oliner, "Propagation in periodically loaded waveguides with higher symmetries," in Proceedings of the IEEE, vol. 61, no. 2, pp. 183-195, Feb. 1973.

[4] A. Kerr, "Surface Impedance of Superconductors and Normal Conductors in EM Simulators," National Radio Astronomy Observatory, Charlottesville, Virginia, Tech. Rep. 245, January 1999.

[5] D. J. Thoen, et al., "Superconducting NbTin Thin Films With Highly Uniform Properties Over a Ø 100 mm Wafer," IEEE Transactions on Applied Superconductivity, vol. 27, no. 4, pp. 1–5, 2017.

[6] F. Mesa, G. Valerio, R. Rodrıguez-Berral, and O. Quevedo-Teruel, "Simulation-assisted efficient computation of the dispersion diagram of periodic structures: A comprehensive overview with applications to filters, leaky-wave antennas and metasurfaces," IEEE Antennas and Propagation Magazine, vol. 63, no. 5, pp. 33–45, 2021.

NOTES:

Design Considerations for a W-band Josephson Junction Travelling Wave Parametric Amplifier

Javier Navarro Montilla[1], Arnaud Barbier[2], Eduard F.C. Driessen[2] and Boon-Kok Tan[1]

***Abstract*—Most Josephson junction Travelling Wave Parametric Amplifiers (JTWPAs) developed so far have been focused on operation below 20 GHz, primarily driven by the choice of the qubit resonance frequency used in quantum computation research. Consequently, there is a lack of effort to extend their operation to higher frequency ranges. However, millimetre (mm)-wave JTWPAs could offer potential significant advantages for astronomy, but their operation in this regime is largely unexplored. In this paper, we describe the design considerations for extending JTWPAs operation to the W-band range. We present two JTWPA designs, one with and one without phase matching elements, and we discuss the design methodology of both approaches, before showing their predicted performance respectively.**

***Keywords*—travelling wave, parametric amplifier, millimetre wave, Josephson junctions, metamaterial, W-band.**

I. Introduction

The design and operation of JTWPAs at frequencies higher than 20 GHz is a relatively unexplored topic, so far. Our interest lies in extending the operation of microwave JTWPAs to the W-band range (75-110 GHz), to pave the way for operation at even higher frequencies into the millimetre (mm) and sub-mm bands. W-band is one of the key windows to astronomical observation, including the Event Horizon Telescopes (EHT). Some of these more demanding observations have driven ALMA (Atacama Large Millimetre/sub-mm Array) to upgrade their band 2 (67-116 GHz) cartridge [1]. However, even state-of-the-art High Electron Mobility Transistors (HEMT) receiver still struggle to achieve the ultimate noise performance [2].

TWPAs exploit the nonlinear reactive properties of superconducting transmission lines for amplification, and have demonstrated broadband high gain with noise approaching the quantum limit in the microwave regime [3, 4]. They can in principle operate at higher frequencies, into the sub-mm wave regime, potentially overhauling the commonly used heterodyne receivers' architecture based on superconductor-insulator-superconductor (SIS) mixers or HEMT amplifiers as first-stage detector. Deployment of an ultra-low noise superconducting pre-amplifier before these detectors would improve the receiver sensitivity [5], especially into the THz region.

While high kinetic inductance films based (K-)TWPAs have shown their potential feasibility in achieving large gain in the W-band [6, 7, 8], a similar demonstration with the Josephson-based metamaterial counterpart has not yet been performed. As JTWPAs generally require much smaller pump powers than KTWPAs, they may be more suitable for operation alongside SIS mixer that require additional high frequency local oscillator source. In this paper, we discuss the design and fabrication considerations of the envisioned W-band JTWPA.

II. W-Band JTWPA with Interdigitated Stubs

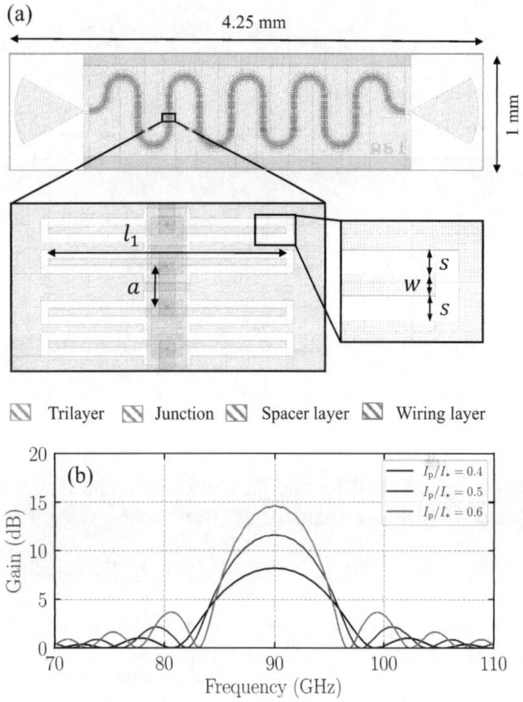

Fig. 1. (a) Layout of the unloaded W-band JTWPA device. Zoom-in shows the structures forming the device, including the junctions and the interdigitated stubs used to match the output impedance of the antennas. The stubs length (l_1), width (w) and gap (s), as well as the distance between the junctions (a) is indicated on the layout. (b) Predicted gain for a pump tone at 90 GHz, with different pump power to critical current ratio (I_p/I_*).

[1]University of Oxford, Oxford, OX1 3RH, United Kingdom; [2]Institut de Radioastronomie Millimétrique (IRAM), Saint Martin d'Hères, 38406, France;

NOTES:

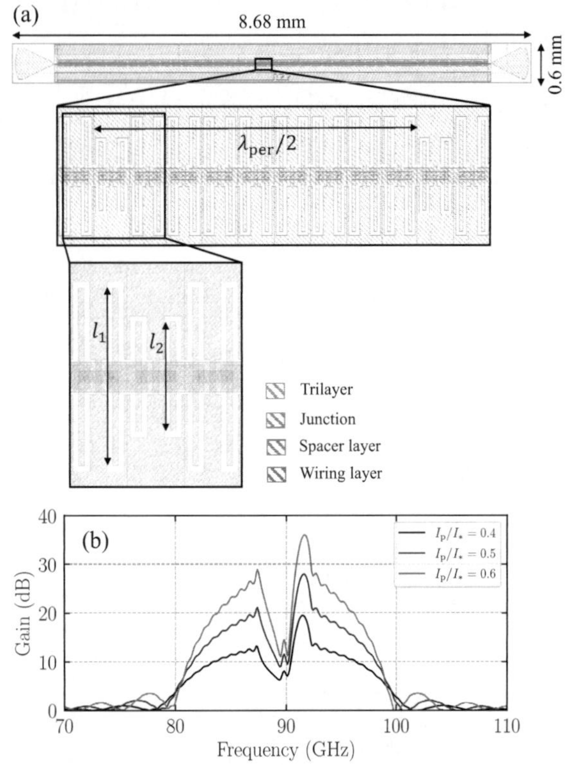

Fig. 2. (a) Layout of the loaded W-band JTWPA device. The zoom-in image shows the variation of the stubs' length (l_1 and l_2) creating the loading section. The loading sections are separated of a distance $\lambda_{per}/2$, where λ_{per} is the wavelength at $f_{per} \approx 92$ GHz. (b) Predicted gain for a pump tone at 89.8 GHz, for different pump power to critical current ratio (I_p/I_*).

While recent JTWPA designs utilise aluminium-based junctions [9, 10], harnessing the maturity of the superconducting qubit fabrication techniques, this technology has an operational limit f < 90 GHz due to the aluminium superconducting gap energy. Our designs are based on niobium/aluminium-aluminium oxide/niobium (Nb/Al-AlO_x/Nb) junctions, comprising a larger superconducting gap energy, extending the operational frequency to f < 680 GHz.

A critical parameter to consider for high frequency operation beyond the microwave regime is the plasma resonance frequency of the junctions, which increases inversely with the junction's inductance. As the parametric gain of a JTWPA is directly proportional to the junction inductance, this implies that we may need more junctions to achieve decent gain. Therefore, careful consideration is required to optimise the

TABLE I. DESIGNS CHARACTERISTICS†

Design	J_c (kA/cm²)	A_{jj} (µm²)	N_{jj}	l_1 (µm)	l_2 (µm)	w (µm)	S (µm)	a (µm)
Unloaded	3.4	0.5	704	60	NA	2	2	10
Loaded	3.4	0.5	704	60	37	1.75	2	10

†Design parameters: critical current density (J_c), junction's area (A_{jj}), number of junctions (N_{jj}), main stubs length (l_1), loading section stub length (l_2), stub width (w), stub gap (S), distance between junctions (a).

design of the mm-wave JTWPA. Furthermore, high frequency devices are generally coupled with waveguide connections instead of microwave connectors, hence different deliberation is required for the JTWPA interfaces, as well as the control of the characteristic impedance of the metamaterial line.

As this endeavor is unprecedented, we first opt for a minimalistic design approach. Specifically, we omit dispersion engineering elements that promote exponential gain to ease fabrication. This approach provides us the flexibility in positioning the pump frequency anywhere within the W-band during the experiments without being constrained by the phase matching conditions. It further allows for greater error limits in the fabrication tolerance, particularly these inaccuracies may be a more pronounced effect at higher frequencies compared to microwave designs.

This design, hereafter the unloaded design, is presented in Fig.1(a). It consists of a coplanar waveguide (CPW) Josephson junction array coupled with interdigitated stubs used to match the output impendence of our probe antennas (Z_{out} = 70 Ω), while ensuring the phase velocity remains high to reduce the number of junctions required for high gain. The critical current of the 704 junctions composing our JTWPA is fixed to J_c = 3.4 kA/cm², resulting in a cut-off frequency about 180 GHz, high enough to suppress the unwanted pump harmonics while not affecting the amplification bandwidth of the device. The design also includes ground bridges to avoid the generation of higher order modes, and a meandered geometry to reduce the size of the device chip, to limit the substrate resonant modes. We simulated the behaviour of the unit cell and cascaded them to estimate the gain-bandwidth product of the entire chip using our established couple-mode equations (CME) framework [11]. Fig.1(b) shows the expected gain profiles when pumping the device with an 89.8 GHz pump tone at different strength.

In a separate design, hereafter the loaded design, we improve the gain-bandwidth product of the unloaded design by incorporating phase-matching elements, namely periodic loading structures, along the transmission line. This results in a transmission line where the length of the stubs is periodically shortened, as shown in Fig.2(a), with a modulation period of $\lambda_{per}/2$, where λ_{per} is the wavelength at $f_{per} \approx 92$ GHz. This modulation creates a stopband around f_{per}, locally distorting the dispersion relation of the device, allowing for phase mismatch

NOTES:

correction, hence facilitating exponential gain [3, 12]. The predicted gain profiles for this loaded design when pumped at f_p = 89.8 GHz is presented in Fig. 2(b), showcasing a significant enhancement in the gain-bandwidth product. The key design parameters of both the unloaded and the loaded designs are summarised in Tab.1.

III. Summary

The operation of JTWPAs at frequencies higher than microwaves is largely unexplored. In this paper, we have presented the design considerations required to extend the operation of JTWPAs to the W-band range. We introduce two JTWPA designs based on Nb/Al-AlO_x/Nb junctions, consisting of a CPW with interdigitated stubs. The two resulting designs, with and without periodic loading modulation of the transmission line, are simulated and their performance calculated for different pump amplitudes. We estimate peak gain values of approximately 15 dB for the unloaded design and 25 dB for the loaded design.

Acknowledgments

This research was supported by the United Kingdom (UK) Science and Technology Facility Council (STFC) under the Quantum for Science Grant (ST/Y004973/1), and the European Research Council (ERC) under the European Union's Horizon 2020 research and innovation programme with grant agreement No. [803862] (Project [SPA4AstroQIT]). J. Navarro Montilla's D.Phil. studentship is supported by the UK STFC and the Foley-Bejar Scholarship from the Balliol College, Oxford. For the purpose of Open Access, the author has applied a CC BY public copyright licence to any Author Accepted Manuscript version arising from this submission.

References

[1] P. Yagoubov et al., "Wideband 67-116 GHz receiver development for ALMA band 2", *Astronomy and Astrophysics*, February 2020.

[2] CC. Chiong et al., "Low noise amplifier for next-generation radio astronomy telescopes: Review of the state-of-the-art cryogenic LNAs in the most challenging applications", *IEEE Microwave Magazine*, December. 2021.

[3] Byeong Ho Eom et al., "A wideband, low-noise superconducting amplifier with high dynamic range", *Nature Physics*, July. 2012.

[4] C. Macklin et al., "A near-quantum-limited Josephson traveling-wave parametric amplifier", *Science*, October. 2015.

[5] O. Noroozian, "Cycle 5 NRAO ALMA development study report technology development of quantum-limited, ultra-wideband RF amplifiers for ALMA: A 65-150 GHz test case", Technical report, April. 2020.

[6] P.K. Day et al., "Millimeter-wave superconducting parametric amplifiers based on kinetic inductance." *Millimeter, Submillimeter, and Far-Infrared Detectors and Instrumentation for Astronomy XI.* SPIE, 2022.

[7] B-K. Tan et al. "Operation of kinetic-inductance travelling wave parametric amplifiers at millimetre wavelengths", *Superconductor Science and Technology*, February. 2024.

[8] Longden, J. C. et al., "Non-degenerate-pump four-wave mixing kinetic inductance travelling-wave parametric amplifiers", *Engineering Research Express*, 2024.

[9] L. Planat et al., "Photonic-crystal josephson traveling-wave parametric amplifier," *Phys. Rev. X*, April. 2020.

[10] A. Ranadive et al., "Kerr reversal in josephson meta-material and traveling wave parametric amplification," *Nature Communications*, April. 2022.

[11] B-K Tan et al., "Engineering the thin film characteristics for optimal performance of superconducting kinetic inductance amplifiers using a rigorous modelling technique", *Open Res Eur*. September. 2023.

[12] K. O'Brien, C. Macklin, I. Siddiqi, and X. Zhang, "Resonant phase matching of Josephson junction traveling wave parametric amplifiers", *Physical review letters*, October. 2014.

NOTES:

Demonstration of a compact high-resolution spectrograph for far-infrared astronomy: silicon-based virtually imaged phased array

Bugao Zou[1*], Thomas Nikola[2], Nicholas F. Cothard[3], Alexander S. Kutyrev[3], Eric Mentzell[3], Karwan Rostem[3], Edward J. Wollack[3], Jake A. Connors[3], Willem Jellema[4], Tsung-Yu Kao[5], Alan W.M. Lee[5] and Gordon J. Stacey[6]

***Abstract*—We present the first measurement of the spectral performance of a compact silicon-based Virtually Imaged Phased Array (VIPA) in the far-infrared regime. We cryogenically cooled the VIPA to 4 K and measured its transmission spectrum using a quantum cascade laser operating near 115.7 μm and a pyroelectric detector. The measured spectral resolving power is about 16,500. The demonstrated performance of the VIPA indicates the suitability of this technology to achieve spectral resolving powers of 100,000 and thus shows its great promise for applications in balloon-borne and space-borne velocity-resolved astrophysics and represents a valuable addition to the existing suite of high-resolution far-infrared astronomical instruments.**

***Keywords*—Spectrometer, virtually imaged phased array, far-infrared astronomy, silicon, nanofabrication.**

I. Introduction

The VIPA, a novel spectral disperser capable of achieving extremely high spectral resolution, distinguishes itself from conventional methods due to its compact structure without moving components and cost-effectiveness [1]. Its versatility spans diverse fields like frequency comb spectroscopy [2] and biological imaging [3]. Recently, the use of a VIPA as the prime spectral element to deliver a resolving power of 100,000 has been proposed for direct detection, far-infrared astronomy projects, e.g. POEMM and FIRSST.

As shown in Fig. 1, the VIPA system channels a line-focused beam through an entrance slit, using two highly reflective surfaces to form a resonating cavity. The entrance side, excluding the slit, has a metal film for near-total reflection, while the exit side has a partially reflective film (usually over 90%) to allow a small portion of the beam to pass through with each reflection. An anti-reflection coating can be applied to the entrance slit to enhance transmittance if needed. The incident light, injected at an angle, resonates within the cavity, with each wavelength following its unique path. After multiple reflections, the light exits through the partially reflective surface, and the transmitted beams are focused to create angular dispersion based on their wavelengths.

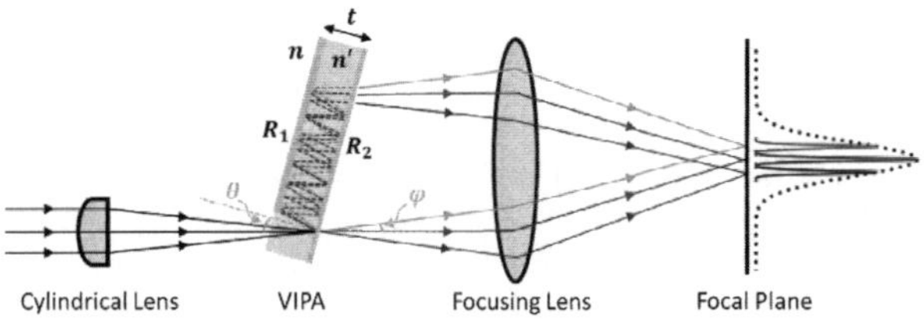

Fig. 1. Schematic of the VIPA system's operation principle. The colored ray traces illustrate how beams of different wavelengths follow distinct paths and disperse on the focal plane.

In this study, we present the development and successful tests of a far-infrared VIPA device, demonstrating the potential of using VIPAs in upcoming high-spectral resolution, velocity-resolved astronomical instruments.

II. Fabrication and Experimental Setup

The VIPA demonstrator was fabricated at the Cornell NanoScale Facility using high-purity silicon from the same boule studied in Wollack et al. [4]. At room temperature, the sample measures 50.020 mm in length, 30.029 mm in width, and 9.907 mm in thickness. This silicon, produced using the float-zone method, has an electrical resistivity of approximately 30-40 kΩ·cm. The entrance side, excluding the slit, is coated with a gold film, while the exit side features gold inductive metal meshes with a pitch of 27.5 μm and an inner opening width of 14.8 μm.

We utilized a cooled THz Quantum Cascade Laser (QCL) coupled with a room-temperature pyroelectric detector to capture the transmitted signal through the VIPA. We tuned the QCL wavelength around 115.7 μm by modifying the operating voltage.

Fig. 2 provides a simplified schematic of the testbed's main components. The QCL is housed in a separate cryostat, maintaining a stable temperature of approximately 52 by using a liquid nitrogen bath. To produce a clean beam, the QCL laser

[1]Department of Applied and Engineering Physics, Cornell University, Ithaca, NY 14853, USA; [2]Cornell Center for Astrophysics and Planetary Sciences, Cornell University, Ithaca, NY 14853, USA; [3]NASA Goddard Space Flight Center, Greenbelt, MD 20771, USA; [4]SRON Netherlands Institute for Space Research, PO Box 800, NL-9700 AV Groningen, the Netherlands; [5]LongWave Photonics, 958 San Leandro Ave, Mountain View, CA 94043, USA; [6]Department of Astronomy, Cornell University, Ithaca, NY 14853, USA. *Corresponding author (email: bz332@cornell.edu).

NOTES:

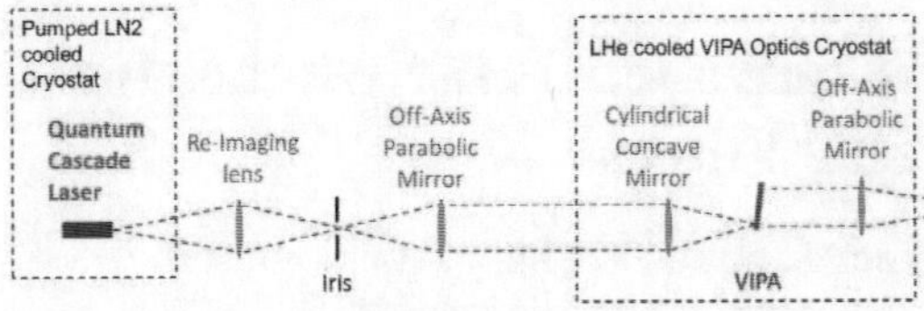

Fig. 2. Simplified schematic of the main components of the testbed

is refocused through a 2 mm diameter iris diaphragm, forming a spatial filter. An off-axis parabolic (OAP) mirror then captures the beam and redirects it as a collimated beam into the liquid helium-cooled cryostat containing the VIPA. Initially, the VIPA stabilized at 26 K due to the thermal background radiation load from room temperature. To reduce this load, subsequent measurements utilized two cooled quartz windows and scatter filters at the entrance and exit of the VIPA cryostat, achieving a VIPA temperature of 4 K. Incoming radiation, upon entering the cryostat, is deflected by a flat mirror and focused onto the VIPA's exit side by a cylindrical concave mirror. The beam is focused on the VIPA's back surface for better radiation coupling to the resonator space. Inside the VIPA, the radiation exits as a system of collimated beams, which are collected by an OAP mirror and directed to a focal plane via two flat mirrors. The radiation then exits the VIPA cryostat and is detected by a single-pixel pyroelectric detector. To record the VIPA's spectral transmission profile, the detector is translated along the dispersion direction in the focal plane.

III. Results

The VIPA was operated at a stable temperature of 4 K. The dispersion measurements obtained from the VIPA are depicted in Fig. 3 as data points with error bars. These measurements are delineated into three discrete sets of data points, each distinguished by distinct colors and acquired at different QCL wavelength settings. The measurements have been normalized to unity to facilitate more effective comparison. The observed wavelength shift of the three recorded spectral peaks in the focal plane aligns with the anticipated VIPA output and QCL tuning.

The solid curves in Fig. 3 represent the normalized model calculations for the three distinct wavelengths. The plot illustrates an impressive consistency between the model predictions and the observed data, with the full width at half maximum (FWHM) measurements differing from the model predictions by less than 10%. The FWHM measurements for the peaks are approximately 7 nm at 115.722 μm and 115.714 μm, and around 8 nm at 115.697 μm. These values correspond to resolving powers of 14,500 and 16,500, and hence, finesses of 25 and 28.5, respectively.

During the optical alignment process, we also measured a single wavelength at 300 K. The resonating peak's resolving

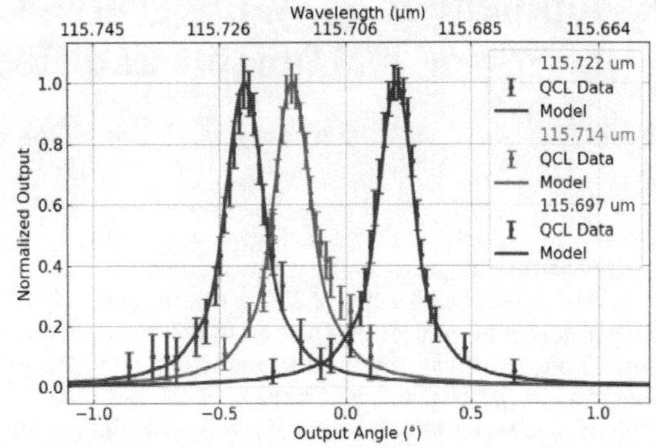

Fig. 3. Measured VIPA transmission profile and corresponding model calculations at three different QCL wavelengths. The measured data is shown as dots with errorbars, and the simulations are shown as solid lines.

power was found to be 5800, with an approximate finesse of 10. As expected, these values are considerably lower than those observed at 4 K, though the signal remains detectable at room temperature.

Differences among the spectral profiles and minor discrepancies in the side wings between measurements and model calculations are likely caused by vignetting in the optical system, non-uniform slit illumination, or stray light. Further investigation into the experimental setup or refining the modeling parameters could provide deeper insights into these small discrepancies.

IV. Conclusions

In this study we presented the tests of a far-infrared VIPA and demonstrated its spectral resolving power of approximately 16,500 at 115.7 μm, 4 K. Our findings establish a robust foundation for future far-infrared VIPA devices. This innovative technology holds potential for diverse scientific applications in velocity-resolved spectroscopy and represents a valuable enhancement to current high-resolution far-infrared astronomical instruments.

References

[1] M. Shirasaki, "Large angular dispersion by a virtually imaged phased array and its application to a wavelength demultiplexer," *Optics Letters*, vol. 21, no. 5, p. 366, Mar. 1996
[2] Scholten S K, Anstie J D, Hébert N B, et al. Complex direct comb spectroscopy with a virtually imaged phased array[J]. Optics Letters, 2016, 41(6): 1277-1280.
[3] Edrei E, Gather M C, Scarcelli G. Integration of spectral coronagraphy within VIPA-based spectrometers for high extinction Brillouin imaging[J]. Optics express, 2017, 25(6): 6895-6903.
[4] Wollack E J, Cataldo G, Miller K H, et al. Infrared properties of high-purity silicon[J]. Optics Letters, 2020, 45(17): 4935-4938.

NOTES:

Development of Fully-Integrated Optically-Controlled THz Switches for Tunable and Reconfigurable Filters

Peizhao Li[1*], Yu Shi[1], Yijing Deng[1], Patrick Fay[1], and Lei Liu[1]

***Abstract*—We report our recent progress toward the development of fully-integrated optically-controlled switches for tunable and reconfigurable terahertz (THz) filters. The switches were designed based on optical modulation of photoconductivity using photo-induced free carriers in semiconductors. For different application scenarios, switches employing two different configurations, i.e., series and shunt CPW, were investigated and initially demonstrated. Encouraged by the promising results including low insertion loss (e.g., 0.3 dB), high isolation (e.g., 70 dB), and broadband operation, fully-integrated G-band switches have been designed and fabricated using Si-on-Sapphire (SoS) wafers for the development of tunable and reconfigurable filters in future THz sensing, imaging and communication systems. The theory, design, simulation, prototype demonstration and fabrication of the proposed switches are presented. On-wafer probe testing of the fully-integrated switches will soon be performed. Finally, a dual-band reconfigurable THz notch filter is initially designed and simulated to highlight the promising applications of the switches.**

***Keywords*—Optically-controlled switches, terahertz filters.**

I. INTRODUCTION

THE THz spectrum has gained increasing interest due to its promising applications in medical imaging, biological sensing, and communications [1]. As a consequence, tunable and reconfigurable THz circuits such as reconfigurable filters have become more desired in advanced THz sensing, imaging and beyond 5G adaptive communications [2]. High performance RF switches operating in the THz region are one of the crucial elements for realizing such circuits with maximum tunability, reconfigurability and multi-functionality [3]. However, conventional RF switches relying on solid-state devices or phase-change materials suffer significant performance degradation as frequencies transition into the mmW-THz range [4]. MEMS-based RF switches designed for THz frequencies thus far have been bulky and challenging to be integrated into sophisticated systems for achieving more advanced circuit functionalities, despite their complex fabrication and potential reliability issues [5].

To address the aforementioned issues, we proposed to develop high performance THz switches based on optical modulation of photoconductivity using photoinduced free carriers in semiconductors. Fig. 1 shows the conceptual switch design and operation mechanism, taking the series configuration in a CPW with interdigital structure as an example. Different from conventional switch designs, an insulation layer is inserted between the gold contact and the active layer using Si or Ge to form a non-contact capacitive coupling architecture. With light illumination, high carrier concentration in the active layer can be maintained and hence high photoconductivity can be achieved, leading to RF switches with extremely low insertion loss (IL) and superior isolation (Iso) in the THz region.

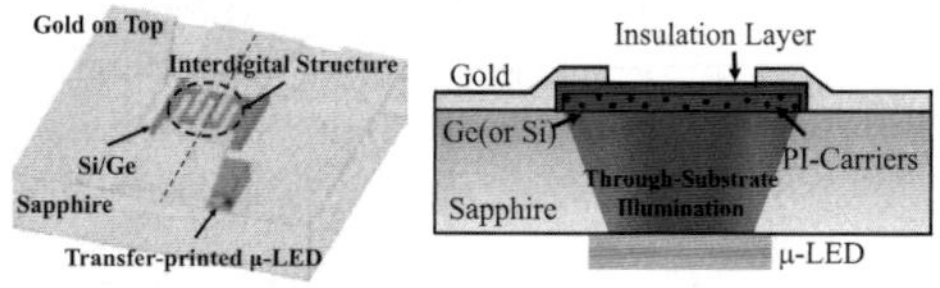

(a) (b)

Fig. 1. The concept of optically-controlled THz switches with (a) transfer-printed μ-LED and (b) its cross-section view along the dashed line.

II. MODELING, SIMULATION AND PROTOTYPE DEMONSTRATION

On the basis of our physics-based model on photoconductivity modulation [6], two types of THz switches with series and shunt configurations, respectively, were designed and simulated using the proposed approach. Fig. 2 illustrates the simulated frequency responses of the proposed fully-integrated optically-controlled THz switches. As depicted, the switch design using series configuration demonstrates an IL of 0.6 dB and Iso of ~35 dB with Ge, while the one with shunt configuration exhibits an IL of 0.3 dB, and Iso of >70 dB. For a proof-of-concept demonstration, D-band and G-band switches assembled using 73 μm-thick Si chips on top of CPW lines were characterized, and the results show promising performance [7]. Specifically, a measured IL of 0.4 dB and Iso of 32 dB have been achieved at 170 GHz (using a light intensity of ~40 W/cm^2) as shown in Fig. 2 (b).

III. FABRICATION

Motivated by the promising results shown above, fully-integrated G-band switches using the proposed approach have

[1]University of Notre Dame, Notre Dame, 46556, US; [*]Corresponding author (email: pli8@nd.edu).

NOTES:

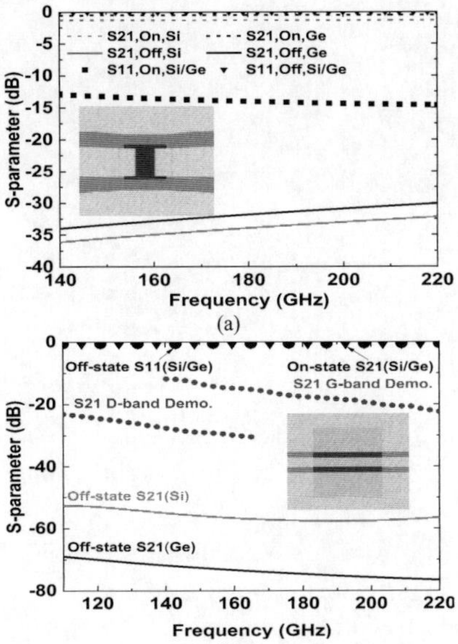

Fig. 2. Simulated and measured frequency responses of the proposed THz switches for (a) series configuration and (b) shunt configuration (with measured data from proof-of-concept demonstration).

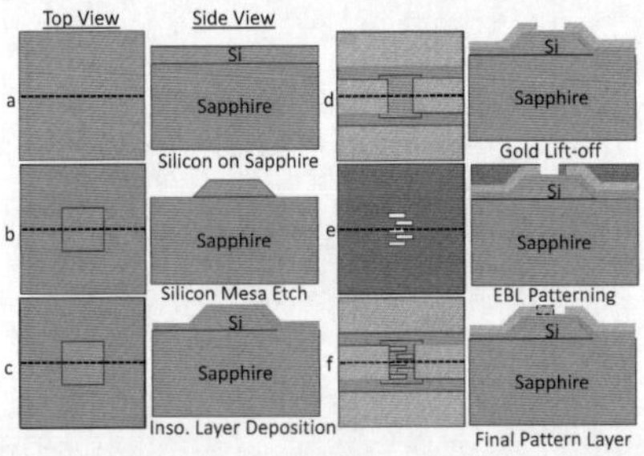

Fig. 3. Process flow diagram for fabricating the proposed optically-controlled THz switches using lift-off on SoS substrates.

been designed and fabricated using Si-on-Sapphire (SoS) substrates for more advanced tunable/reconfigurable THz circuits. The process flow is shown in Fig. 3. The process involves a combination of traditional photolithography and electron-beam lithography (EBL, optional for submicron-scale interdigital structures) followed by lift-off. The initial fabrication results are shown in Fig. 4, and the switches will soon be characterized with on-wafer probe testing.

IV. Applications for Advanced THz Filter

To exploit the potential of this novel switch technology for realizing more advanced tunable/reconfigurable THz circuits, an optically-controlled dual-band reconfigurable THz notch filter based on WR4.3 waveguide platform has been designed and simulated as shown in Fig. 5 (a). The results show that the center frequency of the bandstop filter can be reconfigured between 200 GHz and 230 GHz by individually turning on/off the split-ring resonator groups A/B (see Fig. 5 (b)). In addition, the in-band suppression is greater than 40 dB. The filter will soon be implemented for measurement.

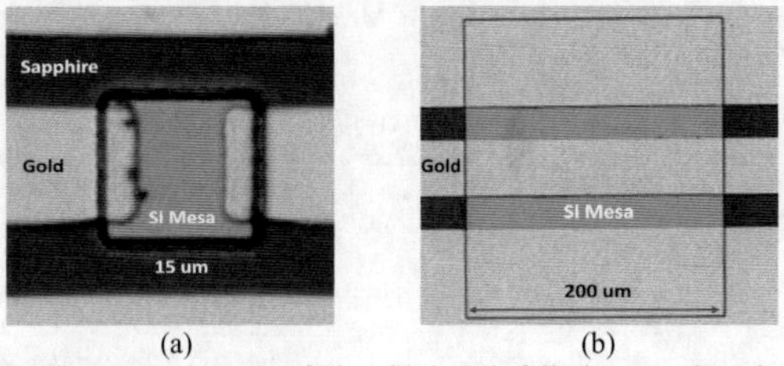

Fig. 4. Microscope images of the fabricated fully-integrated optically-controlled THz switches with (a) series-configuration and (b) shunt-configuration.

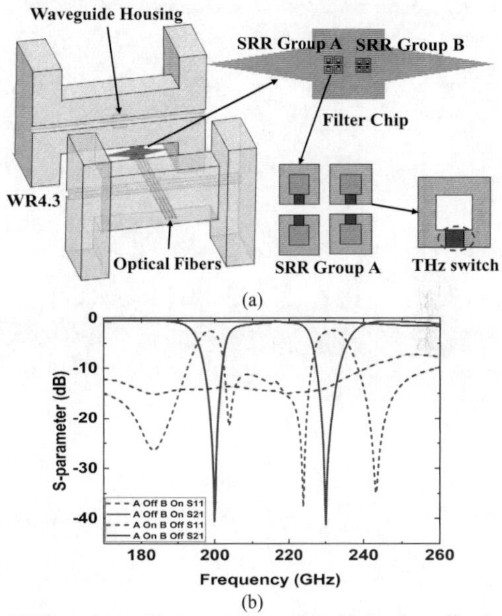

Fig. 5. (a) The concept of the proposed reconfigurable bandstop filter based on WR4.3 waveguide platform and (b) its frequency response.

References

[1] M. Tonouchi, "Cutting-edge terahertz technology," *Nature Photon.*, vol. 1, pp. 97–105, 2007.

[2] V. Sanphuang, N. Ghalichechian, N. K. Nahar, and J. L. Volakis, "Reconfigurable THz filters using phase-change material and integrated heater," *IEEE Trans. THz. Sci. Technol.*, vol. 6, no. 4, pp. 583–591, Jul. 2016.

NOTES:

[3] G. Rebeiz, et al., "Tuning in to RF MEMS," *IEEE Microw*. Mag., vol. 10, no. 6, pp. 55–72, Oct. 2009.

[4] H. Madan et al., "26.5 Terahertz electrically triggered RF switch on epitaxial VO2-on-Sapphire (VOS) wafer," *2015 IEEE International Electron Devices Meeting (IEDM)*, Washington, DC, 2015, 9.3.1-9.3.4.

[5] R. R. Mansour, "RF MEMS-CMOS Device Integration: An Overview of the Potential for RF Researchers," in *IEEE Microw. Mag*., vol. 14, no. 1, 2013. 39-56.

[6] Y. Shi, Y. Deng, J. Ren, P. Li, P. Fay, and L. Liu, "Computational analysis of novel high performance optically controlled RF switches for reconfigurable millimeterwave-to-THz circuits," *OSA Continuum*, vol. 4, no. 10, pp. 2642-2654, Oct 2021.

[7] P. Li, W. Wu, Y. Shi, Y. Deng, P. Fay and L. Liu, "Broadband THz switching with extremely low insertion loss and superior isolation," 2023 *IEEE/MTT-S International Microwave Symposium - IMS 2023*, San Diego, CA, USA, 2023, pp. 688-691.

NOTES:

SESSION 3

Design and Simulation of a Ultra-Wideband 211-375 GHz SIS Mixer based on a Micromachined Metallic Substrate

C. López, V. Desmaris, D. Meledin, A. Pavolotsky and V. Belitsky

Abstract—**In this paper, we describe the design and simulation of a novel ultra-wideband SIS mixer for radio astronomy applications, specifically to cover the ALMA telescope's bands 6 and 7 corresponding to an extensive 56% fractional bandwidth. The design features an electroplated metallic substrate that integrates a finline waveguide-to-substrate transition. Furthermore, the design employs a twin-junction SIS configuration with an Al/AlN barrier. The simulation of the RF matching circuit predicted a 93% coupling efficiency across most of the band. Additionally, this paper details the development and simulation of an integrated IF output circuit, optimized for the 4-16 GHz band. The simulation of the IF output circuit shows the losses in the IF coupling are better than 0.3 dB for most of the band and a reflection coefficient for a 50 Ω output impedance of -15 dB.**

Keywords—**SIS mixer, Broadband, Finline, Simulation.**

I. Introduction

IN the ever-evolving field of mm and sub-mm radio astronomy instrumentation, the technological development of wideband low-noise receivers has been the main focus of research over the last decades. Specifically, the ambitious goals set for the Atacama Large Millimeter/submillimeter Array (ALMA) in the 2030 roadmap [1], striving for a threefold improvement of the IF band, imply a de-facto need for receivers that can cover multiple frequency bands simultaneously, such as bands 6 (211–275 GHz) and 7 (275–373 GHz), i.e., a total a 55.5% fractional bandwidth. The mixers employed at these frequencies rely on superconductor-insulator-superconductor (SIS) technology due to its outstanding sensitivity at millimeter and sub-millimeter-waves [2][3]. Historically, designing SIS mixers, local oscillator sources (LO), and RF components [4-9] with a fractional bandwidth greater than 44% has been challenging. Although various examples in the literature demonstrate SIS mixers exceeding 40% fractional bandwidth [10-13], only a few have surpassed 55% fractional bandwidth, e.g. [14]. The majority of such mixers are based on traditional designs that employ an E-probe waveguide-to-substrate transition [3]. While these designs remain effective for smaller fractional bandwidths, they face limitations for larger bandwidths. Specifically, they require precise machining of waveguide backshorts, reduced height waveguides that increase RF losses [14], or the addition of waveguide capacitive elements in the input waveguide to increase the bandwidth [13]. Additionally, the tolerance for mounting the mixer chips and thus E-probe is a critical parameter for optimal mixer performance.

This paper addresses the design of a wideband SIS mixer in the 211-375 GHz range, i.e., targeting a 56% fractional bandwidth. Our mixer design does not employ a conventional dielectric substrate; instead, the Nb-Al/AlN-Nb SIS junctions [15] and RF matching circuitry are supported by an electroplated metallic substrate. This approach offers multiple advantages over dielectric substrates, e.g., avoiding the excitation of unwanted substrate modes, reducing dielectric losses, and naturally achieving chip grounding, while allowing substrate shaping. The latter facilitates the integration of a metallic finline seamlessly with the substrate. Furthermore, the absence of a dielectric substrate between the fins eases the matching over a wide bandwidth [4]. In addition, finline structures do not use waveguide shorts and demonstrate greater tolerance for the precision of mounting with respect to positioning in the waveguide, simplifying block fabrication requirements. Moreover, our design partially integrates the Intermediate Frequency (IF) output circuit for the 4-16 GHz band, including an extraction pad shaped as a landing capacitor that functions as an IF circuitry tuning component. This integration eliminates the need for additional lumped capacitors, simplifying the IF extraction process.

II. Mixer Chip Overview

The proposed broadband SIS mixer utilizes a metallic finline waveguide-to-substrate transition [4], seamlessly integrated into the substrate. The finline structure is positioned on the split-block plane of a 760 μm x 380 μm rectangular waveguide, as depicted in Fig. 1a. The metallic finline [16], matches the high impedance of the rectangular waveguide to a 100-ohm slotline. However, this impedance is still too high to properly

The authors are with the Group for Advanced Receiver Development (GARD), Department of Space, Earth and Environment, Chalmers University of Technology, Gothenburg, Sweden

NOTES:

match the SIS's low RF impedance, which is close to 11 ohms at the center frequency of 293.5 GHz. Therefore, further impedance transformation is necessary. Fig. 1b illustrates the employed transmission lines in the RF matching network. Slotlines 1 and 2 reduce the impedance from 100 Ohm to 60 Ohm, better suited for a slotline to microstrip transition. The silicon layer, with its high dielectric constant ($\varepsilon r \sim 11.3$) [17], is retained in this chip area to help lower the slotline impedance, thus facilitating the transition to a microstrip. The slotline-microstrip transition uses a 3rd-order Marchand Balun [5], formed by slotlines (3) and (4) and microstrips (5) and (6), as shown in Fig. 1b. Additionally, microstrips (6) and (7) create a 2-step Chebyshev transformer that completes the RF impedance transformation, enabling a broadband impedance match to the SIS junctions where the mixing process occurs. The mixer design, shown in Fig. 1a, employs twin SIS junctions to significantly ease the matching of the RF imaginary part impedance. The IF circuit, integrated into the same chip, is displayed in Figure 1b. The IF signal is extracted from the Marchand Balun transition using an RF filter that presents a high impedance to the RF signal, utilizing suspended microstrip lines over substrate cavities. The IF bandwidth is designed to be 4-16 GHz.

The mixer's various layers are detailed in Fig. 1c. As previously stated, the dielectric substrate is not entirely removed but remains in two areas of the chip: over the slotline-to-microstrip transition and at the suspended microstrip lines of the IF output. The 30 μm thick silicon layer is preserved to assist in impedance matching, provide structural support for the suspended microstrip, and enhance design robustness. Additionally, the chip incorporates two SiO_2 areas with different thicknesses. The first area, 250 nm thick and depicted in green, serves as a dielectric layer for the microstrip inverter of the SIS twin junction [18] configuration and a single section of impedance transformation. The second SiO_2 area, depicted in red and 650 nm thick, is used in both the Marchand balun and RF filter areas. This thicker SiO_2 layer reduces the overall capacitance seen by the IF system, thereby achieving a wider IF band. The electroplated metallic finline, made of 30 μm thick electroplated metal covered by 300 nm of Nb, acts as a counter electrode for the microstrip line. The finline chip's total length is 2420 μm, as shown in Fig. 1c.

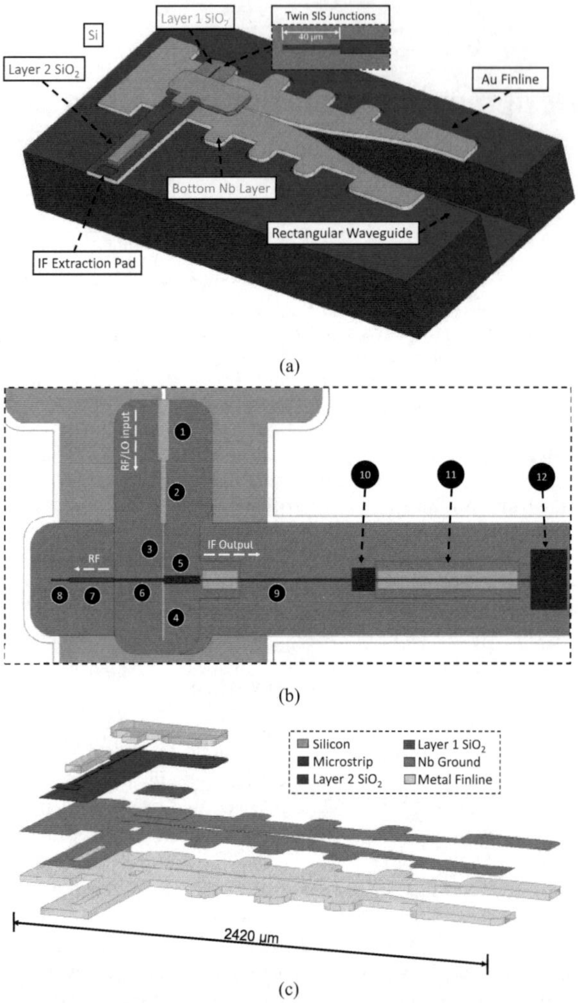

Fig. 1. Proposed broadband finline Mixer. (a) Mixer chip mounted in the split block. The rectangular waveguide is 760 μm x 380 μm. Detail of the different materials employed for the mixer chip fabrication. The twin junction configuration consist on 2 SIS junctions separated by a section of microstrip of 40 μm long and 4 μm width that serves as impedance inverter. (b) Top view of the slotline to microstrip transition and the integrated IF circuitry. For clarity the Si layer and the second SiO_2 layer are transparent. (c) Exploted view of the mixer chip that detail the layer layout and the different materials. The overall chip length is 2420 um.

III. RF Design

The proposed design utilizes a twin SIS junctions configuration [18] with a targeted RnA product of 14 Ohm·μm² and a nominal area of 1 μm². The twin junction configuration offers a broader bandwidth compared to a single junction, as the imaginary impedance of each junction can be offset by an impedance inverter. In our design, this inverter is comprised of a microstrip measuring 40 μm in length and 4 μm in width, placed over a 250 nm layer of SiO2. Additionally, the design incorporates SIS technology with an AlN tunnel barrier, offering lower capacitance at a given current density than Al-AlO_x tunnel barrier junctions [19], crucial for a wide IF response. The specific junction capacitance, calculated from

NOTES:

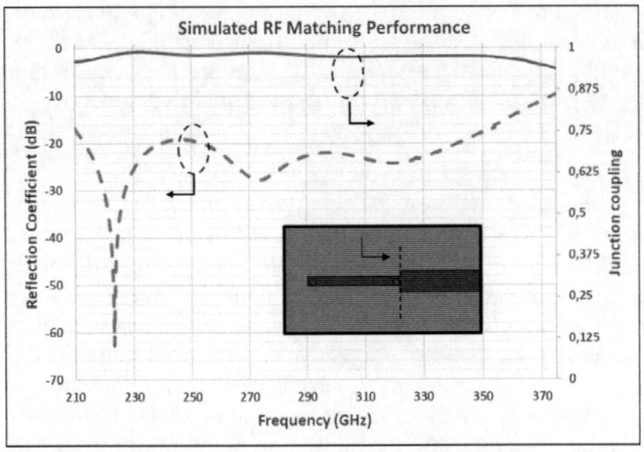

Fig. 2. Simulated performance of the RF matching circuit. For the HFSS simulation a lumped port was located as illustrated by the miniature. The impedance values for the twin junction circuit where calculated and load in the lumped port.

[20], is Cs=64.4 fF/μm². For the chosen area of 1 μm², the total capacitance from the twin junction amounts to 128.8 fF/μm². With the knowledge of Rn and the SIS capacitances, the RF admittance of the SIS junction is calculated using the quantum theory of mixing as outlined in [21]. According to this theory, all higher harmonics of the local oscillator signal are short-circuited by the junction capacitance, leading to the predominance of junction capacitance over the RF susceptance of the SIS. Consequently, the calculation of the quantum conductance provides a comprehensive understanding of the SIS RF impedance for a DSB mixer operation. At the central frequency of the mixer, i.e., 293.5 GHz, the RF resistance for a single SIS is approximately 11.6 Ohm, reducing to half if there's perfect cancellation of the imaginary part. Using a single-section transformer to match this low impedance to the 45 Ohms output of the Marchand balun is insufficient for a broadband performance. Therefore, a 2-section Chebyshev transformer, consisting of lines (6) and (7) as shown in Fig. 1b, was utilized. However, this approach has the drawback of increasing the capacitance seen at IF, effectively limiting the maximum IF bandwidth. To mitigate this issue, the first stage of the transformer is implemented with a thicker 650 nm layer of SiO_2, effectively lowering the total capacitance for the IF circuit. It's important to note that the differences in SiO_2 thickness do not create a vertical step in the superconducting Nb microstrip, as such a step could lead to electromagnetic discontinuities and potentially introduce an unwanted superconducting weak link affecting overall performance. This issue is circumvented as the device is fabricated starting from the silicon layer and concluding with the electroplating of the metallic finline.

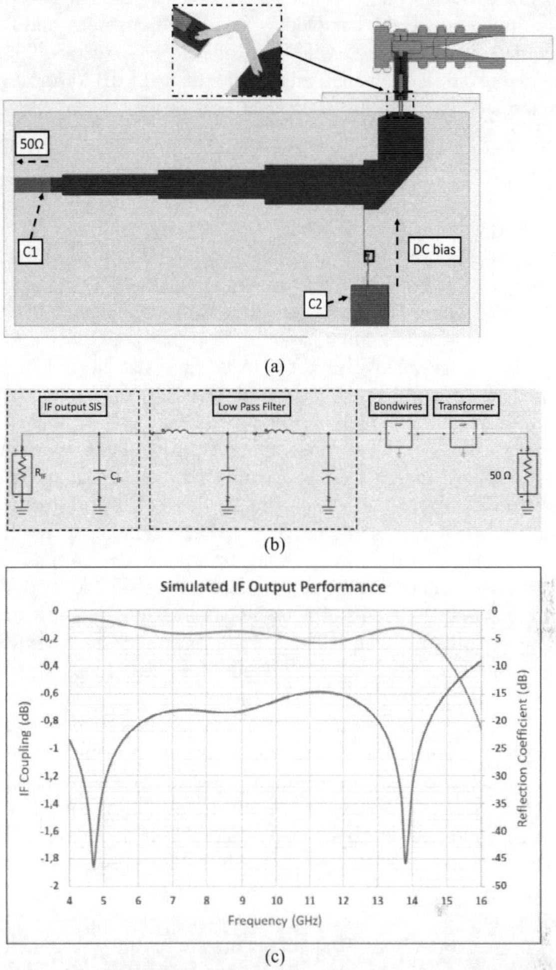

Fig. 3. IF output . (a) IF layout. The IF signal is extracted from the mixer chip with the help a 2 paralel bondwires. A 20 Ω to 50Ω impedance tranformer with integrated Bias T is employed. (b) Schematic of the IF circuitry. (c) Simulated performance for the IF Output circuit. to avoid cluttering.

The simulation of the mixer was conducted in the full wave 3D simulator Ansys HFSS (high-frequency structure simulator) in two phases. Initially, the RF impedance for a single SIS was calculated, followed by a simulation to determine the optimal length and width for the twin-section inverter. Upon identifying these parameters, the input complex impedance of the twin junction for the RF band 211-375 GHz was calculated. These values were then integrated into HFSS using a lumped port at microstrip (7), as depicted in Fig. 2. The model was

NOTES:

subsequently fine-tuned to maximize both the coupling to the twin junction and the bandwidth. Fig. 2 demonstrates that the coupling level to the twin junction circuit exceeds 93%, indicating a reflection coefficient better than -15 dB for most of the targeted bandwidth.

IV. IF Output Circuit Design

The IF output circuit is partially integrated into the mixer chip and was designed to cover the 4-16 GHz band. In Fig. 1b, lines 9 to 12 form an RF/LO filter. Additionally, the IF circuit employs a 20 Ω to 50 Ω transformer, as illustrated in Fig 2.

To calculate the IF response, both the IF impedance of the mixer chip (parallel combination of real and imaginary parts, R_{IF} and C_{IF}, respectively) needs to be known. The IF output impedance of an SIS junction is determined by the slope of the pumped IV curve [21], typically 8-10 times the Rn values. To prevent excessive conversion gain and ensure stable operation of the mixer, the IF load impedance was set 20 Ω, using a transformer from 50 Ω. Meanwhile, C_{IF} was calculated from the junction capacitances and the geometric capacitance of the RF tuning circuitry, resulting in a total capacitance of 376 fF.

The twin junction configuration lacks a natural "cold point" for extracting IF signals. This issue can be addressed by using an LC circuitry realized as a high-impedance line with a landing capacitor in conjunction with bondwires [22]. In our mixer design, the IF signal extraction uses a mix of high and low-impedance lines to form a low-pass filter that effectively rejects the RF/LO signal. The first section of this filter is marked as 9 in Fig. 1b. The electroplated metal's flexibility allows for the creation of small cavities in the substrate to form suspended microstrip lines with high impedance to the RF/LO signal. The filter is completed by a low-impedance line (10), a second section of suspended microstrip line (11), and the IF contact/ extraction pad (12). It's important to note that the filter is situated on top of the 650 nm SiO_2 area to minimize the added capacitance to the IF output. The structure acts as an LC filter for the IF signal, as shown in Fig. 2b. The output performance is influenced by the inductance of the bondwires used in the IF extraction circuitry, necessitating a detailed 3D simulation including the bondwires for accurate performance prediction. Two parallel bondwires, as seen in Fig. 2a, were used to decrease the inductance added to the output LC filter.

The IF extraction is completed with a multisection superconducting transformer on a 254 μm thick alumina substrate. This transformer, shown in Fig. 2a, incorporates a bias T consisting of a DC blocking capacitor (C1) and an IF shunting capacitor (C2). The latter works in tandem with a spiral inductor and a high-impedance line as a stop filter for the IF signal.

The simulated performance of the complete IF output circuit is shown in Fig. 2c. The graph indicates that the losses in the IF coupling is below 0.3 dB for most of the band, and the reflection coefficient is below -15 dB for the majority of the IF band.

V. Conclusion

In this paper, we present a novel broadband SIS mixer design targeting the frequency range of the ALMA telescope's current bands 6 and 7, i.e., 211-375 GHz. While existing SIS mixer designs employing traditional transitions and substrates have achieved significant fractional bandwidths, the research into alternative wideband solutions remains a critical topic for the future of radio astronomy receivers. The design presented in this paper not only addresses current limitations but also opens new opportunities for the development of prospective SIS mixer technology. The presented mixer chip represents a significant shift from traditional design approaches by eliminating the use of a dielectric substrate and replacing it with an electroplated metallic substrate, facilitating the integration of a metallic finline. The suggested design and simulations of RF matching circuits for a twin junction SIS configuration, achieve a simulated coupling of over 93% for most of the frequency band. Additionally, we detail the design of the IF output circuit, which is directly integrated onto the mixer chip. The IF performance, simulated for the 4-16 GHz band, shows a coupling loss better than 0.3 dB for most of the IF bandwidth.

References

[1] Carpenter, J., Iono, D., Kemper, F., & Wootten, A. (2020). The ALMA development program: Roadmap to 2030. *arXiv preprint arXiv:2001.11076*.

[2] V., Belitsky et al. "Alma band 5 receiver cartridge-design, performance, and commissioning," Astron. & Astrophys., vol. 611, p. A98,2018.

[3] D. Meledin, I. Lapkin, M. Fredrixon, et al., "SEPIA345: A 345 GHz dual polarization heterodyne receiver channel for SEPIA at the APEX telescope," Astronomy & Astrophysics, vol. 668, A2, 2022.

[4] C. López, V. Desmaris, D. Meledin, et al., "Waveguide-to-substrate transition based on unilateral substrateless finline structure: Design, fabrication, and characterization," IEEE Transactions on Terahertz Science and Technology, vol. 10, no. 6, pp. 668-676, 2020.

[5] C. D. López, M. A. Mebarki, V. Desmaris, et al., "Wideband Slotline-to-Microstrip Transition for 210–375 GHz Based on Marchand Baluns," IEEE Transactions on Terahertz Science and Technology, vol. 12, no. 3, pp. 307-316, 2022.

[6] C. D. López, D. Montofré, V. Desmaris, et al., "Ultra-Wideband 90° Waveguide Twist for THz Applications," IEEE Transactions on Terahertz Science and Technology, vol. 13, no. 1, pp. 67-73, 2022.

[7] I. Lapkin, C. López, M. Fredrixon, et al., "Vacuum-Seal Waveguide Feedthrough for Extended W-Band 67–116 GHz," IEEE Journal of Microwaves, 2023.

[8] C. Lopez, V. Desmaris, D. Meledin, et al., "Design and implementation of a compact 90° waveguide twist with machining tolerant layout," IEEE Microwave and Wireless Components Letters, vol. 30, no. 8, pp. 741-744, 2020.

NOTES:

[9] A. Gouda, C. D. López, V. Desmaris, et al., "Millimeter-wave wideband waveguide power divider with improved isolation between output ports," IEEE Transactions on Terahertz Science and Technology, vol. 11, no. 4, pp. 408-416, 2021.
[10] J. Y. Chenu et al., "The front-end of the NOEMA interferometer," IEEE Trans. Terahertz Sci. Technol., vol. 6, no. 2, pp. 223–237, Mar. 2016.
[11] J. S. Ward, K. A. Lee, J. Kawamura, et al., "Sensitive broadband SIS receivers for microwave limb sounding," Proc. 33rd Int. Conf. Infrared, Millim. Terahertz Waves, Pasadena, CA, USA, 2008, pp. 1–2.
[12] J. W. Kooi et al., "Performance of the caltech submillimeter observatory dual-color 180–720 GHz balanced SIS receivers," IEEE Trans. Terahertz Sci. Technol., vol. 4, no. 2, pp. 149–164, Mar. 2014.
[13] J. W. Kooi et al., "A 275–425-GHz tunerless waveguide receiver based on AlN-barrier SIS technology," IEEE Trans. Microw. Theory Techn., vol. 55, no. 10, pp. 2086–2096, Oct. 2007.
[14] T. Kojima et al., "275–500-GHz Wideband Waveguide SIS Mixers," IEEE Transactions on Terahertz Science and Technology, vol. 8, no. 6, pp. 638-646, Nov. 2018.
[15] A. Pavolotsky, V. Desmaris, V. Belitsky, "Nb/Al-AlN/Nb superconducting tunnel junctions: fabrication process and characterization results," 29th IEEE International Symposium on Space THz Technology (ISSTT2018), 2018, pp. 107-111.
[16] C. Lopez, V. Desmaris, D. Meledin, et al., "Design and Fabrication of All-metal Micromachined Finline Structures for Millimeter and Sub-millimeter Applications," 32th IEEE International Symposium on Space THz Technology (ISSTT2022), 2022.
[17] J. Krupka, J. Breeze, A. Centeno, et al., "Measurements of permittivity, dielectric loss tangent, and resistivity of float-zone silicon at microwave frequencies," IEEE Transactions on microwave theory and techniques, vol. 54, no. 11, pp. 3995-4001, 2006.
[18] V. Y. Belitsky, S. W. Jacobsson, L. V. Filippenko, E. L. Kollberg, "Broadband twin-junction tuning circuit for submillimeter SIS mixers," Microwave and Optical Technology Letters, vol. 10, no. 2, pp. 74-78, 1995.
[19] P. Y. Aghdam, H. Rashid, A. Pavolotsky, et al., "Specific Capacitance Dependence on the Specific Resistance in Nb/Al–AlOx/Nb Tunnel Junctions," IEEE Transactions on Terahertz Science and Technology, vol. 7, no. 5, pp. 586-592, 2017.
[20] A. Pavolotsky, C. D. López, I. V. Tidekrans, et al., "Specific capacitance of Nb/Al-AlN/Nb superconducting tunnel junctions," ISSTT 2019-30th International Symposium on Space Terahertz Technology, Proceedings Book, 2019, pp. 92-94.
[21] J. R. Tucker, M. J. Feldman, "Quantum detection at millimeter wavelengths," Reviews of Modern Physics, vol. 57, no. 4, pp. 1055, 1985.
[22] B. Billade, A. Pavolotsky, V. Belitsky, "An SIS Mixer With 2hf/k DSB Noise Temperature at 163-211 GHz Band," IEEE Transactions on Terahertz Science and Technology, vol. 3, no. 4, pp. 416-421, 2013.

NOTES:

Improved Process Flow of Heterogeneously Integrated Gallium Arsenide Schottky Diodes

Christopher Moore[1*], Michael Cyberey[2], Dustin Widmann, Richard Zhou[3],
Robert Weikle[4] Scott Barker[4] Arthur Lictenberger[4]

***Abstract*— An improved process flow for fabricating heterogeneously integrated gallium arsenide Schottky diodes on membranes of 15 µm thick high-resistivity silicon microstrip substrate is reported. This effort has three focuses: 1) optimizing the fabrication of devices with high packing density to maximize devices per run, 2) co-fabrication of beam leads and side wall-Au plated vias, and 3) simplifying the process by reducing the number of processing steps.**

***Keywords*—GaAs Schottky Diode fabrication, microstrip, vias, beam leads, Silicon on Insulator, SOI**

I. INTRODUCTION

High-frequency radars, spectrometers, and radiometers employ frequency multiplication chains for local oscillators pumping mixers to receive high-frequency signals [1]. These multiplication chains usually comprise a cascade of doublers, triplers, [2]and mixers implemented in separate waveguide blocks. This modular packaging leads to larger system sizes. The heterogeneous integration of devices onto silicon can significantly miniaturize these systems by implementing cascaded multipliers and mixers on a common substrate, resulting in compact sub-mm wave electronics. The main goal for the resulting chips is to improve thermal dissipation, efficiency, and power output, resulting in smaller, more efficient systems with more output power.

Progress has been reported on the heterogeneous integration of these components onto a common chip; however, the complicated fabrication process limited yields. A new process has been developed utilizing a microstrip substrate with side wall-Au plated vias co-fabricated with beam leads. This new process results in better yields, thermal management, and power handling that may eliminate the need for additional power amplification and cooling systems, making it of particular interest to the space industry and manufacturers of consumer-grade electronics.

The diodes presented in this work were fabricated on a 15 µm thick high resistivity silicon microstrip substrates, with side wall-Au plated vias, and beam leads. This preliminary work is to verify that this process flow can be standardized and used to create highly integrated components on a single chip. This paper reports an enhanced process flow adapted from Nadri's and Xie's work in integrating GaAs Schottky diodes onto thin silicon [3], [4].

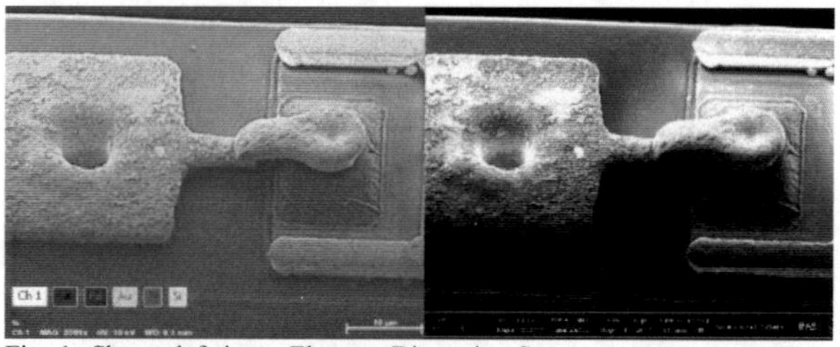

Fig. 1. Shown left is an Electron Dispersive Spectroscopy measurement highlighting the different material sections of the diode structure, asernide is hidden to provide contrast. Shown right is a Scanning Electron Microscopy measurement of the same diode. Both have the same structure, which shows the side wall-Au plated vias and the Schottky Diode.

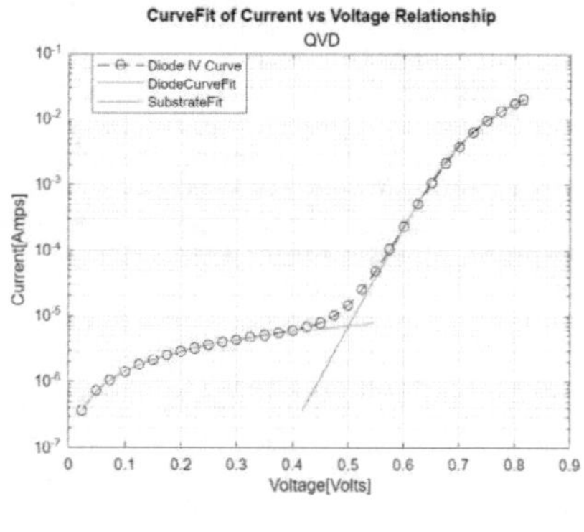

Fig. 2. The diode current-voltage relationship shows approximately Rs=4.6, Ideality=1.13, and Isat=1.4e-13. The ideal series resistance is 1.54 ohms. Measurement was taken with a 2pt probe. Curve fitting was done in Matlab via a nonlinear least square fitting. R-square=0.998 Root Mean Squared Error=.0045

[1]University of Virginia, Charlottesville, 22904, USA.
[*]Christopher Moore (email: cm9ee@virginia.edu).

NOTES:

II. FABRICATION

The fabrication process employed to create these chips involves a unique epitaxy transfer method for bonding GaAs material to a silicon-on-insulator (SOI) substrate to create Schottky diodes [4]. A GaAs epitaxy on a semi-insulating 650 μm GaAs handle featuring an AlGaAs etch stop layer, n-GaAs, n+-GaAs device layers, and a 90 nm highly-doped ($>10^{19}$ cm^{-3}) graded $In_xGa_{1-x}As$ cap layer (x:0→0.6) was used for the diode.The procedure used to create the devices is adapted from the process described in reference [5]. The process is started by first depositing a titanium, palladium, gold, and titanium (Ti/Pd/Au/Ti) (20/40/150/20 nm) ohmic metal layer onto a GaAs wafer with electron beam physical vapor deposition. The 15 μm thick high resistivity SOI wafer is then bonded onto the ohmic metal layer using SU-8. The GaAs handle removal is a 4-step process: Argon ion mill, followed by a nitric-based etch, a citric-based etch step, and a hydrofluoric acid-based etch for the AlGaAs etch stop layer. Photolithography and a citric based etch form the GaAs mesas. The ohmic pedestal was formed using a combination of dry and wet etches. A CF_4/Ar RIE etch, optimized for high selectivity to photoresist, Pd, and Si, was engineered for etching Ti and SU-8. Transene Gold Etch Type A is used to etch Pd and Au. The optimized etch processes developed in this work results in a high resistivity silicon surface with only ohmic pedestals and GaAs Mesas, without etch residue and minimal etching of the silicon substrate.

Lithography and a Bosch etch were used to form 11 μm diameter vias. A sacrificial resist is spun and exposed to define the geometry for the Schottky diodes. A thin layer of Ti (~15 nm) and Au (~30 nm) was deposited by magnetron sputtering to form the diode contact. Then, lithography defines the topside metallization, beam lead, and airbridge features. O_2 plasma cleans the developed features before electroplating. Electroplating coats a 3.5 μm thick Au layer using Technic gold solution. Ultrasonic agitation and a magnetic stir bar ensure a smooth surface finish and side wall metallization of defined vias. A combination of WaferBOND and epoxy was used to create a semi-permanent bond of the wafer to a Si carrier, allowing debonding and release of the final chips. The silicon handle was removed through lapping, plasma etching, and a buffered oxide etch (BOE). The microstrip ground plane is formed on the high-resistivity silicon's exposed backside in a photolithography and electroplating step. Photolithographic and DRIE processes are used to define the chip's outlines, allowing the precise formation of complex chip geometries not possible through traditional dicing techniques [6]. The finished chips are released through a heated solvent-based debonding process. Figure 1 shows an SEM image of the fabricated diode structure. The measured and curve-fitted current and voltage relationship of the diode is shown in Figure 2. Figure 3 shows the design drawing for comparison.

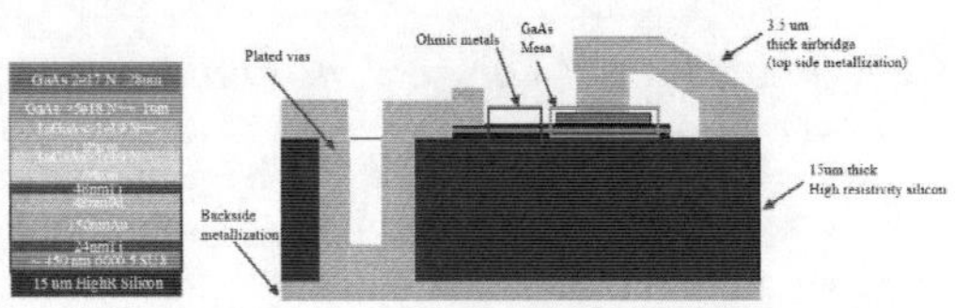

Fig. 3. Side profile of quasi-vertical fabricated for this work

III. CONCLUSION

This paper reports an enhanced and novel process flow optimizing the fabrication of quasi-vertical Schottky diodes on a thin microstrip substrate. In comparison to previously reported processes, this eliminates the need for multiple GaAs etches, reduces the amount of etches required for the ohmic metal stack, selectively etches the ohmic layers without significantly etching the resist and silicon, and significantly reduces residual etch byproducts of fluorine-based etches. This work also reports a heterogeneously integrated Schottky diode fabricated on a thin microstrip incorporating 11 μm side wall-Au plated vias alongside suspended beam leads, accomplished by sputtering and electroplating.

REFERENCES

[1] T. J. Reck, S. Durant and J. L. Hesler, "Design of a 2.5 THz Schottky-Diode Fourth-Harmonic Mixer," in IEEE Transactions on Terahertz Science and Technology, vol. 13, no. 6, pp. 580-586, Nov. 2023, doi: 10.1109/TTHZ.2023.3307566.

[2] Penfeld and Rafuse. (P. Penfield, Jr. and R. P. Rafuse, Varactor Applications. Cambridge, MA, USA: MIT Press, 1962.)

[3] S. Nadri et al., "A 160 GHz Frequency Quadrupler based on Heterogeneous Integration of GaAs Schottky Diodes onto Silicon using SU-8 for Epitaxy Transfer," in 2018 IEEE/MTT-S International Microwave Symposium - IMS, Philadelphia, PA: IEEE, Jun. 2018, pp. 769–772. doi: 10.1109/MWSYM.2018.8439536.

[4] N. Alijabbari, M. F. Bauwens, and R. M. Weikle, "160 GHz Balanced Frequency Quadruplers Based on Quasi-Vertical Schottky Varactors Integrated on Micromachined Silicon," IEEE Trans. THz Sci. Technol., vol. 4, no. 6, pp. 678–685, Nov. 2014, doi: 10.1109/TTHZ.2014.2360983.

[5] L. Xie, "Submillimeter-wave Schottky Diodes Integrated on Micromachined Silicon Probes," PhD Thesis, University of Virginia, Charlottesville, Virginia, 2019.

[6] R. Bass et al., "Ultra-Thin Silicon Chips for Submillimeter-Wave Applications," Fifteenth International Symposium on Space THz Technology, Jan. 2004, Accessed: Jan. 23, 2024. [Online]. Available: https://pdxscholar.library.pdx.edu/ece_fac/566/

NOTES:

A Planar RF-LO Coupler Design for Heterodyne Receiver at 220 GHz

Yan-Jun, Wang[1,2], Chun-Lun Wang[2], Chao-Ching Wang[2], Yu-Chieh Huang[2] and Ming-Jye, Wang[2*]

[1] *National Taiwan University, Taipei, 10617, Taiwan;*

[2] *Institute of Astronomy and Astrophysics, Academia Sinica, Taipei 10617, Taiwan*

[*] *Corresponding author: Ming-Jye Wang, mingjye@asiaa.sincia.edu.tw*

Abstract

A planar 220 GHz directional coupler, comprising two distinct layers of microstrip lines (MLs) with an overlapping section, has been designed. The length of this overlapped section allows for the adjustment of coupling efficiency between the two MLs. This planar directional coupler, intended for coupling radio frequency (RF) and local oscillator (LO) signals, can be integrated with mixers to create a compact heterodyne receiver. Such a receiver is highly desirable for constructing large-format multi-pixel array receivers at millimeter and sub-millimeter wavelengths. The detailed design of the 220 GHz directional coupler and its scaled model, operating at 10 GHz, are presented. Preliminary measurement results of the scaled model confirm the feasibility of this concept, although some external issues still need to be resolved.

Keywords—Heterodyne receiver, Directional coupler, Planar circuit

Introduction

Integrating electrical components on a chip is a modern approach to building complex circuits with optimized space utilization. This concept can also be implemented in heterodyne receivers at sub-millimeter wavelengths. A heterodyne receiver typically requires many waveguide components, such as directional couplers, local oscillator power distributors, and orthogonal mode transducers. Optimized components are selected to assemble single-pixel receivers for the best performance. These high-performance single-pixel receivers are deployed on the elements of interferometers. However, there is a high demand for multi-pixel heterodyne receivers for large aperture telescopes. A few multi-pixel heterodyne receivers have been developed and deployed on telescopes, such as Nobeyama-BEARS (25 pixels), Delinha (9 pixels), JCMT-HARP (16 pixels), and SuperCAM (64 pixels), APEX-CHAMP+ (14 pixels), and IRAM-HERA (9 pixels). Some new multi-pixel receivers are currently under development. The architecture of these multi-pixel receivers is similar to that of single-pixel receivers, which makes scaling the pixel number up to 100 very challenging for 50-meter class telescope in the future.

Recently, Shan et al. demonstrated a two-pixel heterodyne receiver chip operating at 140-170 GHz using planarized circuits to replace most waveguide components [1]. This compact design makes constructing large-format multi-pixel receivers more feasible, although the receiver's performance is not yet optimal. However, in their design, the local oscillator (LO) power is injected into each mixer through waveguide coupling, requiring four waveguides for LO distribution. By designing a planar LO power distribution circuit that incorporates an LO/RF coupler for each mixer, the receiver would require only a single waveguide for LO power, making the receiver even more compact.

In this paper, we present the design of a planar 220 GHz RF/LO directional coupler with two overlapped microstrip lines (MLs), inspired by the work of Tan et al. [2], who reported a directional coupler design with an overlapped section of coplanar-waveguide (CPW) and ML. Our design features an ML with a Z-bent section overlapping another ML on the bottom. The coupling coefficient is controlled by the length of the overlap. As a proof of concept, a scaled model chip operating at 10 GHz was designed, fabricated, and tested. The details of this design and its performance will be presented and discussed.

Results and Discussion

A. Design of 200 GHz coupler

The basic design concept involves overlapping two microstrip lines (MLs) for a certain length. The electromagnetic (EM) fields of the two MLs in the overlapped region will couple to each other, and the coupling efficiency (CE) can be tuned by adjusting the coupling length (CL). An exploded view of the device structure is shown in Fig. 1(a). The upper ML, which is 6 μm in width, is bent to align with the lower ML, which is 3 μm in width. These three metal layers, made of 0.3 μm niobium (Nb), are separated by two dielectric layers of 0.3 μm silicon dioxide (SiO2). The impedances of the two MLs are designed to be 16 ohms. Fig. 1(b) shows the top view of the directional coupler and the definition of ports.

Figure 2 depicts the simulated S-parameters of the designed directional coupler using HFSS with six different coupling lengths (CLs). The return losses are consistently below -15 dB. Specifically, S41 represents the coupling efficiency (CE) from the LO to the RF port, ranging between -35 dB and -15 dB for coupling lengths from 5 μm to 30 μm. This relationship is illustrated in the inset of Fig. 1 (b). In mixer applications, the coupling efficiency of a directional coupler for LO/RF coupling typically hovers around -17 dB. For the 200 GHz model, a coupling length close to 20 μm achieves this coupling efficiency.

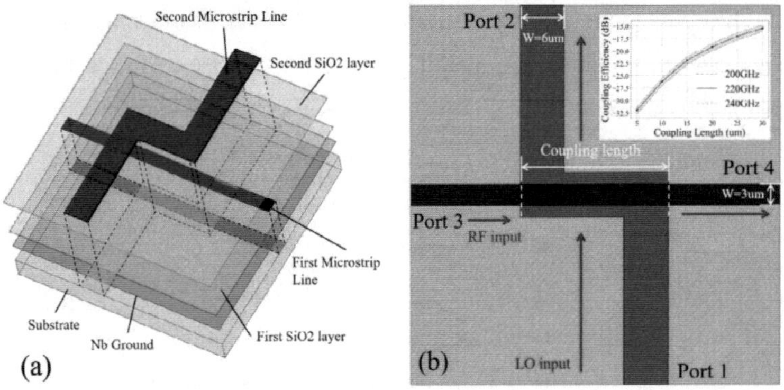

Figure 1 (a) Exploded view of the planar RF/LO coupler. (b) The top view of the planar RF/LO coupler and the definition of each port. The inset shows the relation between coupling efficient and coupler length.

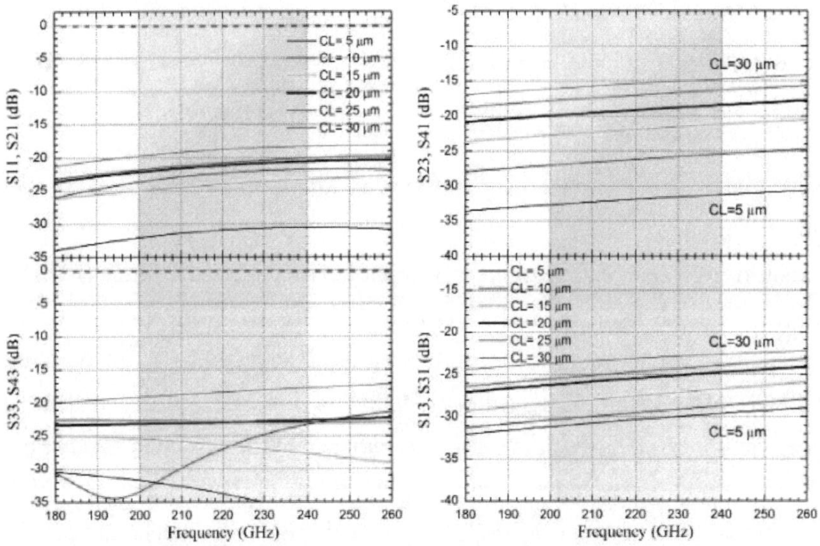

Figure 2 The simulated S-parameters of planar RF/LO coupler at 220 GHz central frequency. The operation frequency is set between 200 and 240 GHz (Blue area). S41 & S23 is the coupling efficiency

B. Scaled Model

To verify the feasibility of our concept, a scaled model, adjusting the central frequency from 220 GHz to 10 GHz, was designed. The configuration of the coupler region (overlapped microstrip lines) is identical to that of the 220 GHz version, but the coupling lengths are scaled according to the wavelength. This scaled chip has been successfully designed and fabricated. In addition to the coupling region, matching circuits from MLs to coplanar waveguides (CPWs) and 50 Ω input/output ports were also designed, depicted on the left side of Figure 3. The comparison of coupling coefficients between the 220 GHz design (black line) and the scaled model (blue line) with varying lengths is shown in the middle of Figure 3. The simulated coupling efficiencies (CEs) of the 220 GHz coupler and its scaled models are displayed in Figure 1(b), revealing a CE difference of less than 1 dB between the two models. The scaled chip was mounted in a test block and connected to signal lines and ground using aluminum bonded wires, as illustrated on the right side of Figure 3.

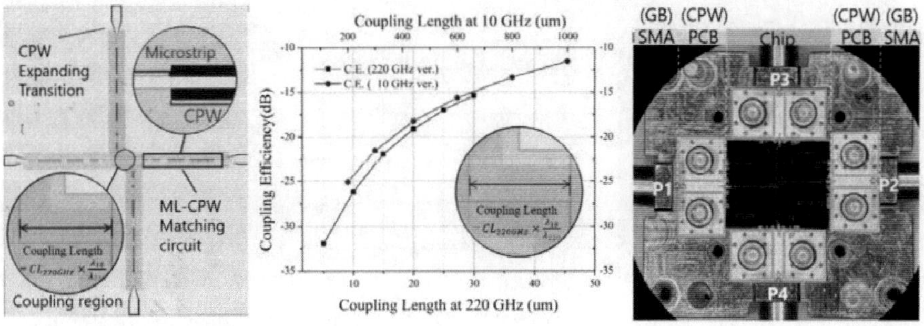

Figure 3 Left: The scaled chip with impedance matching circuits and input/output ports. Middle: The comparison of simulated coupling coefficient between the scaled model and 220 GHz chips. The coupling length was scaled by the wavelength. Right: The scaled chip mounted in the test block.

The performance of the scaled chip was evaluated in a cryostat operating at 2K, as depicted in the left side of Figure 4. The four ports of the scaled-model chip were connected to the feedthroughs on the chamber wall via 40.5 cm long coaxial cables. A vector network analyzer was employed to measure the directional coupler's performance. The simulated S-parameters are shown in blue lines, while the measured results are indicated by black lines in the right side of Figure 4. Generally, the LO and RF transmissions (S21, S43) exhibit values ranging from -1 to -3 dB across the 8-12 GHz frequency band. The LO and RF reflections (S11, S33) are notably worse than the anticipated values, suggesting an impedance mismatch within the system. The coupling (S23, S41) closely approximates the designed specifications, while the isolation (S13, S31) measures approximately 5 dB higher than simulated values.

In the raw data (black lines), two types of ripples, approximately 200 MHz and 3 GHz in frequency, suggest possible impedance mismatches within the circuit. To investigate this issue further, time-domain measurements were conducted for each port using a CL=600 μm chip. Figure 5 illustrates the results for Port 1 and Port 2 (upper panel) as well as Port 3 and Port 4 (lower panel). The impedance responses of Port 1 and Port 2 are nearly identical, and a similar consistency is observed between Port 3 and Port 4.

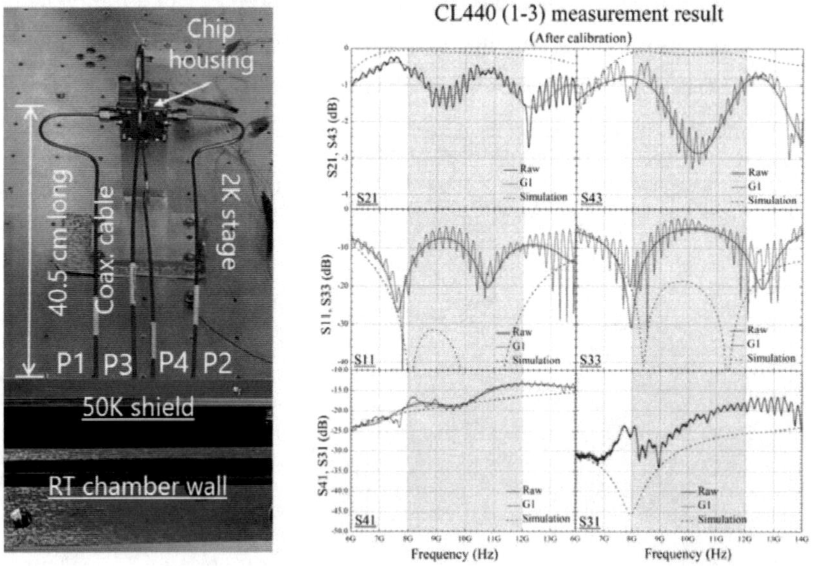

Figure 4 Left: The measurement setup of the scaled directional coupler. The chip housing was mounted on the cold plate (2 K) and it four ports were connected to the chamber wall by 40.5 cm long coaxial cables. Right: The comparison of simulated and measured S-parameters of scaled directional coupler with a coupling length of 440 μm.

To investigate the impedance mismatching issue further, time-domain measurements were conducted for each port. Figure 5 illustrates the results for Port 1 and Port 2 (upper panel) as well as Port 3 and Port 4 (lower panel). The impedance responses of Port 1 and Port 2 are nearly identical, as are those of Port 3 and Port 4. The longer transmission line in the design results in a longer time delay along Port 1(2) compared to Port 3(4). The responses in the coaxial connector (light red) and transition (light blue) regions are similar across all ports. The impedance of the metal layers for Port 1(2) and Port 3(4) measures approximately 21 Ω and 20 Ω, respectively, slightly higher than the intended design value of 16 Ω. A distinct step is evident in the chip region for Port 1(2), marked by a circle, but not observed for Port 3(4). Detailed time-gating measurements may shed light on the factors contributing to the poorer performance of S43.

After identifying impedance variations at different locations, the S-parameters were re-measured using time gating mode. The red lines on the right side of Figure 4 were obtained by gating between the connectors of two ports on the housing, denoted as "G1". The ~200 MHz ripple noticeably disappeared upon gating, indicating it was likely caused by slight impedance mismatches between the housing and the four 40.5 cm long cables. However, gating at the location with the highest impedance did not eliminate the ~3 GHz ripple observed in S21 and S43.

The test results of the scaled chip successfully demonstrated our concept of using two overlapped metal layers (MLs) for the RF/LO directional coupler. The measured coupling strength closely matched the designed values. Fortunately, the absence of wire bonds in the 220 GHz chip may have mitigated impedance mismatch issues. Furthermore, it should be noted that the losses in the transmission lines were not accounted for in our initial simulations. These losses will be incorporated in future designs using a surface impedance model.

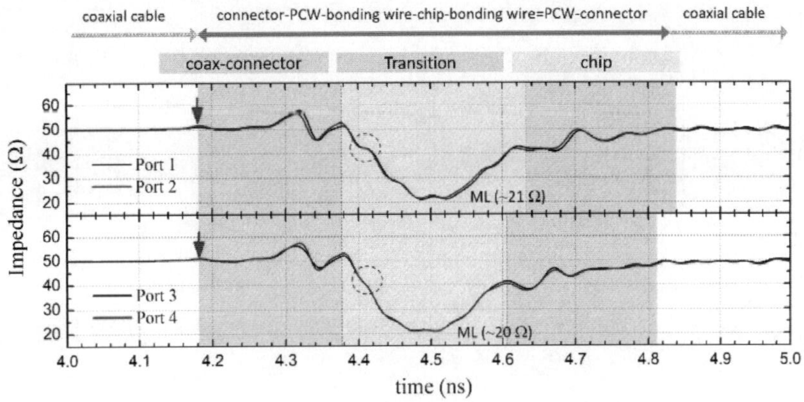

Fig. 5. The time domain measurement for Port 1&2 (upper panel) and Port 3&4 (lower panel). The red circle points out the main difference between two panels.

Summary and Future Work

We presented a design for a planarized RF/LO coupler operating at 220 GHz, along with its scaled model. The coupling coefficient is adjustable through the overlapped length between two metal layers (MLs). Preliminary measurements indicate a reasonable performance of the scaled-model chip. However, the measured data was notably affected by multiple reflections due to impedance mismatches at the interfaces of the coaxial cables, housing, and bonding wires to the chip. The 220 GHz LO/RF coupler chip will be to fabricated and tested in the near future.

Acknowledgement

This work is support by the National Science and Technology Council of Taiwan, grant no. NSTC 112-2112-M-001-050.

Reference

1. Shan, Wenlei, et al. "Planar superconductor-insulator-superconductor mixer array receivers for wide field of view astronomical observation." Millimeter, Submillimeter, and Far-Infrared Detectors and Instrumentation for Astronomy IX. Vol. 10708. SPIE, 2018.
2. Tan, Boon-Kok, and Ghassan Yassin. "A planar beam splitter for millimeter and submillimeter heterodyne mixer array." IEEE Transactions on Terahertz Science and Technology 7.6 (2017): 664-668.

Silicon Micromachined 400-600-GHz Orthomode Transducer

Zahraa Rizk[1*], Christopher Moore[1], Michael E. Cyberey[1], Arthur W. Lichtenberger[1], Robert M. Weikle[1], and N. Scott Barker[1]

***Abstract*— This paper presents the design, fabrication, and performance of a silicon-micromachined orthomode transducer (OMT) that operates in the 400-600 GHz band (WR-1.9 band). The simulation results show an insertion loss of 0.5 dB, a return loss >10 dB, isolation >35 dB and a cross polarization >35 dB over 85% of the frequency band.**

***Keywords*— Orthomode transducer, silicon micromachined, ICP-CVD, THz.**

I. Introduction

TERAHERTZ systems are becoming increasingly important in many fields such as astronomy [1,2], astrophysics spectroscopy [3], remote sensing, communication [4], and security applications [5,6].

Orthomode transducer (OMT) is a three-port passive device that has been used in receivers to discriminate between two orthogonal polarizations in the receiving antenna within the same frequency band [7]. At high frequencies, fabricating OMTs using the traditional metal machining becomes a challenge because of the small dimensions which require high precision and uniformity. Silicon micromachining is also more advantageous than CNC machining since entire batches of complex devices can be fabricated with lower costs and less time.

A multi-step deep reactive ion etching (DRIE)-based silicon micromachining is used to fabricate the OMT shown in Fig.1 [8]. The OMT consists of seven steps of different depths ranging from 22 μm to 284 μm. To obtain the stepped-shape structure in the Si, our fabrication process involves depositing SiO_2 masks with different step thicknesses that are proportional to the Si step depths. The SiO_2 layers are deposited using Inductively Coupled Plasma Chemical Vapor Deposition (ICP-CVD) technique which results in good surface uniformity (± 5 nm) compared to other deposition methods. Next, we run the Bosch process so that the SiO_2 layers and the Si are etched cumulatively until the whole SiO_2 is etched and the Si features are obtained. The Si step depths are measured using an optical profilometer. Finally, gold electroless plating is used to make the surface conductive.

II. Results

The simulation results of the OMT design presented in Fig.1 show a return loss better than 10 dB over 85% of the bandwidth for both vertical and horizontal polarizations as shown in Fig.2. Moreover, the insertion loss and the cross-polarization level are

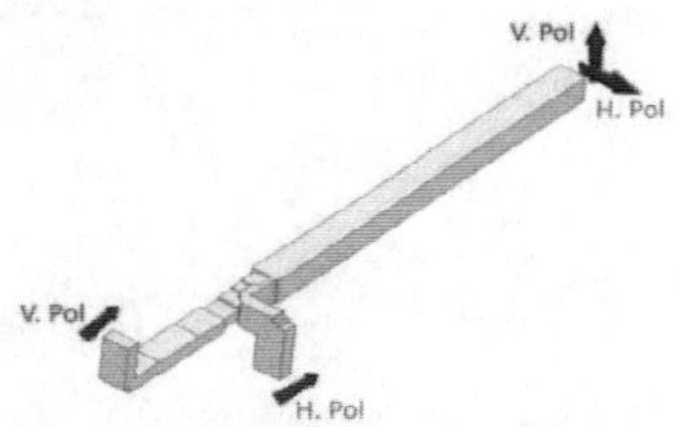

Fig.1. Three- dimensional OMT design and polarizations. V. Pol: Vertical polarization, H. Pol: Horizontal polarization.

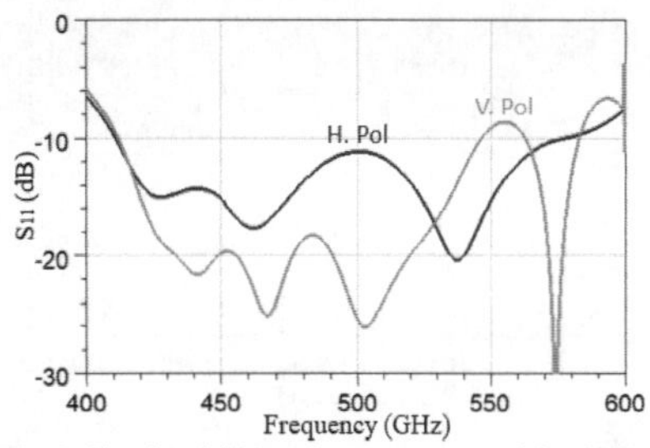

Fig. 2. Simulated S11 at the output ports of the OMT.

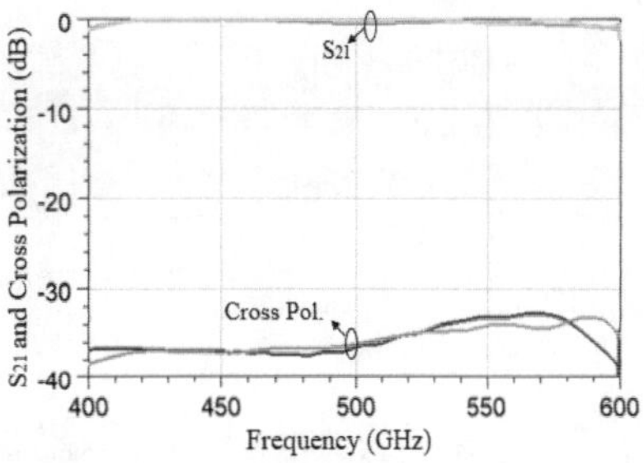

Fig.3. Simulated S_{21} and cross polarization of the OMT.

demonstrated in Fig.3 where the insertion loss is less than 0.7 dB, and the cross polarization is better than 35 dB over the 85% of the frequency range. Additionally, the port isolation is

[1]Electircal and Computer Engineering Department, University of Virginia, Charlottesville, 22904, USA. *Zahraa Rizk (email: zar8z@virginia.edu).

NOTES:

better than 30 dB over the whole range.

Additional simulations with step depths tolerance of ± 5 μm showed similar results with a return loss better than 10 dB over 80% of the bandwidth for both vertical and horizontal polarizations, insertion loss less than 1 dB, and a cross polarization better than 29 dB over the whole frequency band as shown in figures 4 and 5 respectively.

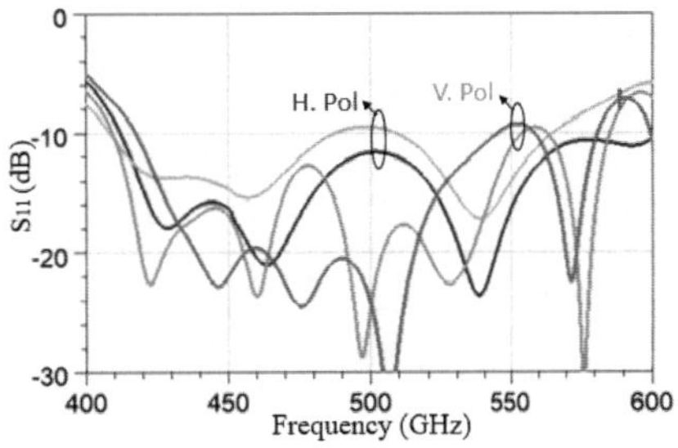

Fig. 4. Simulated S11 at the output ports of the OMT with step depths tolerance of ± 5 μm.

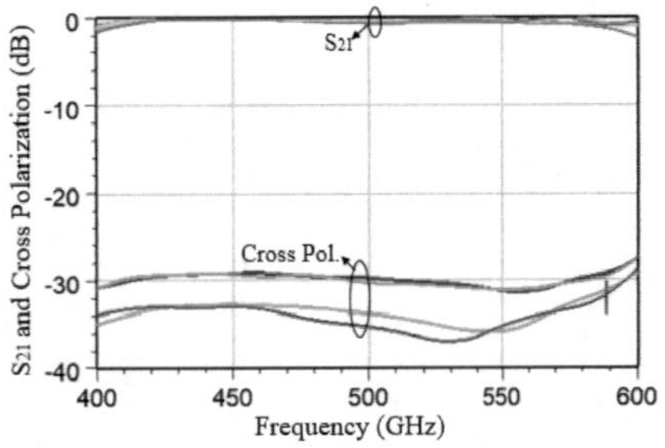

Fig.5. Simulated S_{21} and cross polarization of the OMT with step depths tolerance of ± 5 μm.

REFERENCES

[1] H.W. Hubers, R. Eichholz, S. G. Pavlov and H. Richter, "High Resolution Terahertz Spectroscopy with Quantum Cascade Lasers", *Journal of Infrared, Millimeter, and Terahertz Waves* 34(5) pp. 325-341, 2013.

[2] C. Kulesa, "Terahertz Spectroscopy for Astronomy: From Comets to Cosmology." *IEEE Transactions on Terahertz Science and Technology*, 1(1), pp.232-240. 2011.

[3] S. Withington, "Terahertz Astronomical Telescopes and Instrumentation," *Philosophical Transactions of the Royal Society of London. Series A: Mathematical, Physical and Engineering Sciences*, 362(1815), pp.395-402, 2004.

[4] P. Rodríguez-Vázquez, J. Grzyb, N. Sarmah, B. Heinemann and U.R. Pfeiffer, "A 65 Gbps QPSK One Meter Wireless Link Operating at a 225-255 GHz Tunable Carrier in a SiGe HBT Technology," *In 2018 IEEE Radio and Wireless Symposium (RWS)*, pp.146-149, January, 2018.

[5] K.B. Cooper, R.J. Dengler, N. Llombart, B. Thomas, G. Chattopadhyay and P.H. Siegel, "THz Imaging Radar for Standoff Personnel Screening," *IEEE transactions on terahertz science and technology*, 1(1), pp.169-182, 2011.

[6] H. Quast, and T. Loffler, "Towards Real-Time Active THz Range Imaging for Security Applications," *In 2009 International Conference on Electromagnetics in Advanced Applications*, pp. 501-504, IEEE, September, 2009

[7] A. Navarrini, C. Groppi, and G. Chattopadhyay, "A Waveguide Orthomode Transducer for 385-500 GHz," *In 21st International Symposium on Space Terahertz Technology 2010, ISSTT 2010*, pp. 281-290, December, 2010.

[8] C. Jung-Kubiak, T.J. Reck, J.V. Siles, R. Lin, C. Lee, J. Gill, K. Cooper, I. Mehdi and G. Chattopadhyay, "A multistep DRIE process for complex terahertz waveguide components." *IEEE Transactions on Terahertz Science and Technology*, 6(5), pp. 690-695, 2016.

NOTES:

SESSION 4

Demonstration of Multi-Layer Antireflective Treatments for Gradient Index Silicon Optics at THz frequencies

C. Jung-Kubiak[1*], F. DeFrance[1], J. Gill[1], S. Rahiminejad[1], T. Macioce[2], J. Sayers[2], G. Chattopadhyay[1], and S. Golwala[2]

***Abstract*— Future instruments for millimeter- and submillimeter-wave astronomy applications can benefit greatly from ilicon optics with broadband antireflection treatment. Silicon is an ideal material at these wavelengths, but its high index (n = 3.42) requires antireflection (AR) coatings to avoid reflections. We report on the development of multi-layer flat gradient index (GRIN) silicon optics for wide spectral bandwidths.**

***Keywords*— Antireflective treatments, Deep reactive ion etching, Vacuum window, Silicon optics.**

I. Introduction

Numerous branches within the fields of astronomy and cosmology, such as studies of CMB polarization, the Sunyaev-Zeldovich effect, dusty sources, and millimeter-wave transients would benefit significantly from the utilization of low-loss, wide bandwidth, flat optics operating at submillimeter and terahertz (THz) frequencies. Silicon has favorable characteristics such as high refractive index (3.42), achromaticity, absence of birefringence, mechanical strength, while being low loss; and as such emerges as an optimal optical material for these frequencies. However, the challenge lies in the antireflection (AR) treatment due to silicon's high refractive index. Over the past few years, we have successfully developed multi-layer integrated AR treatments that we are now combining with our flat gradient index (GRIN) silicon optics.

II. Design, Simulations and Fabrication

We use a multi-step deep reactive ion etching (DRIE) [1] process to etch subwavelength features (posts and/or holes) into 100 mm diameter high-resistivity silicon wafers to locally vary silicon's effective refractive index: by creating multiple layers with different refractive indices, we obtain a very wide-bandwidth antireflection (AR) treatment. And by varying the index of a silicon wafer radially we can create flat low-loss gradient index (GRIN) lenses [2]. Our two-layers AR structure has been shown to yield reflections of < 1% over the frequency range of 180 – 310 GHz (1.6:1 bandwidth) [3]. Our three-layer AR surfaces are designed for a 2.3:1 bandwidth (190-350 GHz) and < 1% reflectance and currently we are combining those wafers with flat-faced gradient index (GRIN) optics, see Fig. 1a. Finally, our latest four-layer AR structure is optimized to give −20 dB reflection between 130 – 400 GHz (3.1:1 bandwidth), see Fig. 1b and 1c. Complete results will be presented and discussed along with future designs for five-layer and six-layer AR treatments to cover bandwidths up to 5.7:1 (73-415 GHz and 23-130 GHz).

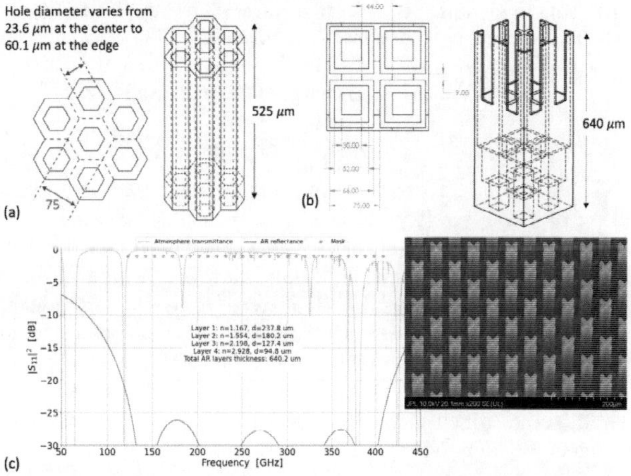

Fig 1: (a) and (b) Representations of our GRIN pattern and our four-layer AR pattern. (c) Simulation for our 4-layer AR surface, with an SEM image of the fabricated wafer.

References

[1] C. Jung-Kubiak, and al., "A Multi-Step DRIE Process for Complex Terahertz Waveguide Components," *IEEE Transactions on Terahertz Science and Technology*, vol. 6, no. 5, pp. 690-695, Sep. 2016

[2] F. Defrance, and al, "Flat Low-Loss Silicon Gradient Index Lens for Millimeter and Submillimeter Wavelengths", *Journal of Low Temperature Physics,* vol. 199, pp. 376-383, May 2020

[3] F. Defrance, and al., "A 1.6:1 Bandwidth Two-Layer Antireflection Structure for Silicon Matched to the 190-310 GHz Atmospheric Window," *Applied Optics*, vol. 57, no. 18, pp. 5196-5209, Jun. 2018

[1]California Institute of Technology, Jet Propulsion Laboratory, 4800 Oak Grove, Pasadena, CA, USA, 91109; [2]California Institute of Technology, Pasadena, CA, USA, 91125. [*]cecile.d.jung@jpl.nasa.gov

NOTES:

Si metalens for quasi-optical THz HEB mixer arrays

Dingding Ren[1*], Wenye Ji[1], Sofia Cremasco[1], Jan de Graaff[1], Tsung-Yu Kao[3], Alan Lee[3], Aurèle J. L. Adam[1], and Jian-Rong Gao[1,2]

***Abstract*—We present the development of a Si metalens for large THz hot electron bolometer (HEB) mixer arrays. The metalens of 1×1 cm² is based on a metasurface consisting of 100 µm long rectangular Si pillars with variable area, arranged in a 50×50 µm² periodic pattern. As proof of the concept, we designed a metalens with a focal distance of 1 cm. A prototype of the metalens has been fabricated on a Si wafer using a deep reactive ion etching process. The metalens was optically characterized at 2.1 THz and was demonstrated successfully by focusing a collimated beam at its designed focal distance.**

***Keywords*—Metalens, metasurface, THz, HEB mixer array**

I. Introduction

HEB mixer arrays have already been used for THz observatories, such as SOFIA and GUSTO, due to the fast-scanning speed, which is roughly scaled to the number of pixels. GUSTO has arrays of 8 pixels using lens-antenna quasi-optical HEB mixers [1]. Using individual elliptical lenses, as for the GUSTO arrays, will be extremely challenging to extend to future arrays of ≥100 pixels. Here we present a metalens concept, that can lead to an flat lens array on a Si wafer, which can be integrated with an HEB array on another Si wafer to form a large HEB array of ≥100 pixels.

II. Results

Fig. 1a shows a photograph of a fabricated metalens on a 525 µm thick Si wafer. The metalens is designed to have a focal distance ~1 cm as a proof of the concept and is composed of meta-atoms of rectangular pillars, with a periodicity of 50 µm. The lateral dimension of each Si pillar varies from 4 to 40 µm to cover the full phase shift from zero to 2π, which is the relative phase change of the transmitted light with respect to the original light. A similar approach was reported [2][3]. A 2D finite element simulation is used to estimate the focusing effect of the designed metalens shown in Fig. 1c. A laser writer was used for the lithography in the fabrication, while 100 iterations of a Bosch process were performed in an ICP-RIE system to etch and to form Si pillars of 100 µm in depth, shown in Fig. 1b. With an incident Gaussian beam of ~ 1 cm in diameter at 2.1 THz normal to the metalens, the focal distance is expected to be ~ 1cm, as illustrated by the red region of high intensity in Fig.

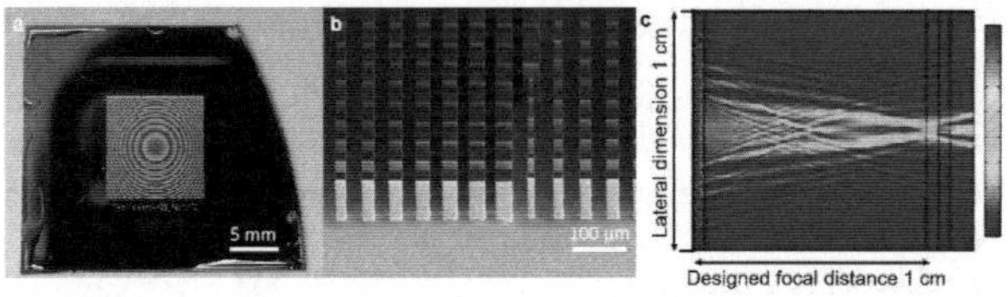

Fig. 1. a) Photograph of the metalens. **b)** 45° tilted view of the Si nanopillars after cleaving the middle of the sample from **a)**. **c)** Finite element simulation of the focusing effect of the metalens shown in **a)**.

1c. To optically characterize the metalens, we used the 2.1 THz radiation from a QCL operated in a Stirling cooler. The QCL beam was collimated using an off-axis parabola. The collimated beam measured in front of the metalens and focused beam at the designed focal point are shown in Fig. 2**a** and **b**, respectively, confirming the focusing effect.

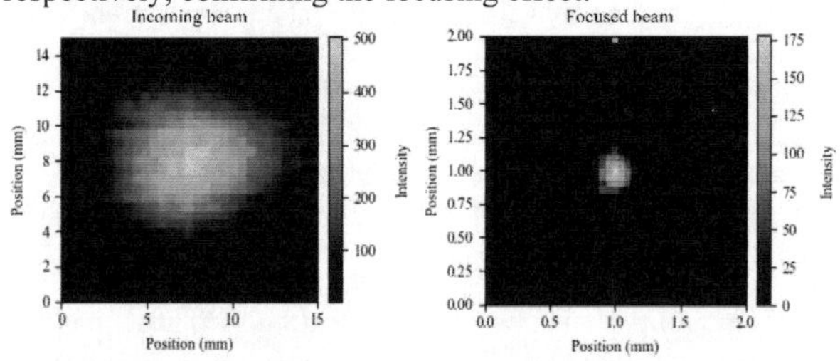

Fig. 2. a) Collimated incident beam measured in front of the metalens and **b)** beam measured at the designed focal point of the metalens.

III. Conclusion

We successfully designed, fabricated, and demonstrated a flat lens at 2.1 THz using a meta-surface, aiming for large HEB arrays. Currently, we are working on a new design of metalens at 2.5 THz, which can combine with a twin-slot antenna HEB to demonstrate the lens-antenna system.

References

[1] J. R. G. Silva et al, "High Accuracy Pointing for Quasi-Optical THz Mixer Arrays," in IEEE Transactions on TST, 12, 53-62 (2022).
[2] Qing Yu et al, "All-Dielectric Meta-lens Designed for Photoconductive Terahertz Antennas". IEEE Photonics Journal, 9, 5900609 (2017).
[3] M. Yang et al.," High focusing efficiency metalens with large numerical aperture at terahertz frequency", Optics Letters, 48, 4677(2023)

[1]Department of Imaging Physics, Delft University of Technology, Delft, the Netherlands; [2]SRON Netherlands Institute for Space Research, Leiden/Groningen, the Netherlands; [3]LongWave Photonics LLC, Mountain View CA 94043, USA, *Corresponding author (email: d.ren-1@tudelft.nl).

NOTES:

Design and Measurements of a 480 GHz Metamaterial Flat Lens

Cassandra N. Whitton, Daniel Lu, Adhitya B. Sriram, Philip D. Mauskopf, Christopher E. Groppi, Michael Marrs, Paul F. Goldsmith, Georgios C. Trichopoulos

***Abstract*—There exist scientifically interesting molecular lines, such as the ground state transitions of water, that cannot be observed except from space. Observations of these lines can be made more cost-effective by lightweighting observation components, such as the primary optical aperture. This is particularly important for SmallSats and CubeSats which have highly limited weight budgets. Here we present a flat lightweight metamaterial lens which operates at 480 GHz, close to the 557 GHz ground state transition of ortho–H_2O. The lens is composed of alternating layers of spin-coated polyimide and patterned aluminum. The aluminum patterning was generated by optimization to a specific phase pattern. We have manufactured and tested the lens. The lens has an optical diameter of 124 mm. It weighs 3 grams and is less than 150 microns thick. It is also flexible. We have demonstrated using a near-field scan that the optical performance of the lens is nearly diffraction- limited. We have found the loss of the lens using radiometric techniques to be 2.5dB. This loss is roughly 1.5dB higher than expected, and we investigate possible reasons for this discrepancy.**

***Index Terms*—Terahertz, cubesat, metamaterial, lens, optics.**

I. Introduction

THIS work focuses primarily on the development of lightweight terahertz optics enabled by the use of metamaterial design. This is of particular interest for Smallsat applications where SWaP restrictions are of paramount importance. ASTRO 2020 recommends the expansion of Smallsat based astrophysics through NASA's Astrophysics Research and Analysis (APRA), Pioneer and Explorer opportunities [1]. For this development to include THz applications, low mass, large aperture antennas are essential. The metamaterial method used in this paper is one possible path to this goal.

The main characteristic of this lens that results in dramatic improvements in SWaP resource requirements is the thin geometry of the metamaterial structure. In a traditional bulk lens, focusing is achieved through the transformation of a planar wavefront to a spherical wavefront with a dielectric of varying thickness. In the sub-mm and THz, this thickness is at a minimum several mm and increases when a lens with a small f/D ratio is required. For example, the state-of-the-art low focal ratio, f/D=0.27, silicon lens in [2] is more than 10 mm thick for a lens diameter of 74 mm.

There have also been lenses and reflectors created that use metamaterials and metasurfaces to generate the same phase shift as a conventional optic [3]–[8]. The lens presented here is based on techniques developed in [3]. Using metamaterials consisting of metal layers embedded in dielectric layers, this group has developed lenses up to 300 GHz in frequency [4]. Here we present a 124mm-diameter lens which performs at 480 GHz, an operation frequency which is close to the 557 GHz line of water without being close enough to incur significant atmospheric losses in a lab environment. In addition, we here use a different fabrication method from previous works; while most previous work utilized polypropylene or polyethylene sheets that were first patterned and then were pressed together [3], our design builds up polyimide using spin-casting followed by patterned aluminum in alternating layers, allowing for very precise top-to-bottom alignment. Our lens has also reached a larger diameter or higher operation frequency than most comparable alternatives lenses.

C. Whitton, D. Lu, A. Sriram, P. Mauskopf, C. Groppi, M. Marrs, and G. Trichopoulous are with Arizona State University, Tempe, AZ, United States.

P. Goldsmith is with the Jet Propulsion Laboratory, California Institute of Technology, Pasadena, CA, United States.

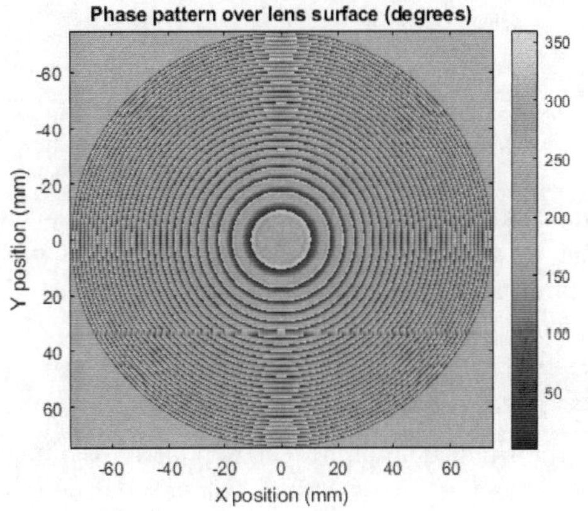

Fig. 1. The ideal phase pattern of the lens.

II. Lens Design

The design of the lens proceeds in stages. First, an ideal lens phase pattern is created depending on the operation frequency, focal length, and diameter of the lens. The phase pattern is based on gaussian quasioptics [9], and is shown in figure 1. This phase pattern is then subdivided into pixels sized according to the metamaterial structure employed in the lens. A library of meta-atoms is then created and simulated: each meta-atom in this structure is a single metal square. Then, using this simulated library of meta-atoms, optimizations are performed to create pixels in the lens structure which approximate the ideal phase pattern as closely as possible. Each pixel then consists of alternating layers of dielectric and meta-atom,

Fig. 2. The lens mounted in its aluminum ring holder. The phase-wrap boundaries can be seen quite clearly.

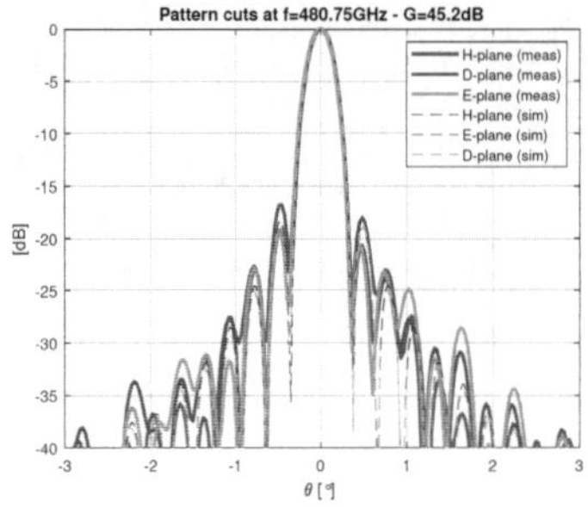

Fig. 3. The measured far-field beam pattern of the lens-horn system, compared to simulation

and the entire lens structure is alternating layers of dielectric and metal patterned grids. This lens structure is then converted into manufacturing files, and the metamaterial lens is created by building up alternating layers of solid polyimide dielectric and patterned metal using photolithography. An image of the complete lens in a ringholder is shown in 2.

III. Lens Characterization

Theoretical simulations of the lens were performed in order to provide a baseline of comparison. These simulations were based on the as-optimized model of the lens, and thus included the phase errors, reflection losses, and material losses of the optimized pixels.

Two primary methods of measurement were used to characterize the lens performance: a 2d near-field scan and a radiometric loss measurement.

To perform the near-field scan, the lens was mounted one focal distance from a near-field probe, acting as the lens feed. The near-field probe illuminated the lens nearly uniformly. A second near-field probe acted as a receiver. The receiver was scanned through the x-y plane in the lens's nearfield, obtaining both frequency and phase information.

The radiometric loss measurement used a diagonal horn connected to a 1100K receiver designed and built at JPL that operates in the frequency range of our lens. A horn feeding the receiver was placed at approximately the focus of the lens. A Y-factor measurement was then performed, alternating between a room-temperature absorber and an absorber submersed in liquid nitrogen on the side of the lens opposite the horn [10]. This Y-factor measurement was then repeated with the lens removed. Given both of these measurements, the loss of the lens alone could be deembedded from the loss of the horn-receiver system.

A. Results

The data obtained from the near-field scan were processed to obtain a far-field beam pattern using standard complex near-to-farfield transformations. The resulting far-field pattern at 480 GHz can be seen in Figure 3, which compares the measured near-field-to-farfield transformed pattern to the theoretically modeled results. It is clear that the far-field beam pattern measured in the lab very closely matches the theoretical expectation. The largest discrepancies are deviations of a few dB in the sidelobe levels.

The results of the radiometric loss measurements were analyzed as well. The combined reflective and transmissive loss is roughly 2.5 dB at the design frequency and only varies by about ± 0.2 dB over the measured band. This is significantly higher than the expected loss of the lens, which our simulations predicted to be 1.0. Of this 1.0 dB of simulated loss, approximately 0.2 dB is reflective loss, 0.5 dB is conductive loss, and 0.3 dB is dielectric loss.

It is possible that the effective conductivity of the aluminum was lower than estimated due to its thinness, an effect that can arise at high frequencies due to defects in the metal surface [11].

IV. Conclusion

We have designed, fabricated, and tested a flat metamaterial 480 GHz lens with a diameter of 124 mm. The design employed optimization of 10 layers of meta-atoms to match a desired phase output. The manufacturing process made use of spin-casted polyimide in combination with photolitographically etched aluminum. The lens weighs only 3.0 grams, significantly less than a comparable conventional lens.

Our lens exhibits near diffraction–limited directivity at the design frequency. Future work will focus on increasing the bandwidth through various means, including using a longer focal length, and using broadband optimizations rather than single-frequency optimizations. We also plan to manufacture 300mm diameter lenses in the future using identical techniques with a larger wafer as a base.

The lens did exhibit 1.5 dB more loss than expected, having 2.5 dB of loss as opposed to the 1.0 dB simulated loss. It is unclear what is the source of this loss. Simulations suggest it is highly unlikely that the loss is due to excess dielectric loss, increased dielectric thickness, or skin depth effects due to decreased metal thickness. Our current expectations are that the loss is due to reflections as a result of under- or over-etching the patterned surfaces, or due to decreased conductivity due to the thinness of the metal. Future work will focus on understanding these losses and designing around them.

Acknowledgments

We would like to thank Alain Maestrini of JPL for conceiving and performing the radiometric loss test of the lens. We would also like to thank Sven van Berkel, also of JPL, for performing the near-field measurement of the lens and for providing critique on this work. This material is based upon work supported by the National Aeronautics and Space Administration under Grant No. 80NSSC20K0650 issued through the APRA Program. This research was conducted in part at the Jet Propulsion Laboratory, which is operated by the California Institute of Technology under contract with the National Aeronautics and Space Administration (NASA).

References

[1] National Academies of Sciences, Engineering, and Medicine, *Pathways to Discovery in Astronomy and Astrophysics for the 2020s*. Washington, DC: The National Academies Press, 2021.

[2] S. van Berkel, M. Alonso-delPino, C. Jung-Kubiak, and G. Chattopadhyay, "An f/0.27 high-gain lens antenna for ultrasmall platforms at thz frequencies," *IEEE Transactions on Terahertz Science and Technology*, vol. 13, no. 5, pp. 549–560, 2023.

[3] G. Pisano, M. W. Ng, F. Ozturk, B. Maffei, and V. Haynes, "Dielectrically embedded flat mesh lens for millimeter waves applications," *Applied optics*, vol. 52, no. 11, pp. 2218–2225, 2013.

[4] G. Pisano, A. Shitvov, P. Moseley, C. Tucker, G. Savini, and P. Ade, "Development of large-diameter flat mesh-lenses for millimetre wave instrumentation," in *Millimeter, Submillimeter, and Far-Infrared Detectors and Instrumentation for Astronomy IX*, vol. 10708. SPIE, 2018, pp. 16–27.

[5] Z.-W. Miao, Z.-C. Hao, Y. Wang, B.-B. Jin, J.-B. Wu, and W. Hong, "A 400-ghz high-gain quartz-based single layered folded reflectarray antenna for terahertz applications," *IEEE Transactions on Terahertz Science and Technology*, vol. 9, no. 1, pp. 78–88, 2019.

[6] T. S. Nowack, Y. D. Shah, J. P. Grant, I. E. Carranza, M. G. Kenney, D. Faccio, E. Wasige, and D. R. S. Cumming, "Metasurface optics with on-axis polarization control for terahertz sensing applications," *IEEE Transactions on Terahertz Science and Technology*, vol. 13, no. 4, pp. 373–380, 2023.

[7] B. Nie, H. Lu, T. Skaik, Y. Liu, and Y. Wang, "A 3d-printed subterahertz metallic surface-wave luneburg lens multibeam antenna," *IEEE Transactions on Terahertz Science and Technology*, vol. 13, no. 3, pp. 297–301, 2023.

[8] Z. Wani, M. P. Abegaonkar, and S. K. Koul, "Thin planar metasurface lens for millimeter-wave mimo applications," *IEEE Transactions on Antennas and Propagation*, vol. 70, no. 1, pp. 692–696, 2022.

[9] P. F. Goldsmith, *Quasioptical systems: Gaussian Beam Quasioptical Propagation and Applications*. New York: IEEE Press/Chapman & Hall, 1998.

[10] D. M. Pozar, "Microwave engineering," *Fourth Editions, University of Massachusetts at Amherst, John Wiley & Sons, Inc*, pp. 26–30, 2012.

[11] N. Laman and D. Grischkowsky, "Terahertz conductivity of thin metal films," *Applied Physics Letters*, vol. 93, no. 5, 2008.

Inverse-Designed Volumetric and Multi-Layer Silicon Metaoptics

Conner Ballew[1*], Sven van Berkel[1], Subash Khanal[1], Goutam Chattopadhyay[1]

***Abstract*—Metaoptics made of near- or sub-wavelength patterned dielectric elements are a low loss platform for realizing advanced optical components which can control electromagnetic radiation based on its fundamental spatial, spectral, and polarization properties. Fundamentally this behavior arises from controlling the interference of multiple modes within the structure. 3D-patterned and multi-layer devices hold many modes, but controlling them requires complex geometries that presently must be identified with advanced algorithms. This talk will highlight the design, fabrication, and measurement results of metaoptics components being designed for space instrumentation, including interferometry systems and filters.**

***Keywords*— Metaoptics, inverse-design, volumetric**

I. Introduction

OPTICAL design has long relied on modular, linear arrangements of traditional optical elements like lenses, gratings, polarizers, waveplates, etc. A collection of these components can control all of the fundamental properties of light (frequency, k-vector, and polarization), but the physical size of the system grows immensely with the complexity of the system function. The reliance on modular system can be mitigated by encoding customizable and multi-functional behavior into individual components, effectively combining multiple basic components into one advanced component.

Broadly speaking, *metaoptics* are optical components with near- or sub-wavelength patterned features whose geometrical parameters substantially affect the optical modes within their volume [1]. Among their most appealing attributes is the multifunctional capabilities, such as the ability to prescribe independent high-transmission phase masks to orthogonal polarizations [2]. Most of metaoptics development has focused on thin and flat elements like transmit/reflect arrays and metasurfaces, which can only support a few modes within the volume of their individual elements. The effect is a fundamental limitation on the degrees of freedom in the design and the achievable complexity [3]. While thinness is often an appealing aspect of these devices, it is also the fundamental limitation on their efficiency and functionality.

Ideally, meta-optical component should be thick enough to contain as many modes as is necessary to achieve its prescribed task. However, the existence of enough modes is not the only requirement, since additionally these modes must be *controlled* via the geometry. To do this, the device typically must be patterned in all three dimensions which poses two main difficulties: 1) identifying the optimal 3D shape, and 2) fabricating the resulting 3D shape at the microscale.

II. Results and Applications

To identify a 3D shape that is both optically efficient and fabricable we use a adjoint-based inverse-design that incorporates fabrication constraints. The adjoint method offers an extremely efficient method for calculating the gradient of a figure-of-merit with respect to the permittivity of a design region, enabling gradient-based optimization of the device.

At terahertz frequencies, high-resistivity Si is an excellent material because of its low absorption loss and its ability to be fabricated with high precision. To fabricate 3D devices, we pattern the individual layers across a single wafer, separate the dies, and then stack them. Both layer-to-layer alignment and minimization of the air gap between layers is critical for realizing efficient devices, and this talk will describe the methods used to achieve this.

Some applications of volumetric Si metaoptics are extremely compact interferometer systems for spectroscopy, and components like orthomode transducers and filter, which can be used to substantially reduce size, weight, and power of Earth, planetary, and astrophysics instrumentation. In all cases, the volumetric nature of these metaoptics allows for much broader band performance than thin metasurfaces typically do.

References

1] N. Yu and F. Capasso, "Flat optics with designer metasurfaces," *Nature Materials*, vol. 13, no. 2. Nature Publishing Group, pp. 139–150, Jan. 23, 2014

[2] A. Arbabi, Y. Horie, M. Bagheri, and A. Faraon, "Dielectric metasurfaces for complete control of phase and polarization with subwavelength spatial resolution and high transmission," *Nat. Nanotechnol.*, vol. 10, no. 11, pp. 937–943, Nov. 2015

[3] David A. B. Miller, "Why optics needs thickness".Science 379, 41-45, Jan. 2023

[1]Jet Propulsion Laboratory, Pasadena, CA 91109, USA
[*]Corresponding author (email: conner.ballew@jpl.nasa.gov).

NOTES:

A Low Loss Dual-polarization Optical Diplexing Scheme for Millimeter to Terahertz Waves.

Keara J. Carter, Cheuk-Yu Edward Tong *Member IEEE*, Lingzhen Zeng, Paul Grimes

***Abstract*—In this paper, we present an optical diplexer composed of periodic stacks of High Resistivity (HR) Silicon and Low Density Polyethylene (LDPE) disks. Operating at an angle of incidence of 30 degrees, this diplexer is characterized by a reflection band centered at 130 GHz and it acts as a band-pass transmission filter around 300 GHz. It offers a very low measured average transmission loss of 3.2% in the transmission band. Similar diplexers created using this technique will be employed with next generation ground and space-based telescopes to support simultaneous dual polarization, multiband observations.**

***Index Terms*—dichroic, diplexer, radio astronomy, submillimeter wave**

I. Introduction

THE addition of simultaneous dual polarization observation capabilities at submillimeter wavelengths offers numerous advantages. From an astronomy standpoint, simultaneous multi-band observations provide fresh insights into various astrophysical processes and it also enhances spectral line surveys. From an engineering perspective, the throughput of radio telescopes is increased and the technique of Frequency Phase Transfer (FPT)[1] is enabled. The operation of radio interferometers in the submillimeter regime is limited by atmospheric phase fluctuations, which has a particularly strong impact on Very Long Baseline Interferometry (VLBI). FPT improves sensitivity by pairing low-frequency receivers with high-frequency ones, using the strong signal-to-noise ratio of the low-frequency data to calibrate the high-frequency data. Next-generation instruments, such as the wideband Submillimeter Array (wSMA)[2], the next-generation Event Horizon Telescope (ngEHT)[3], and the Black Hole Explorer (BHEX)[4] will all require simultaneous dual polarization observation capabilities.

The key to simultaneous dual-polarization multi-band receiver operation is a low-loss multiplexing scheme, typically constructed from optical diplexers, or dichroics. These diplexers direct the telescope beam towards different receivers centered at varying frequency ranges. The performance of optical diplexers can be assessed by a number of metrics, including insertion loss, bandwidth of operation, angle of incidence, and clear aperture size. This paper presents a novel diplexer design that features simplicity, robustness, exceptionally low insertion loss, operation at high angles of incidence, wide bandwidth, and moderate beam size.

All authors are with the Center for Astrophysics | Harvard & Smithsonian, Cambridge 02138, MA. (e-mail: keara.carter@cfa.harvard.edu).

II. Theory

The optical diplexer we have created uses a periodic stack of dielectric materials to define a transmission and a reflection band for the incoming signal beam. The equations governing the propagation of electromagnetic waves in a stack of dielectric disks are well known [5] and can be used to define a characteristic matrix for each layer of the stack. For a stack of dielectric disks, matrix multiplication of the characteristic matrix of each layer in the stack yields the equivalent characteristic matrix of the entire stack, from which the transmission and reflection of the stack can be derived. For dual polarization operation, the responses of the stack for both Transverse Electric (TE) and Transverse Magnetic (TM) waves have to be considered. At normal incidence, the two polarizations are degenerate, but as the angle of incidence increases, the two waves demonstrate increasingly different propagation characteristics. We have designed our diplexers to yield relatively similar TE and TM responses up to and including an angle of incidence of 30 degrees. Please refer to [5] for a full derivation of the equations used.

III. Design

The dielectric materials utilized in our dielectric stacks are selected based on their refractive indices and loss tangents. The thickness of the disks are chosen to effectively reflect and transmit specified frequency range. In the design discussed below, High Resistivity Silicon (HR-Si) and Low Density Polyethylene (LDPE) are used for these reasons. Specified resistivity of the HR-Si is 10 kΩ-cm, and at such resistivity level, silicon is known to be one of the lowest loss materials in the THz regime. HR-Si forms the basis of the design by defining the half-wave and full-wave transmission frequencies. The refractive index of HR-Si and LDPE are 3.42 and 1.52 respectively. The high contrast in refractive indices of the two materials is exploited to produce a strong reflection at frequencies close to quarter-wave or three quarter-wave in the HR-Si layers. In our design, we adjust the thicknesses of the two material such that the diplexer's transmission and reflection bands can be tailored to the desired frequency range.

A nine-layer dielectric stack with alternating 0.11mm HR-Si and 0.18mm LDPE layers which has a transmission band centered around 300 GHz and a reflection band centered around 220 GHz was presented elsewhere [6]. Here, we present a similar design with 0.15mm HR-Si and 0.18mm LDPE layers (Fig. 1) which behaves as a 270 GHz band pass filter in transmission, while reflecting a band around 130 GHz. Both diplexers are designed to operate with an incident angle of 30 degrees. The stack is consolidated by baking in an oven at a temperature of $\approx 110^\circ$ Celsius with weights on top. This acts to fuse the LDPE to the HR-Si wafers and eliminate any air gaps.

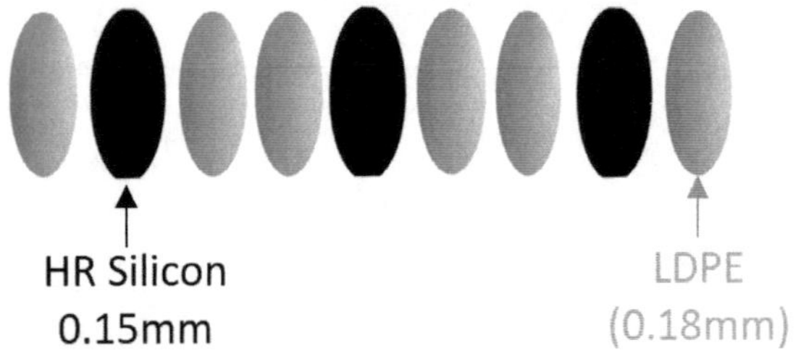

Fig. 1: The dielectric stack used to create an optical diplexer with a transmission band centered at 270 GHz and a reflection band centered at 130 GHz. The inner layers of LDPE are two stacked pieces of 0.18mm (0.36mm in total) while the outer layers are single pieces of 0.18mm. The thickness of the High Resistivity (HR) Silicon disks are 0.15mm. The clear aperture size is 100mm.

IV. Measurement Results

A. Transmission

Transmission measurements of the diplexer were performed using a Quasi-Optical Vector analyzer (QO-VNA) [7]. The diplexer was placed in a vertical holder located near the beam waist of the setup, with the holder rotated by an angle of 30° about the vertical axis of the holder. When the incoming wave is vertically polarized, the diplexer is seeing a TE incident wave; and for horizontally polarized input wave, the diplexer is looking at a TM incident wave. To switch between TE and TM measurements, we simply introduce waveguide twists to both the source and the receiver to rotate the plane of polarization by 90°. Although the QO-VNA employs a WR-3.4 source module and harmonic mixer, by working with weaker harmonics from the Amplifier-Multiplier-Chain based source, we are able to perform measurements, in a single sweep, from 195 to 370 GHz with a signal-to-noise ratio of generally better than -35 dB. Lower frequency TE transmission measurements were performed with a separate WR-6.5 VNA setup.

The results of the transmission measurements of the 130/270 GHz diplexer are plotted in Fig 2. The measured data matches the theory very well, showcasing an advantage of this type of diplexer over the more commonly employed Frequency Selective Surfaces (FSS); no complex electomagnetic simulations are required to model the stack.

B. Loss

We have performed an experimental determination of the transmission loss of the optical diplexer using an SIS receiver designed for the wideband Submillimeter Array project[2]. The method of Intersecting Lines[8], which has been proven to be able to measure optical losses accurately at short millimeter wavelengths[9], was adopted.

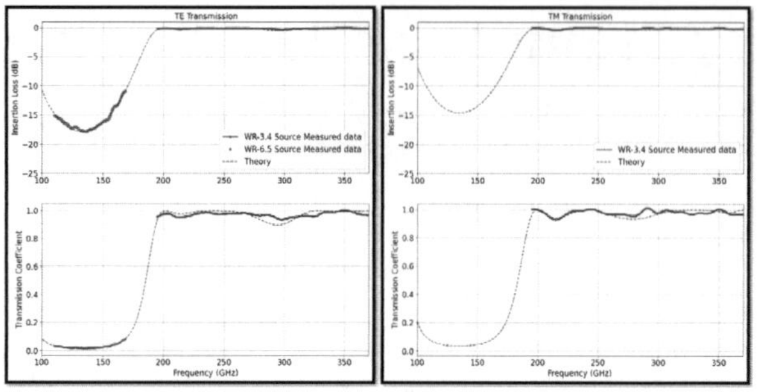

Fig. 2: Transmission curves for both Transverse Electric (TE) and Transverse Magnetic (TM) waves. The solid red curves are the predicted transmission value from our model. The measured data from both the WR-3.4 QO-VNA and WR-6.5 VNA are plotted on the same plot.

In our experiment, we placed the 130/270 GHz optical diplexer in front of the vacuum window of a 350 GHz wSMA receiver, with the diplexer inclined at an angle of 30° to the optical axis of the receiver. A series of Y-factor measurements were made with and without the diplexer. The noise temperature of the diplexer was inferred and converted to an insertion loss value, the compilation of which are included in Table I. Average loss, using two central frequencies in the passband with a bandwidth of around 1 GHz at an incident angle of 30 degrees was 3.2%.

TABLE I: Results of Transmission Loss Measurements

Center Frequency (GHz)	V-Pol Loss	H-Pol Loss
288	(3.8 ± 0.2) %	(3.4 ± 0.2) %
324	(2.5 ± 0.5) %	

V. Conclusion

We have presented an optical diplexer design, which spatially separates two bands of frequencies centered at 130 and 270 GHz, using a periodic stack of High Resistivity Silicon and Low Density Polyethylene. The simplicity, robustness, adaptability, and low insertion loss nature of this design, together with the possibility to operate at an angle of incidence of up to 30 degrees, make this class of diplexer an enabling tool for simultaneous dual-polarization multi-band receiver operation.

VI. Acknowledgements

This work is made possible by the generous support of the Submillimeter Array (SMA), the Black Hole Explorer (BHEX), and the next generation Event Horizon Telescope (ngEHT) through a grant from the Gordon and Betty Moore Foundation (GBMF-10423).

References

[1] M. J. Rioja, R. Dodson, and Y. Asaki, "The Transformational Power of Frequency Phase Transfer Methods for ngEHT," *Galaxies*, vol. 11, no. 1, p. 16, Jan. 2023.

[2] P. Grimes, R. Blundell, S. Paine, E. Tong, R. Wilson, and L. Zeng, "Receivers for the wideband Submillimeter Array," in *Space Telescopes and Instrumentation 2024: Optical, Infrared, and Millimeter Wave*. Proc 31th Int. Symp. Space THz Technology, March 2020, pp. 60–66.

[3] S. S. Doeleman *et al.*, "Reference Array and Design Consideration for the Next-Generation Event Horizon Telescope," *Galaxies*, vol. 11, no. 5, 2023.

[4] M. Johnson *et al.*, "The Black Hole Explorer: Motivation and Vision," in *Space Telescopes and Instrumentation 2024: Optical, Infrared, and Millimeter Wave*, vol. 13092, International Society for Optics and Photonics. SPIE, Jun 2024.

[5] M. Born and E. Wolf, *Principles of Optics: Electromagnetic Theory of Propagation, Interference, and Diffraction of Light*. London: Cambridge University Press, 1959.

[6] K. Carter, E. Tong, L. Zeng, and P. Grimes, "A Low Loss Silicon Optical Diplexer for Millimeter and Submillimeter Radio Astronomy," in *Space Telescopes and Instrumentation 2024: Optical, Infrared, and Millimeter Wave*, vol. 13102, International Society for Optics and Photonics. SPIE, Jun 2024.

[7] C.-Y. E. Tong, K. Carter, and J. Connors, "Quasi-optical characterization of low-loss polymers at 300 ghz for vacuum window applications," *IEEE Transactions on Terahertz Science and Technology*, vol. 10, no. 6, pp. 713–720, 2020.

[8] E. Tong, A. Hedden, and R. Blundell, "An Empirical Probe to the Operation of SIS Receivers- Revisiting the Technique of Intersecting Lines," in *Proc 19th Int. Symp. Space THz Technology*, April 2008, pp. 314–318.

[9] W. Shan, "Inconstant output noise of sis mixers and its implication in input noise measurement," *IEEE Transactions on Applied Superconductivity*, vol. 34, no. 3, pp. 1–7, 2024.

SESSION 5

On the sensitivity limitation in the HEB mixers

Boris S. Karasik[1*] and Chang Yoo[1]

Abstract—**While the superconducting HEB mixer remains the most sensitive heterodyne detector beyond 1.3 THz, its sensitivity is still well below the known quantum and thermal limits. We analyze possible limitations imposed by the biasing scheme as well as the parasitic effects distorting the noise temperature measurements.**

Keywords—**hot-electron, mixer, terahertz.**

THE Hot-Electron Bolometer (HEB) introduced in [1] is the detector of choice for heterodyne spectroscopy in the THz range above the frequency where the SIS mixer stops working (≈ 1.3 THz). Still, a gap between the SOA noise temperature in HEB mixers and the quantum limit T_{QL}^{DSB} = $hv/2k_B$ is several times greater than that for the SIS mixers and there has been no good explanation of why this is the case.

Meanwhile, aggressive attempts to beat the sensitivity records sometimes led to reporting extraordinarily low noise temperature values that could not be reproduced by others (see, e.g., [2]. Another issue has been the use of the Y-factor technique in a way producing a receiver noise response not related to the mixing process. The latter issue is specific to the HEB which is a total power detector. Since most of the measurements have been done using HEB devices integrated with an ultra-broadband (several THz bandwidth) log-spiral antenna, the THz power from the 295K calibration target impinging an HEB detector can easily be several nW. This is a large power for a typical NbN HEB that causes the so-called "direct detection," a change of the output noise due to the shift of the bias point or of the standing wave between the mixer and the LNA (in the absence of the microwave isolator), or both.

By mixing two 2.5 THz monochromatic sources, we investigated the behavior of the mixer conversion gain in an NbN HEB device as a function of the LO power and dc bias. We found that the conversion gain increases as the LO power decreases. However, when the IVC develops an N-shape, the gain abruptly drops (Fig. 1). At the same time, the IF spectrum changes from the single tone into an intermodulation distorted spectrum thus indicating the presence of 10-20 MHz oscillations in the dc bias and/or IF circuit [3]. Even though the IVC appears to be continuous in this regime it is actually a superposition of the "real" dc N-shaped IVC and the time-averaged rectified contribution of fast MHz oscillations. Y-factor measurements performed under such conditions would have a contribution of the electrical noise associated with the oscillations triggered by the "direct detection."

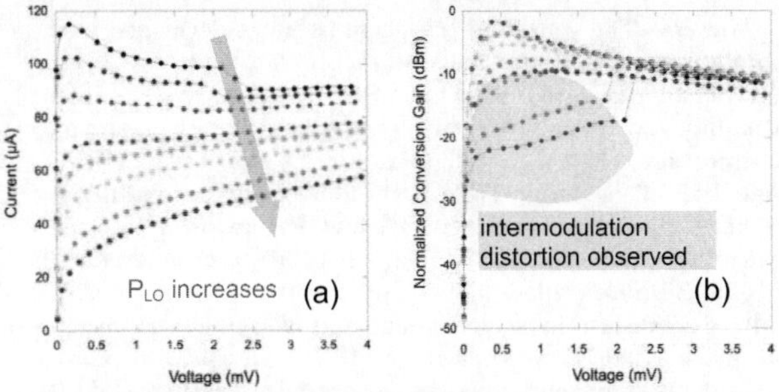

Fig. 1. (a) IVCs vs LO power in an NbN HEB at 4.3 K. (b) conversion gain vs LO power. The color of the symbols matches to that in Fig. 1a.

It is obvious that the regimes when the "direct detection" and/or the intermodulation distortion situations should be carefully avoided in the Y-factor measurements. The former regime can be mitigated by the use of a cold narrowband bandpass THz filter. On the other hand, the bolometric mixing model describes the transition to the N-shaped IVC with the decrease of the LO power continuously and predicts the gain under those hypothetical conditions to be much larger than that for the monotonic IVC. The operation of an HEB mixer in this regime has never been attempted. However, the counterpart direct detector TES always operates in this "voltage-biased" regime. The drastic difference in the approaches is due to the circuit instability issue which is much more severe in the case of the multi-GHz HEB IF circuit compared to the 10-100 kHz TES readout circuit. A solution allowing stable operation of the HEB mixer with the oscillations critically damped would lead to a much higher conversion gain followed by an improved noise temperature.

REFERENCES

[1] E. M. Gershenzon, G. N. Gol'tsman, I. G. Gogidze *et al.*, "Mixer of millimeter and submillimeter wave range based on heat-up of electrons in resistive state superconducting films," *Sverkhprovodimost': Fizika, Khimiya, Tekhnika*, vol. 3, pp. 2143-2160, 1990.

[2] I. Tretyakov, S. Ryabchun, M. Finkel *et al.*, "Low noise and wide bandwidth of NbN hot-electron bolometer mixers," *Appl. Phys. Lett.*, vol. 98, p. 033507, 2011.

[3] R. F. Su, Y. D. Zhang, X. C. Tu *et al.*, "Microwave probing of relaxation oscillations related to terahertz power detection in superconducting hot electron bolometers," *Supercond. Sci. Technol.*, vol. 32, p. 105002, 2019.

[1]Jet Propulsion Laboratory, California Institute of Technology, Pasadena, CA 91109, USA *Corresponding author (email: boris.s.karasik@jpl.nasa.gov).

NOTES:

Wideband Cryogenic Isolators for Sideband-separating Receivers

Lingzhen Zeng[1*], Edward Tong and Robert Kimberk

***Abstract*— The wideband cryogenic isolators developed in the Submillimeter Receiver Laboratory at the CfA have been deployed in the double-sideband (DSB) receiver system in the Submillimeter Array (SMA) for several years. These isolators play an important role in expanding the Intermediate Frequency (IF) bandwidth of the array. To further enhance our capabilities, we are developing phase-matched isolators for use in a sideband-separating (2SB) receiver system. This abstract outlines the essential requirements of the 2SB receiver system and the challenges inherent in isolator design and fabrication. Finally, we will present the measurement results of an isolator pair to demonstrate that our isolators successfully meet the design requirements for 2SB operations.**

***Keywords*— Calcium Vanadium, Ferrite devices, isolators, millimeter wave and terahertz components, YIG.**

I. Introduction

CRYOGENIC isolators are crucial components in low-noise heterodyne receiver systems extensively utilized in millimeter and submillimeter telescopes. To meet the increasing need for wider Intermediate Frequency (IF) bandwidths, there is a growing demand for a wideband, low-loss isolator in modern facilities such as the Submillimeter Array (SMA) and the Atacama Large Millimeter/submillimeter Array (ALMA).

The cryogenic isolators developed for the double-sideband (DSB) receivers have been demonstrated to be able to deliver very wideband performance, covering a frequency range from 4 GHz to 22 GHz [1, 2]. They exhibit insertion loss of less than 1 dB, together with return loss below -15 dB. They provide isolation better than 17 dB across most of the band.

In addition to the wideband performance for individual isolator mentioned above, the amplitude and phase of the S_{21} of an isolator pair used in a sideband-separating (2SB) receiver has to be well matched, if the IF quadrature hybrid is to be placed after the low noise IF amplifiers. This is the case for the ALMA Band6v2 development. To achieve a good image rejection ratio in this configuration, the amplitude and phase matching requirement are set to be 0.3 dB and 3 deg for an isolator pair.

In this presentation, we will report on the design, fabrication, and assembly of the isolator pairs to meet the requirements for such a 2SB receiver applications. We will provide measurement results to demonstrate that our isolator design meets the requirements for the ALMA Band6v2 project.

References

[1] L. Zeng, C. E. Tong, R. Blundell, P. K. Grimes and S. N. Paine, "A Low-Loss Edge-Mode Isolator With Improved Bandwidth for Cryogenic Operation," in *IEEE Transactions on Microwave Theory and Techniques*, vol. 66, no. 5, pp. 2154-2160, May 2018, doi: 10.1109/TMTT.2018.2799574.

[2] L. Zeng, C. -Y. E. Tong and S. N. Paine, "A Low-Insertion Loss Cryogenic Edge-Mode Isolator With 18 GHz Bandwidth," *in IEEE Journal of Microwaves*, vol. 3, no. 4, pp. 1258-1266, Oct. 2023, doi: 10.1109/JMW.2023.3307297.

[1]All authors are with: Center for Astrophysics | Harvard & Smithsonian, Cambridge, MA, USA.

[*]Corresponding author (email: lingzhen.zeng@cfa.harvard.edu).

NOTES:

Design of Reverse-Coupler Orthomode Transducer for 209-281 GHz

A. Navarrini[1*], P. Dindo[1], A. R. Kerr[1], J. Lambert[1], P. Mena[1], G. Morris[1], B. Casto[1], J. Effland[1] and K. Saini[1]

***Abstract*— We describe the design of a waveguide orthomode transducer (OMT) for the 209–281 GHz frequency band. The device is one of three candidates being considered for deployment in the upgraded ALMA Band 6 receiver, called "Band 6v2," currently under development by NRAO. The OMT is based on a symmetric reverse coupler structure. It has a circular waveguide input port (diameter 1.29 mm) and two single-mode oval waveguide output ports with full-radius corners matched to WR3.7 rectangular waveguide (0.94×0.47 mm²). A circular-to-square waveguide transition is used on the input side. The two oval waveguide outputs have E-plane orientations parallel to each other and are located on opposite sides of the OMT module. A commercial 3D electro-magnetic simulator was used to optimize the design's performance.**

***Keywords*— Branch-line coupler, hybrid coupler, orthomode transducer, polarimetry, polarization splitter, power combiner, reverse coupler, waveguide transitions, radio astronomy.**

I. Introduction

An orthomode transducer is a passive device that separates two orthogonal linearly polarized signals propagating in a common circular or square waveguide into two independent single-mode output ports. Highly symmetric waveguide structures are required to avoid the excitation of higher-order modes in the common OMT waveguide input and to achieve a broad fractional frequency bandwidth (40% or wider). Since the small dimensions and tight tolerances pose a significant challenge for fabrication and assembly of the parts, only a few broadband OMT designs have been demonstrated to work well at mm- and sub-millimeter wavelengths [1-17].

The broadband Bøifot junction orthomode transducer [1] currently used in the ALMA Band 6 cartridges [2] utilizes a thin septum in the common waveguide arm and wires in the waveguide side-arms, whose positioning in the split-blocks is critical to performance, particularly to its polarization properties. As part of the ALMA Band 6v2 receiver upgrade project [18], NRAO is evaluating three types of OMT for the 209-281 GHz frequency band that do not use pins or septa: two double-ridge designs [12], including one from NAOJ [19], and a symmetric reverse-coupler design [20,21].

II. Specifications of Band 6v2 OMT

Due to the complexities of performing precise electrical measurements at cryogenic temperatures (4 K), the Band 6v2 OMT's electrical specifications, detailed in Table I, are to be met at room temperature (293 K). This requires accurate room-temperature testing of the device before its integration into the cryogenically cooled receiver where it will ultimately operate.

The Band 6v2 OMT is specifically designed to integrate with the ALMA Band 6v2 receiver system. At its input, a circular waveguide with a diameter of 1.29 mm ensures compatibility with the Band 6v2 feed horn. At its outputs, two oval waveguides (matching 0.94 x 0.47 mm² WR-3.7 rectangular waveguide) connect to the Band 6v2 sideband separating (2SB) SIS mixer modules [22, 23]. These outputs are positioned on opposite sides of the OMT module, with their E-planes (electric field polarization planes) parallel to each other to ease integration with the other receiver cartridge components. In the following section, we present the design of the NRAO symmetric reverse-coupler waveguide OMT for potential application in the Band 6v2 receiver cartridge.

TABLE I. Band 6v2 OMT electrical specifications to be met at room temperature, 293 K

Parameter	Specification
Frequency Band	209-281 GHz
Input Return Losses	≥15 dB
Output Return Losses	≥15 dB
Insertion Losses	≤0.5 dB
Isolation	≥30 dB
Cross-polarization	≥30 dB

III. Design of Polarization-Splitting Symmetric Reverse Coupler Waveguide Structure

The symmetrical reverse-coupler waveguide structure illustrated in Fig. 1 is at the core of the broadband NRAO OMT candidate for Band 6v2. This design comprises four key elements:

a) Square waveguide input (0.94 x 0.94 mm²): This serves as the entry point for the incoming signal. Across the 209-281 GHz frequency range, it supports the two fundamental orthogonal modes TE_{10} (horizontal polarization, Pol 0) and TE_{01} (vertical polarization, Pol 1), as well as the higher order modes TE_{11} and TM_{11}, whose excitation is avoided by the twofold symmetry of the structure;

b) Two-section transformer for Pol 1: This component transitions the TE_{01} signal (vertical polarization) from the square waveguide to a rectangular waveguide (0.94 x 0.486 mm²), enabling efficient forward coupling;

c) Sidearm hybrid couplers: Two 90-degree hybrid waveguide couplers are positioned on the sidearms. These couplers split the incoming TE_{10} signal (horizontal polarization) into two equal outputs. The hybrids utilize branch-line coupling structures with four branches and reduced-height coupled waveguide (0.940×0.379 mm²). The separation between the main arm and the side arms (the width of the branch-lines) is 0.332 mm;

d) Reactively loaded terminations: Reactive terminations are placed at each hybrid coupler port. These terminations reflect the coupled Pol 0 signals back (reverse coupling) in anti-phase with equal power distribution (-3 dB).

[1] NRAO Central Development Laboratory, Charlottesville, VA 22903, USA. *Corresponding author (email: anavarri@nrao.edu).

The National Radio Astronomy Observatory is a facility of the National Science Foundation operated under cooperative agreement by Associated Universities, Inc

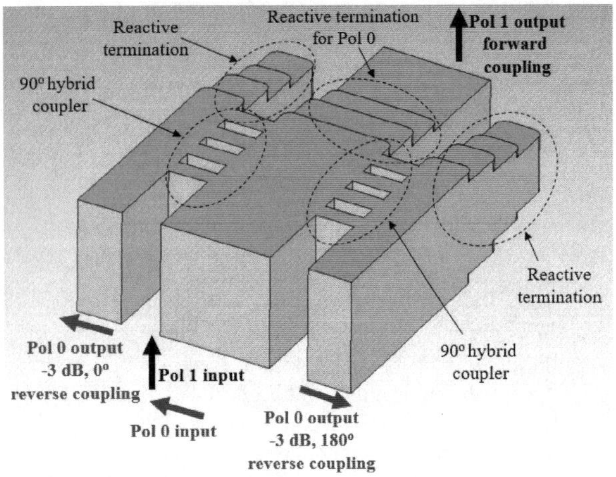

Fig. 1. Internal view of the symmetric dual-side backward coupler of our OMT with input square waveguide in common with two 3-dB branch-line coupling structures utilizing reactive terminations for Pol 0. The device has four physical ports and five electrical ports.

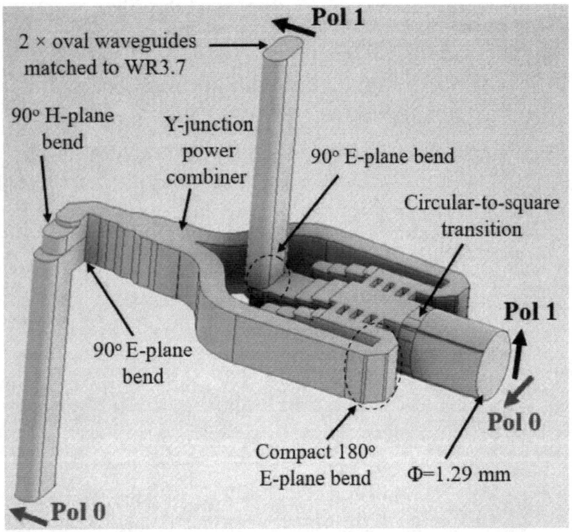

Fig. 2. Full OMT showing the circular waveguide input and the two oval waveguide outputs (matched to WR-3.7 waveguide).

IV. Complete Symmetric Reverse-Coupler OMT

While the previous section focused on the core functionality of the signal splitter (refer to Fig. 1), Figure 2 reveals the complete Band 6v2 OMT design. It includes various components working together to process the signals:

a) Quarter-wave circular-to-square waveguide transition: This section smoothly transforms the incoming signal from a circular waveguide (1.29 mm diameter, used for feed-horn compatibility) to the square waveguide (0.94 x 0.94 mm^2) of the OMT reverse-coupler polarization-splitting structure. To minimize higher-order mode excitation caused by rotational misalignment, an OMT with a circular waveguide input is preferable to one with a square waveguide input. This approach contrasts with the Band 6v1 design, which features a circular-to-square waveguide transition within the feedhorn and a square waveguide interface, potentially increasing susceptibility to mode excitation due to misalignment;

b) Polarization-splitting structure: The symmetric branch-line coupling structure with reactive terminations, as explained earlier, separates the incoming horizontal polarization signal (Pol 0) and reflects it using branch lines and special terminations;

c) Pol 0 Signal Routing: Two compact 180° E-plane waveguide bends change the direction of the Pol 0 signal paths by 180 degrees within the plane of the electric field (E-plane), re-orienting them in the forward direction;

d) An E-plane 180° Y-junction combiner [24] merges the two separate Pol 0 signals into a single waveguide output;

e) 90° bends in the E-plane and H-plane: The cascade of these waveguide bends guides the signal through 90-degree turns in the designated planes to achieve the routing of Pol 0 to the OMT oval waveguide output. It is noted that the E-plane bend twists the polarization plane of Pol 0, achieving the desired orientation at the OMT output (this E-plane bend functionality is not required in the Bøifot OMT design used in ALMA Band 6v1. There, the Y-junction power combiner can be directly cascaded with a 90-degree H-plane bend due to a different requirement for the final E-plane orientation of the Pol 0 signal at the Bøifot OMT output);

f) A 90° waveguide bend in the E-plane cascaded to the forward-coupled port of the polarization-splitting structure routes the Pol 1 signal to the desired OMT oval waveguide output.

CST Studio Suite [25] was employed to perform electromagnetic field simulations and optimizations for the OMT design. Assuming perfectly aligned block halves and a conductivity of $\sigma = 2.23 \times 10^7$ S/m, simulations predict the complete structure (as shown in Fig. 2) will meet the specifications outlined in Table 1. Notably, simulated input and output reflections are below -20 dB, isolation is below -79 dB, cross-polarization is below -61 dB, and transmission surpasses -0.5 dB for both polarizations across the entire 209-281 GHz band. It's important to acknowledge that these exceptional isolation and cross-polarization values are predicted for an ideal OMT with perfect machining and alignment. Any misalignment of the block halves in a real-world OMT degrades this performance.

V. Conclusion

We presented the design of a symmetrical reverse-coupler orthomode transducer (OMT) for the 209-281 GHz frequency band, one among three candidate designs being considered for the ALMA Band 6v2 cryogenic receiver cartridge. The OMT successfully meets the electromagnetic requirements set for this application.

References

[1] A. M. Boifot, E. Lier, and T. Schaug-Pettersen, "Simple and broadband orthomode transducer," *IEEE Proc.*, vol. 137, no. 6, pp. 396–400, Dec. 1990.

[2] E. J. Wollack and W. Grammer, "Symmetric waveguide orthomode junctions," *Proc. 14th Int. Symp. Space THz Technol.*, 2003, pp. 169–176.

[3] G. Narayanan and N. R. Erickson, "Full-waveguide band orthomode transducer for the 3 mm and 1 mm bands," *Proc. 14th Int. Symp. Space THz Technol.*, 2003, pp. 508–512.

[4] A. Navarrini and M. Carter, "Design of a dual polarization SIS sideband separating receiver based on waveguide OMT for the 275–370GHz frequency band," *Proc. 14th Int. Symp. Space THz Technol.*, pp. 159-168, 2003-Apr.

[5] A. Dunning, S. Srikanth, and A. R. Kerr, "A simple orthomode transducer for centimeter to submillimeter wavelengths," Proc. 20th Int. Symp. Space THz Technol., 2009, pp. 191–194.

[6] A. Navarrini and R. L. Plambeck, "A turnstile junction waveguide orthomode transducer", *IEEE Trans. Microw. Theory Tech.*, vol. 54, no. 1, pp. 272-277, Jan. 2006.

[7] A. Navarrini, A. Bolatto and R. L. Plambeck, "Test of 1 mm band turnstile junction waveguide orthomode transducer", *Proc. 17th Int. Space Terahertz Technol. Symp.*, 2006-May-10–12.

[8] G. Pisano, L. Pietranera, K. Isaak, L. Piccirillo, B. Johnson, B. Maffei, et al., "A broadband WR10 turnstile junction orthomode transducer," *IEEE Microw. Wireless Compon. Lett.*, vol. 17, no. 4, pp. 286-288, Apr. 2007.

[9] D. Henke and S. Claude, "Design of a 70–116 GHz W-band turnstile OMT," *2014 44th European Microwave Conference*, Rome, Italy, 2014, pp. 456-459, doi: 10.1109/EuMC.2014.6986469.

[10] P. Yagoubov et al., "Wideband 67-116 GHz receiver development for ALMA Band 2," *A&A*, Vol. 634, Feb. 2024, https://doi.org/10.1051/0004-6361/201936777

[11] A. Gomez-Torrent, U. Shah and J. Oberhammer, "Compact Silicon-Micromachined Wideband 220–330-GHz Turnstile Orthomode Transducer," *IEEE THz Sci. Technol.*, vol. 9, no. 1, pp. 38-46, Jan. 2019, doi: 10.1109/TTHZ.2018.2882745.

[12] A. Dunning, "Double ridged orthogonal mode transducer for the 16-26 GHz microwave band," *Proc. Workshop Appl. Radio Sci.*, 2002.

[13] G. Moorey et al., "A 77-117 GHz cryogenically cooled receiver for radio astronomy," *Proc. Workshop Appl. Radio Sci.*, 2006, pp. 1–7.

[14] A. Gonzalez and S. Asayama, "Double-ridged waveguide orthomode transducer (OMT) for the 67-116-GHz band," *J. Infrared, Millimeter, THz Waves*, vol. 39, no. 8, pp. 723–737, Aug. 2018.

[15] S. Asayama and M. Kamikura, "Development of doubled-ridged waveguide orthomode transducer for the 2mm band," *J. Infrared Millimeter THz Waves*, vol. 30, pp. 573–579, Jun. 2009.

[16] M. Kamikura, M. Naruse, S. Asayama, N. Satou, W. Shan, Y. Sekimoto "Development of a 385-500 GHz Orthomode Transducer (OMT)," *19th Int. Symp. on Space THz Tech.*, Groningen, 28-30 April 2008.

[17] G. Valente and A. Navarrini, "Design of a Superconducting Planar Orthomode Transducer for the 84-116 GHz Band," *2024 4th URSI Atlantic Radio Science Meeting (AT-RASC)*, Meloneras, Spain, 2024, pp. 1-4, doi: 10.46620/URSIATRASC24/YPKO4796.

[18] A. Navarrini et al., "ALMA Band 6v2 receiver development status," *2023 XXXVth Gen. Assem. Sci. Symp. Int. Union Radio Sci. (URSI GASS)*, vol. 00, pp. 1–4, 2023, doi: 10.23919/ursigass57860.2023.10265563.

[19] A. Gonzalez and K. Kaneko, "Practical Aspects of the Design and Fabrication of High-Performance (sub)mm-Wave Dual-Ridged Waveguide Orthomode Transducers, and Application to a 205–280 GHz Design," *IEEE THz Sci. Technol.*, vol. 13, no. 6, pp. 587-593, Nov. 2023, doi: 10.1109/TTHZ.2023.3308057.

[20] A. Navarrini and R. Nesti, "Symmetric Reverse-Coupling Waveguide Orthomode Transducer for the 3 mm Band," *IEEE Trans. Microwave Theory Tech.*, Vol. 57, Issue 1, pp. 80-88, Jan. 2009.

[21] A. Navarrini, C. Groppi, R. Lin, G. Chattopadhyay, "Test of a Waveguide Orthomode Transducer for the 385-500 GHz Band," *Proc. 22nd Int. Symp. on Space THz Tech.*, Tucson, 26-28 April 2011.

[22] J. G. Lambert et al., "Development Status of the ALMA Band 6v2 SIS Mixers," submitted to *33nd IEEE Int. Symp. on Space THz Tech.*, Charlottesville, VA, USA, April 7-11, 2024.

[23] P. Dindo et al., "Design and Analysis of the Waveguide Circuitry for the ALMA Band 6v2 Sideband Separating Mixer," submitted to *33nd IEEE Int. Symp. on Space THz Tech.*, Charlottesville, VA, USA, April 7-11, 2024.

[24] A. R. Kerr, "Elements for -plane split-blocks waveguide circuits," National Radio Astronomy Observatory, *ALMA Memo 381*, 5 July 2001. [Online]. Available: https://library.nrao.edu/public/memos/alma/memo381.pdf.

[25] CST Studio Suite. [Online]. Available: https://www.3ds.com/products/simulia

Comprehensive laboratory characterization of the AMKID instrument

Nicolás Reyes[1], Andrey Baryshev[1-3], Ivan Cámara-Mayorga[1], Gerrit Grutzeck[1],
Christopher Heiter[1], Sebastian Jorquera[4], Axel Weiss[1], Stephen Yates[2], J.A. Baselman[2], Bernd Klein[1]

***Abstract*—Deployment of a complex instrument requires extensive and precise laboratory measuremnt to asses on instrument quality. In this talk we will review the case of the AMKID instrument, a large field of view incoherent detector camera designed for the APEX telescope. We will present different experimental techniques developed to evaluate the performance of the optical setup, the camera sensitivity, and the inference of external magnetic fields on measuremnt quality. Finnally we will show the current status of the commissioning of AMKID, which is on track for scientific operations in 2024, preliminary telescope results show a good agreement with the laboratory measuremnts, providing a prove of quality of the experimental system described in this talk**

***Keywords*—Incoherent detectors, astronomical instruments, experimenta methods.**

I. Introduction

THE APEX telescope Microwave Kinetic Inductor (AMKID) instrument is a wide-field camera operating in the sub-millimeter window. It is a dual-color instrument composed of two arrays at 870 and 350 µm. Both arrays are diplexed in polarization and can operate simultaneously. The instrument is designed to be installed at the APEX telescope [1], making use of the superb observation conditions of the Chajnantor plateu. This fact is especially relevant for the high frequency band, as regular operation at 350um requires precipitable water vapour to be lower than 0.5mm. The instrument covers a field of view of 15' x 15' with an unprecedent number of pixels: 2.800 pixels at the low frequency band and 13.800 pixels at the high frequency band. Before deployment at the telescope, the instrument was extensively tested in laboratory enviroment. Performed measurements and analysis include optical quality and alignment, optical performance, instrument sensitivty, and detector pick-up from external magnetic fields, between others. To precisely assest the instrument performance a dedicated experimental setup was built to mimic the instrument (telescope) operational conditions. The setup includes, among others, a wire-scanner: a device designed to perform fast and reliable instrument characterization at the telescope focal plane.

In this talk we will present the design of the laboratory enviroment, the main test and simulatiuons we performed, and the achieved results. Finnally we will show the current status of the commissioning of AMKID at the APEX telescope, which is on track for scientific operations in 2024.

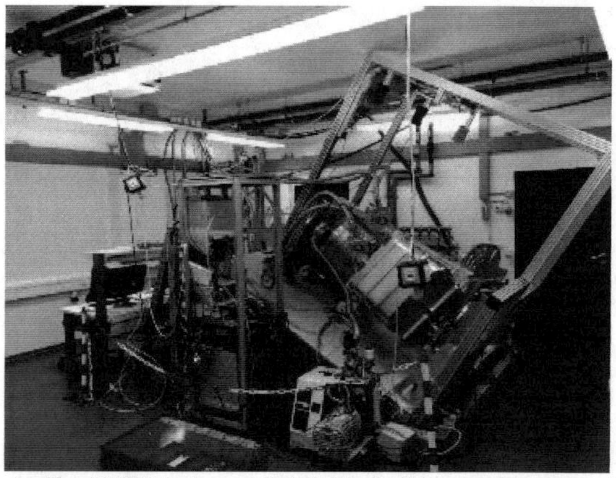

Fig. 1. The AMKID instrument installed in the laboratory test system. The setup allows to tilt the instrument to its normal operation conditions. A large Liquid Nitrogen load provide a background loading similar to the expected atmosphere load.

References

[1] R. Güsten, et al., "The Atacama pathfinder experiment (APEX), a new submillimeter facility for southern skies" A&A, 454(2): L13–L16, 2006.

[1]Max Planck Institute for Radio-Astronomy, Auf dem Hugel 69, 53121, Bonn.
[2]SRON, Netherland Institute for Space Research, Groningen, The Netherlands.
[3] Kapteyn Astronomical Institute, University of Groningen, The Netherlands
[4] Pontificai Universidad Catolica de Chile, San Joaquin, Chile
*Corresponding author (email: nireyes@mpifr-bonn.mpg.de)

NOTES:

Installation and testing of the wSMA prototype receiver system

Paul K. Grimes[1*], Robert Christensen, Steve Leiker, Edward Tong, Robert W. Wilson and Lingzhen Zeng

***Abstract*—For the past several years, the Submillimeter Array has been developing the wideband Submillimeter Array (wSMA) upgrade. This upgrade includes the complete replacement of the aging SMA receiver systems with new receiver systems consisting of a new cryostat, receiver cartridges, SIS receiver front ends, receiver optics, local oscillators and IF processor. Following extensive laboratory testing, the prototype receiver has recently been installed in an SMA antenna for on-sky testing. We describe the prototype receiver, the deployment and commissioning process and the results of on-sky testing.**

***Keywords*—Heterodyne receivers, astronomical instrumentation.**

I. Introduction

The Submillimeter Array[2] has operated with its current receiver systems since its commissioning in 2003. In 2015, we began the design of the wideband Submillimeter Array upgrade, a complete redesign of the receiver and other systems with the aim of improving the sensitivity, and increasing the throughput, of the SMA, while also replacing the aging cryogenic systems.

The wSMA upgrade consists of new cryostats for each of the SMA antennas, containing cooled receiver selection optics and two receiver cartridges, each of which house new integrated dual polarization receivers based on DSB SIS mixers optimized for wide IF bandwidth. New receiver control electronics and local oscillator modules and an IF processor complete the wSMA receiver system.

The wSMA receivers will provide two frequency bands for the SMA, with LO frequencies from 210-270 GHz and 280-360 GHz, and IF bandwidths from 4-20 GHz. By using wire grid polarizers and optical diplexers [1] in the cooled receiver optics, dual frequency observing in either single or dual polarization modes will be possible. The two polarization channels of each receiver band can be independently tuned, allowing for "split" LO tunings that provide greater on-sky instantaneous bandwidth.

The first prototypes of the new cryostats and receiver cartridges were delivered during the pandemic in 2020 and 2021. After extensive laboratory testing of the cryostats, integration of receiver frontends to the cartridges, testing of new receiver electronics and local oscillator modules, and performance testing of the full receiver system, one prototype receiver was delivered to the SMA site in early 2023.

Fig. 1. Photo of the prototype wSMA receiver installed in SMA Antenna 7.

We are currently installing this prototype receiver into an SMA antenna, for commissioning checks and on-sky testing. In this paper we will describe the design of the prototype receiver, discuss the results of laboratory performance testing, the receiver commissioning, and early results of on-sky testing.

[1] Carter et al, "A Low Loss Optical Diplexing Scheme for Millimeter to Terahertz Waves," to be presented at ISSTT2024.

[1]All authors are with: Center for Astrophysics | Harvard & Smithsonian, Cambridge, MA 02138, USA. *Corresponding author (email: pgrimes@cfa.harvard.edu).

[2]The Submillimeter Array is a joint project between the Smithsonian Astrophysical Observatory and the Academia Sinica Institute of Astronomy and Astrophysics and is funded by the Smithsonian Institution and the Academia Sinica.

NOTES:

SESSION 6

Estimating the Sensitivity of Ultra-Wideband Cryogenic IF-LNAs to Input Mismatch by Noise Wave Measurement

R. I. Amils[1,2*], I. Malo-Gómez[1], M. C. Diez-González[1], I. López-Fernández[1], A. García-Merino[1], and J. D. Gallego[1]

Abstract— **The noise performance of cryogenic LNAs is usually evaluated using a single parameter: its noise temperature. In this work we present a simple experimental method to measure a complete set of noise parameters, expressed as noise waves, which allow an intuitive evaluation of the effect on the sensitivity of the mismatch at the input of cryogenic LNAs. This method is applicable to broadband LNAs and only requires knowledge of the input return loss of the amplifier and the noise power measured with an electrically long short-circuited line connected at its input. This is particularly useful for the case of LNAs used in the IF of cryogenic heterodyne receivers with not perfectly matched SIS or HEB mixers.**

Keywords— **Cryogenic Low Noise Amplifiers, High Sensitivity, Noise Parameters, Noise Waves, Millimeter Receivers, SIS Mixers.**

I. Introduction

THE trend of modern radio astronomy receivers is to achieve larger instantaneous bandwidths and better sensitivities [1]. In such wide bands, matching of the mixer and the LNA becomes complicated. Traditionally, noise temperature has been the main indicator used to evaluate the quality of an LNA at cryogenic temperature. However, this parameter does not completely characterize the sensitivity degradation produced by the mismatch of the impedance presented at its input. To evaluate this, we have selected a full set of four noise parameters using the input noise wave representation proposed in [2]. This representation has the advantage of providing an easy and intuitive interpretation of the noise degradation as a function of the reflection coefficient presented at the input by direct inspection of its parameters. This formulation can be considered the noise analogue of the familiar S-parameters.

In this communication we propose a method to measure this set of noise parameters which is applied to a 4-20 GHz cryogenic LNA developed in Yebes. This method requires a simpler experimental setup and less computation than other attempts to measure the noise parameters of an LNA at cryogenic temperature [3], [4].

II. Input Noise Waves

The input noise wave representation was proposed by Meys [2] in 1978 and considers two partially correlated noise waves at the input of the LNA, one going into the amplifier, A_n, and another emerging from it, B_n, as depicted in Fig. 1. Wave B_n gets reflected at the source and interacts with A_n to produce the complete noise contribution at the input of the LNA, A_{ns}, given by

$$A_{ns} = A_n + \Gamma_S B_n \quad (1)$$

where Γ_S represents the reflection coefficient of the source.

Taking the square modulus of (1) and following the notation proposed by Meys, the total noise power at the input of the LNA can be expressed in units of temperature as

$$T_{ns} = T_a + |\Gamma_S|^2 T_b + 2T_c|\Gamma_S|\cos(\phi_S + \phi_c) \quad (2)$$

where $|\Gamma_S|$ and ϕ_S are the modulus and phase of the reflection coefficient of the source impedance, and T_a, T_b, T_c and ϕ_c are the four noise parameters of the Meys noise wave representation, which we will refer to as noise wave parameters. T_a is the noise temperature measured using a matched load at the source and represents the non-correlated part of the wave going into the amplifier, commonly referred to as the effective noise temperature. T_b represents the non-correlated part of the noise wave emitted from the LNA towards the source. And finally, the correlated part of the noise waves is characterized by parameters T_c and ϕ_c. Parameters T_b, T_c and ϕ_c contribute to produce a change in the noise temperature of the LNA, modulated by the input mismatch as expressed in (2). This set of noise parameters is completely equivalent to other noise representations into which they can be easily transformed [5].

III. Experimental Procedure

We present a method to measure the noise wave parameters T_a, T_b, T_c and ϕ_c of an LNA at cryogenic temperature. This method is inspired by [2] and based on the simplification of expression (2) when a lossless short-circuit ($|\Gamma_S| = 1$) is used as the source of the amplifier depicted in the schematic of Fig. 1

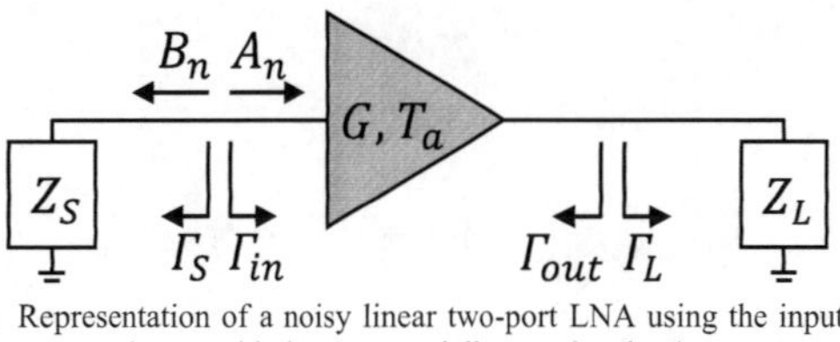

Fig. 1. Representation of a noisy linear two-port LNA using the input noise wave representation considering two partially correlated noise waves.

[1]Yebes Observatory, CDT (IGN), Yebes, 19141, Spain; [2]University of Alcalá, Alcalá de Henares, 28871, Spain. *Corresponding author (email: r.amils@oan.es).

This work was supported in part by the RADIOBLOCKS project within the Horizon Europe programme under grant agreement No 101093934, and the YDALGO project No ICTS-2018-01-CNIG-11 within the Multiregional Operational Programme for Spain 2014-20 ERDF.

NOTES:

$$T_{ns} = T_a + T_b + 2T_c\cos(\phi_s + \phi_c) \quad (3)$$

where T_{ns} is a sinusoidal function of frequency with an average value of $T_a + T_b$ and a peak-to-peak value of $4T_c$. To experimentally implement this procedure, the short-circuit presented at the input of the LNA must be low loss and sufficiently electrically long to allow the correct evaluation of the envelope of the sinusoid described by (3) which will be used to determine the average and peak-to-peak values of T_{ns}.

The electrically long short-circuited line was fabricated in-house by soldering a fitted brass cylinder to one end of an 80 mm piece of 0.141 mil low loss coaxial line (Microcoax UT-141C-LL) connecting the internal and external conductors (Fig. 2). The other end of this coaxial line was terminated in a male 2.92mm connector. The physical length of this line was selected to produce a period in T_{ns} of approximately 1 cycle/GHz in the 4-20 GHz band of the cryogenic LNA used to demonstrate this method (Section IV). This periodicity proved to be adequate to allow the correct evaluation of the envelope of T_{ns}.

Three cryogenic measurements are required to obtain T_{ns} and therefore the four noise wave parameters: a) the gain (G) and noise temperature of the LNA measured with a matched source (T_a) using a variable temperature load as described in [6], b) the S_{11} of the amplifier, and c) the total output noise power of the LNA when loaded at its input with the electrically long short-circuited line (T_{out}). This output power measurement is transformed into noise temperature at the input of the amplifier, T_{ns}, using

$$T_{ns} = \frac{|1 - \Gamma_S\Gamma_{in}|^2}{G} T_{out} \quad (4)$$

where Γ_S is the reflection coefficient of the short-circuited line and Γ_{in} is the input reflection coefficient of the LNA which is assumed equals to its S_{11}. The term $|1 - \Gamma_S\Gamma_{in}|^2$ in (4) accounts for the mismatch between the source and the input of the LNA. As described, the average value of the envelopes of T_{ns} allows to obtain T_b by subtraction of the previously measured value of T_a, while T_c is one fourth of the peak-to-peak value of the envelopes of T_{ns}. The fourth parameter, ϕ_c, can be obtained by feeding the results for T_a, T_b and T_c into (3) and fitting the value of ϕ_c that minimizes its difference with the measured value of T_{ns}.

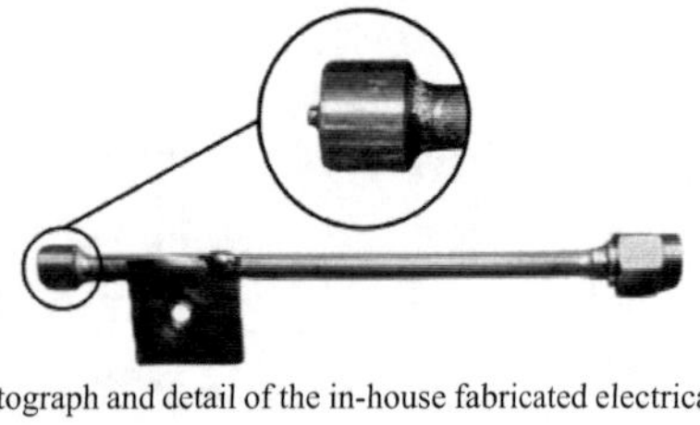

Fig. 2. Photograph and detail of the in-house fabricated electrically long short-circuited line.

The experimental setup used for the proposed method consists of a noise-receiver-equipped (opt. 029) Keysight N5247A PNA-X interfaced with an experimental cryostat and the previously described short-circuited line.

The PNA-X is used to measure cryogenic noise power and cryogenic one-port S-parameters of the LNA and the electrically long short-circuited line. A SOLT calibration of this instrument for S-parameter measurement is done using an electronic calibration module (Agilent N4693-6001) at room temperature. The length and loss variation of the stainless-steel line used to interface the inside and the outside of the cryostat is calibrated by measuring a cryogenically cooled flush short located inside the cryostat. The S_{11} of the LNA is characterized by terminating its output with a matched load and measuring the reflection at its input. The reflection coefficient of the short-circuited line, Γ_S, was carefully measured obtaining a value of the magnitude better than 0.4 dB within most of the 4-20 GHz band.

All measurements are conducted consecutively in the same calibration plane with the same experimental setup and using a PID temperature controller (Lake Shore 336) to reduce temperature variation of DUTs.

IV. RESULTS

To illustrate the proposed method, a 4-20 GHz cryogenic LNA[1] designed and fabricated at Yebes was characterized experimentally at 7 K. Fig. 3 illustrates the measured results for the total output power, T_{out}, and noise temperature, T_{ns}, of the LNA loaded at its input with the electrically long short-circuited line. The results for the four noise wave parameters are depicted in Fig. 4, together with their modeled results. These simulated values were obtained from the complete CAD model of the cryogenic LNA, generated using Keysight ADS software in the same way as described in [6], where the noise contribution of the transistors is included using Pospieszalski's model [7].

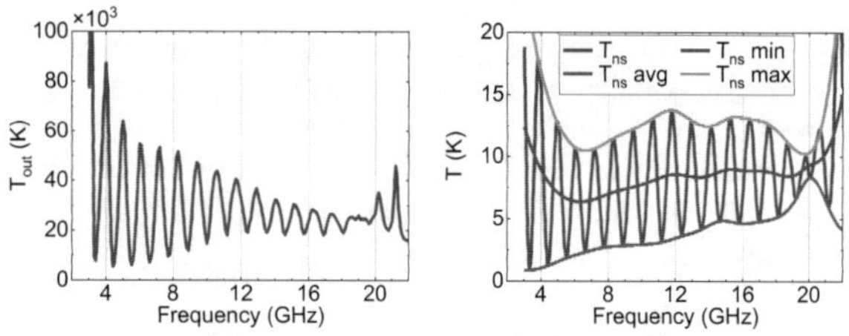

Fig. 3. Experimental results at 7 K for (a) T_{out} and (b) T_{ns} of the 4-20 GHz cryogenic LNA developed at Yebes when terminated with the electrically long short-circuited line. T_{ns} is plotted together with the extracted values of its average and envelopes.

[1]This LNA was developed under the ESO Technology Development Programme aiming at fulfilling the requirements of the ALMA Wideband Sensitivity Upgrade.

NOTES:

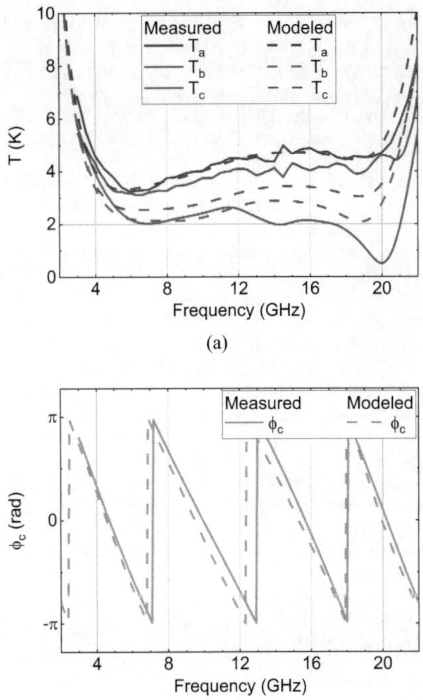

Fig. 4. Experimental (solid) and modeled (dashed) results for (a) T_a, T_b, T_c and (b) ϕ_c of the 4-20 GHz cryogenic LNA from Yebes at 7 K. Note that T_a is the noise temperature measured using a matched input load.

Agreement between measured and modeled noise wave parameters is very good for T_a in the whole band, and up to 12 GHz in the case of T_c. Parameter T_b is offset from modeled values, which could be justified due to loss in the experimental short-circuited line.

Parameters T_a and T_b have a similar flat evolution within the frequency band. The parameter responsible for the description of the magnitude of the correlation, T_c, presents an evident minimum at 20 GHz caused by the low ripple amplitude of T_{ns} at this frequency (see Fig. 3(b)). It is interesting to point out that, for this experiment, the noise temperature that emerges from the input of the LNA, T_b, is lower than its 7 K physical temperature.

V. Method Limitations

The proposed method is only applicable if the LNA complies with some restrictions. First, the LNA must be unconditionally stable for it to be possible to measure its output power without breaking into oscillation when loaded with a short-circuit. Second, it must be wideband in order for a ripple to form in T_{ns} that allows the correct evaluation of its envelope without requiring an excessively long short-circuited line. And third, the LNA gain should be high enough to ensure that the noise contribution of the PNA-X noise receiver is negligible relative to the value of T_{out}. Additionally, gain must also be constant between the three consecutive cooldowns required for the evaluation of the noise parameters. This gain variation has been estimated to be lower than 0.15 dB for the measurement setup used, and according with simulations, this may produce errors lower than 10% on T_b and lower than 5% on T_c.

Finally, another identified potential limitation of the method is the unknown reflection coefficient of the PNA-X noise receiver (Γ_L in the circuit depicted in Fig. 1) which has repercussions on the approximation of Γ_{in} by the S_{11} of the LNA and on the possible formation of standing waves at the output of the amplifier. For $\Gamma_{in} \approx S_{11}$ to be true, the S_{12} of the amplifier and/or the Γ_L of the PNA-X must be approximately zero. Regarding the formation of output standing waves, it should be noted that in the proposed experimental setup Γ_{out} adopts its worst possible value for a given LNA, since the input load is a short-circuit and therefore $\Gamma_S \approx$ -1. In this scenario any possible deviation of Γ_L from a matched load would have the effect of producing a standing wave pattern on the output of the LNA. Both contributions, $\Gamma_{in} \approx S_{11}$ and standing wave formation, have been studied concluding that even a poor $|\Gamma_L|$ of -15 dB would only translate into a variation of about 10% in T_b and 5% in T_c, for the values of $|S_{21}| > 30$ dB and $|S_{12}| < -50$ dB of the tested LNA.

VI. Alternative Method

An alternative method to obtain T_b, T_c and ϕ_c for a known T_a has also been developed. In this case, instead of using the envelope of T_{ns}, three instances of non-linear equation (3), corresponding to different lengths of short-circuited line (0, 80 and 160 mm), are numerically solved simultaneously at each frequency point. This can be expressed as

$$\begin{aligned} T_{ns,k} &= T_a + T_b + 2T_c\cos(\phi_{s,k} + \phi_c) \\ k &= 1,2,3 \end{aligned} \tag{5}$$

where $T_{ns,k}$ is the input noise of the LNA loaded with the corresponding short-circuited line and $\phi_{s,k}$ the phase of each line. The results obtained using this method are shown in Fig. 5 as individual points. A considerable level of dispersion can be appreciated at frequencies where all instances of T_{ns} in (5) have similar values. Fig. 5 also compares these results with the ones obtained with the previous method (solid lines), showing an overall good agreement. Although this new method has the advantage of allowing a direct extraction of parameter ϕ_c, it requires more measurements and produces noisier results making it less experimentally appealing and therefore only used in this study to corroborate the proposed method.

VII. Conclusions

A simple method for measuring the noise parameters of LNAs at cryogenic temperature has been presented and demonstrated. The measured noise parameters are based on the

NOTES:

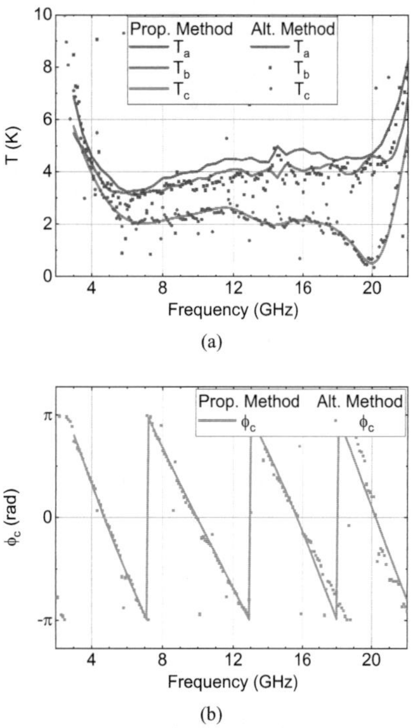

Fig. 5. Results for (a) T_b, T_c and (b) ϕ_c obtained using the alternative (discrete points) and proposed (solid lines) methods at 7 K.

input noise wave representation proposed by Meys. These parameters provide a direct and intuitive interpretation of the sensitivity degradation produced by the source mismatch, making them a useful tool for the evaluation and selection of amplifiers for a wide range of low noise applications such as the IF chain of radio astronomy receivers. This set of experimental noise parameters can also be transformed into any other representation, as for example T_{min}, Γ_{opt} and r_n, which allow to generate a complete experimental Touchstone S-parameter file of a cryogenic LNA. The proposed method has the advantage of requiring a simple experimental setup and only one additional cryogenic measurement beyond those usually made to characterize this kind of devices, i.e. noise temperature and S-parameters.

This method has been applied to a broadband 4-20 GHz cryogenic LNA showing good agreement between experimental and modeled results. Confidence in these results has been provided through corroboration by an alternative method. The limitations of the proposed experimental procedure, along with possible sources of measurement error, have been analyzed. These uncertainties have been studied in the context of this amplifier and have been found to be similar to those obtained in the measurement of the noise temperature of cryogenic LNAs [6], [8].

References

[1] J. Carpenter, D. Iono, L. Testi, N. Whyborn, A. Wootten, and N. Evans, "The ALMA Development Roadmap," 2018. [Online]. Available: http://www.almaobservatory.org/wp-content/uploads/2018/07/20180712-alma-development-roadmap.pdf. [Accessed Jul. 5, 2024].

[2] R. P. Meys, "A wave approach to the noise properties of linear microwave devices," *IEEE Trans. Microw. Theory Tech.*, vol. 26, no. 1, Jan., pp. 34-37, 1978.

[3] D. Russell and S. Weinreb, "Cryogenic Self-Calibrating Noise Parameter Measurement System." *IEEE Trans. Microw. Theory Tech.*, vol. 60, no. 5, May., pp. 1456-1467, 2012.

[4] R. Hu and S. Weinreb, "A novel wideband noise-parameter measurement method and its cryogenic application", *IEEE Trans. Microw. Theory Tech.*, vol. 52, no. 5, May., pp. 1498-1507, 2004.

[5] J. A. Dobrowolski, *Introduction to Computer Methods for Microwave Circuit Analysis and Design*. Boston, MA: Artech House, 1991.

[6] I. López-Fernández, J. D. Gallego-Puyol, C. Diez, I. Malo-Gomez, R. I. Amils, R. Flückiger, D. Marti, and R. Hesper, "A 16-GHz Bandwidth Cryogenic IF Amplifier With 4-K Noise Temperature for Sub-mm Radio-Astronomy Receivers," *IEEE Transactions on Terahertz Science and Technology*, vol. 14, no. 3, May., pp. 336-345, 2024.

[7] M. W. Pospieszalski, "Modeling of noise parameters of MESFETs and MODFETs and their frequency and temperature dependence," *IEEE Trans. Microw. Theory Techn.*, vol. 37, no. 9, Sep., pp. 1340-1350, 1989.

[8] J. D. Gallego and J. L. Cano, "Estimation of uncertainty in noise measurements using Monte Carlo analysis," presented at Radionet FP7 1st Eng. Forum Workshop, Gothenburg, Sweden, 2009, doi: 10.5281/zenodo.8203244. [Online]. Available: https://zenodo.org/records/8203244. [Accessed Jul. 5, 2024].

NOTES:

345 GHz SIS Junction development for the ngEHT

Jacob Kooi[1*], Bruce Bumble[1], Henry Leduc[1], Alan Kleinsasser[1], Darren Hayton[1], Katie Bouman[2], Gopal Narayanan[3], Sheperd "Shep" S. Doeleman[4]

Abstract—**The Microdevices laboratory (MDL) at the Jet Propulsion Laboratory (JPL), California, with a long history of innovative SIS mixer technology development and fabrication was contracted in 2021 to deliver quantum limited ALMA band 7 (275 – 370 GHz) SIS mixing devices for the ngEHT. Added goals were device uniformity and high yield for 2SB mixing operation and up to 20 GHz of supported intermediate frequency (IF).**

Keywords— **ALMA Band 7, ngEHT, SIS Tunnel Junction, 20 GHz IF, SOI Frame.**

I. Introduction

In 2018 the Event Horizon telescope (EHT) obtained the first ever image of a super massive black hole in M87 in the center of a giant elliptical galaxy in the constellation Virgo, ~55 million light years away from Earth. Then in May 2022 the EHT revealed a first ever image of the supermassive black hole at the center of the Milky Way galaxy: Sagittarius A*, approximately 27,000 light-years away from Earth. These images synthesized observational data from eight submillimeter telescopes around the world in a very long baseline interferometer (VLBI) network. To further enhance the imaging resolution of the EHT and create high-definition black hole images and movies the next generation Event Horizon Telescope (ngEHT) [1] endeavors to double the number of antennas in the existing telescope array, utilize tri-color observations (85-, 230-, 345 GHz), and increase the intermediate frequency (IF) continuum throughput.

The Microdevices laboratory (MDL) at the Jet Propulsion Laboratory (JPL), California, with a long history of innovative SIS mixer technology development and fabrication has taken up the challenge to deliver a new generation of quantum limited high current density SIS mixing devices (ALMA band 7, 275 – 370 GHz) with an Intermediate Frequency (IF) up to 20 GHz and high yield/uniformity. To do so the SIS devices under development utilize:

1. *8 um Silicon on Insulator (SOI) substrate.*
2. *Use a novel Si frame concept with ~ 75 frames / 4" wafer, each frame holding 16 SIS devices (Fig. 1), each junction on the wafer being uniquely identified.*
3. *Nb/AlNx/Nb high current density (~25 kA/cm²).*
4. *IF bandwidth 2 - 20 GHz.*
5. *High yield and uniformity*

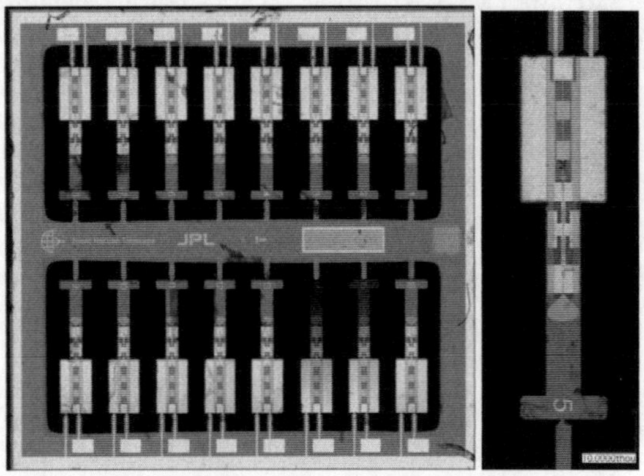

Fig. 1. Silicon Frame with 16 SIS Tunnel Junctions. Each device/Frame has a unique ID.

In addition to the development of the 345 GHz 'Ultra broad IF bandwidth' SOI devices, two double sideband (DSB) mixer blocks have been acquired to verify the RF and IF performance of the set forth SIS mixers.

The talk will outline concept, frame and junction design, and finally measured RF and IF performance.

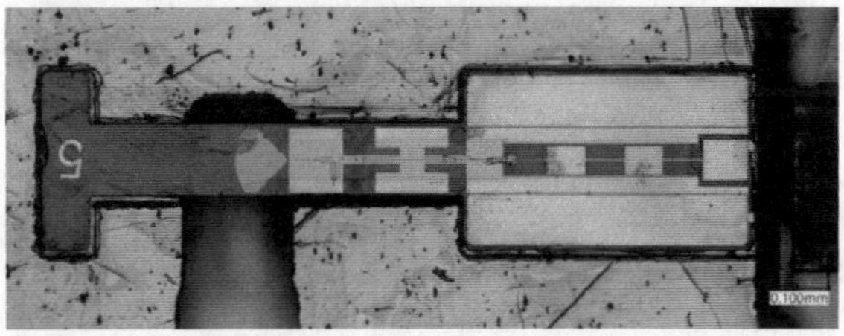

Fig. 2. Device from Wafer Run B221308 mounted in a DSB test block.

References

[1] "Next generation Event Horizon Telescope", [Online]. Available: https://www.ngeht.org/

1. NASA Jet Propulsion Laboratory California Institute of technology, Pasadena, CA 91109-8099 USA; [2]Caltech, Pasadena, 9125, USA; [3]University of Massachusetts, MA 01003-8270, USA; [4]Smithsonian Astrophysical Observatory, Cambridge, MA 02138, Country. *Corresponding author (email: jacob.w.kooi@jpl.nasa.gov).

NOTES:

Lumped-element aluminum KIDs with hierarchical phased-array antennas

F. Defrance[1*], A. Beyer[1], S. Shu[2], J-M. Martin[2], J. Kim[3], C. Frez[1], J. Sayers[2], P. Day[1], and S. Golwala[2]

***Abstract*—We have developed lumped-element aluminum kinetic inductance detectors (KIDs) for millimeter and submillimeter wavelengths. The incoming light is received with a hierarchical phased-array of slot-dipole antennas, split into four frequency bands (between 125 GHz and 365 GHz) with on-chip lumped-element band-pass filters, and routed to different KIDs using microstriplines. The antenna far-field beam-pattern, band-pass filter spectral response, and noise equivalent power were characterized and we report on these results.**

***Keywords*— aluminum, hierarchical antenna, KID, kinetic inductance detector, superconductor, THz.**

I. Introduction

For studies of the Sunyaev-Zeldovich effect, the cosmic microwave background, and dusty sources, future large-aperture ground-based telescopes would benefit from detectors able to observe simultaneously several spectral bands between 75 GHz and 415 GHz. By coherently summing signals from multiple individual pixels, hierarchical antennas are able to scale the pixel size with the observed frequency, allowing the coverage of a very wide frequency range while keeping a constant beam size, as demonstrated by Cukierman et al. [1]. The use of hierarchical antennas in combination with band-pass filters would allow the simultaneous observation of multiple frequency bands within the antennas frequency range.

II. Design

We initially designed a three-scale hierarchical phased-array slot-dipole antenna concept, designed to be coupled to titanium nitride KIDs, covering 75–415 GHz [2], and we designed and fabricated a two-scale version with four bands covering 125 GHz to 365 GHz, coupled with aluminum lumped-element KIDs [3]. The incoming light received through the two-scale hierarchical phased array of slot-dipole antennas is summed via superconducting microstriplines, and routed through superconductive microstripline and banks of band-defining filters to lumped-element aluminum KIDs. The microstripline uses hydrogenated amorphous silicon dielectric to ensure good transmission into the submillimeter, and the same dielectric is used in parallel-plate capacitors for the KIDs to achieve low loss, low noise, and low susceptibility to direct absorption. A schematic of the hierarchical phased-array antennas with the band-pass filters is shown in Fig. 1.

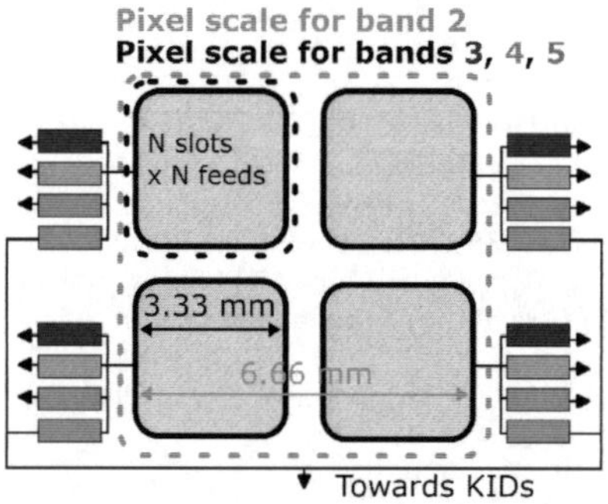

Fig. 1. Schematic of the hierarchical phased-array antennas. The N slots of each pixel are coherently summed using microstriplines. Each pixel is connected to lumped-element band-pass filters corresponding to bands 3, 4, and 5, while the four pixels are coherently summed together and connected to the band-pass filters defining band 2.

We report on the characterization of the 2-scale hierarchical phased-array antennas, on the spectral response of the 4-band lumped-element band-pass filters using Fourier Transform Spectroscopy, and on the noise-equivalent power of the KIDs.

References

[1] A. Cukierman, A. T. Lee, C. Raum, A. Suzuki, B. Westbrook, "Hierarchical sinuous-antenna phased array for millimeter wavelengths," *Applied Physics Letters*, vol 112, issue 13, pp. 132601, 2018. Available: https://doi.org/10.1063/1.5021962. [Accessed: Jan. 26, 2024].

[2] C. Ji, A. Beyer, S. Golwala, J. Sayers, "Design of antenna-coupled lumped-element titanium nitride KIDs for long-wavelength multi-band continuum imaging," *Proc. SPIE*, vol 9153, 2014. Available: https://doi.org/10.1117/12.2056777. [Accessed: Jan. 26, 2024].

[3] S. Shu, A. Beyer, P. Day, F. Defrance, J. Sayers, S. Golwala, "A multi-chroic kinetic inductance detectors array using hierarchical phased array antenna," vol. 209, pp. 330–336, 2022. Available: https://doi.org/10.1007/s10909-022-02890-x. [Accessed: Jan. 26, 2024].

[1]Jet Propulsion Laboratory, California Institute of Technology, Pasadena, CA, 91109, USA; [2]California Institute of Technology, Pasadena, CA, 91125, USA; [3]Korea Advanced Institute of Science and Technology, Daejeon, South Korea. *Corresponding author (email: fabien.m.defrance@jpl.nasa.gov).

NOTES:

A Wideband RF and Wideband IF DSB SIS Mixer

Igor Lapkin, Victor Belitsky*, Alexey Pavolotsky, Cristian López, Denis Meledin, Erik Sundin, Mohamed Aniss Mebarki, Leif Helldner, Magnus Strandberg, Mathias Fredrixon, Sven-Erik Ferm, and Vincent Desmaris

***Abstract*— The ALMA 2030 Roadmap document calls for a new generation of receivers for ALMA. Such Wideband Sensitivity Upgrade, WSU, aims to improve the existing receiver bands of ALMA with prospectively 4-times wider IF bandwidth and possibly a wider RF bandwidth. In order to explore a possibility to make wideband RF and simultaneously wideband IF receiver, we constructed and built a DSB SIS mixer covering the RF band 210-380 GHz. The mixer exhibits a 4 – 20 GHz IF band and produces non-corrected nose temperature between 3hf/k and 4 hf/k averaged over the IF band.**

***Index Terms*— ALMA receivers, wideband, SIS DSB mixer**

I. INTRODUCTION

The ALMA observatory is the most powerful instrument for modern radio astronomy and has been operational for over 10 years, while the original fundamental science goals of ALMA have been essentially achieved [1]. A new generation of receivers is needed to "significantly expand ALMA's capabilities and enable it to produce even more exciting science in the coming decades" [1]. The frequency band between 200 and 400 GHz is suggested as the most scientifically significant for upgrading with such new receivers.

Earlier works on wideband receivers have demonstrated a possibility to achieve a twice wider IF bandwidth, e.g. [2] as compared to the counter-part of ALMA. In [3], a wideband RF SIS mixer covering RF band width of 275 – 500 GHz have been presented. In this paper, we present our work on a DSB SIS mixer that covers the RF band 210 – 380 GHz, fractional band 56%, with IF bandwidth 4 - 20 GHz.

II. SIS MIXER DESIGN

The SIS mixer design takes advantage of an improved process for SIS junction fabrication [4] yielding high quality superconducting tunnel junctions with AlN tunnel barrier. Besides featuring a lower specific capacitance as compared to the tunnel junctions using AlOx as the tunnel barrier, this type of junctions also allows the fabrication of higher current density SIS junctions (or low RnA product, where Rn - junction normal resistance and A is junction area) without compromising the junction quality. Fig. 1 illustrates improvements in the specific SIS junction capacitance vs. RnA product, where Rn is SIS junction normal resistance and A is junction area in square micrometers.

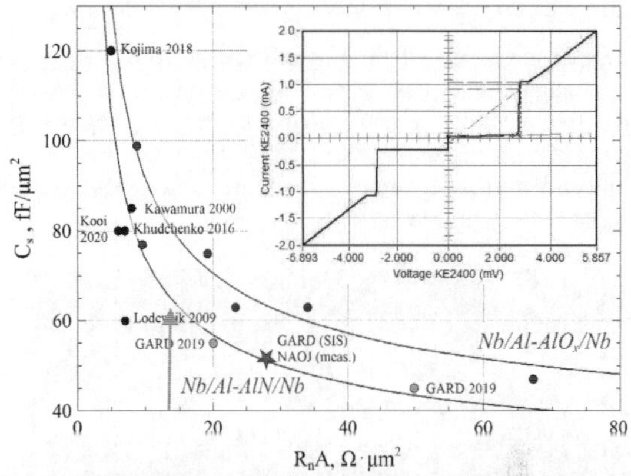

Fig. 1. Diagram connecting the specific capacitance per unit of area of SIS junctions with AlxOy and AlN tunnel barriers with product of its normal resistance, Rn, and area, A. The green arrow/dot shows the choice of the RnA for the SIS junctions used for the mixer in the current project. The junction area is 2 µm^2. The insert shows IVC of the fabricated SIS junctions (twin configuration, the DC measurements performed in the "current-source" mode) with measured Rj/Rn~30.

The designed mixer chip has twin-junction tuning circuitry [5] complemented with 2-step transformer connected to a broadband E-probe. The virtual ground employs hammer-type low-pass filter planar structure. Similar hammer-type filters structure is used to preclude leaking the RF into the IF/DC circuitry. The mixer block, Fig. 2, was milled from Tellurium copper, the waveguide dimensions are 380x760 µm.

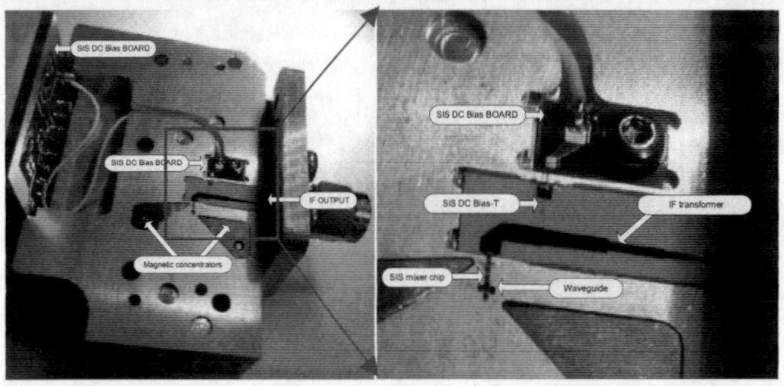

Fig. 2. The mixer block, Fig. 2, was milled from Tellurium copper, the waveguide dimensions are 380x760 µm. The IF circuitry includes 50-to-20 Ohm 3-step transformer with integrated DC bias-T. The latter uses spiral inductor and high-impedance microstrip line with termination capacitor to effectively isolate IF/DC. The entire IF circuitry is fabricated on alumina substrate with Nb sputtered lines.

The IF circuitry includes 50-to-20 Ohm 3-step transformer with integrated DC bias-T. The latter uses spiral inductor and high-impedance microstrip line with termination capacitor to

All authors are with the **Group for Advanced Receiver Development**, Onsala Space Observatory, Department of Space, Earth and Environmental Sciences, Chalmers University of Technology, Gothenburg, SE-41296, Sweden. E-mail of all authors is: name.surname@chalmers.se.

**Corresponding author: Victor Belitsky, victor.belitsky@chalmers.se.*

effectively isolate IF/DC. The entire IF circuitry is fabricated on alumina substrate with Nb sputtered lines. DC blocking capacitors are fabricated together with the IF circuitry and use SiO_2 insulator of about 300 nm thick. All DC and IF interconnections between the SIS mixer chip, IF board and DC bias circuitry was performed via wire-bonding with 16 μm diameter gold wires.

III. SIS Mixer Measurements and Performance

The measurements of the mixer were performed in a test cryostat using cold optics identical to [2], while for the ALMA Band 6 frequencies, the corrugated horn was replaced with a diagonal horn while using the same cold mirrors, Fig. 3. The LO injection was performed via 30 μm thick Mylar beam splitter at room temperature.

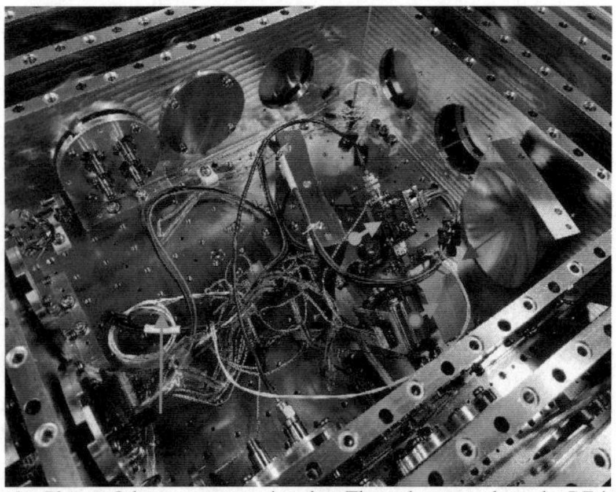

Fig. 3. Photo of the test cryostat interior. The red arrows show the RF input beam path. The blue dot indicates the mixer block position while the orange dot points to the cold 4-20 GHz IF amplifier, produced by Yebes [6], that directly attached to the mixer IF output. No IF isolator was used.

At the time of measurements, we had to use 4-18 GHz room temperature amplifier that have noticeable degradation right after 18 GHz with some noisy feature around 18.4 GHz, which affects the measured noise temperature. Fig. 4 displays the measured noise temperature at 4-20 GHz IF band at LO frequency 285 GHz and have quite typical behavior for all measured LO frequencies. The hot and cold temperature IF traces also presented in the

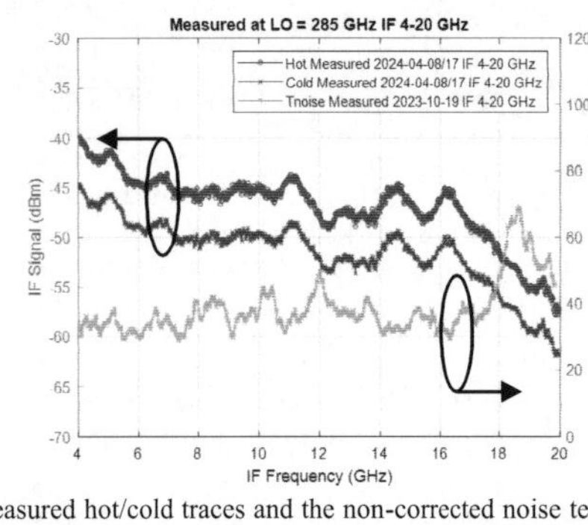

Fig. 4. Measured hot/cold traces and the non-corrected noise temperature for LO frequency 285 GHz, DSB SIS mixer, IF 4-20 GHz. The room temperature IF amplifier specified for 4-18 GHz band is the likely the reason of higher noise above 18 GHz.

Fig. 4 shows a slope about -20 dB across 3-20 GHz IF band that is very repeatable for all LO frequencies and thus could be equalized. Our measurements show that about 11,5 dB of the slope are connected to the RF insertion loss in the coaxial cables connecting cryogenic LNA, room temperature LNA and a spectrum analyzer used to characterize the receiver. The non-corrected measured noise temperature vs. LO frequency is presented in Fig. 5. It should be noted that for LO frequencies at ALMA Band 6, we used quartz window with Teflon matching layer produced by QMC and a diagonal horn while for LO frequencies in the ALMA Band 7, an HDPE plastic window with anti-reflection ridges were fabricated and used during the measurements with the corrugated horn from the SEPIA345 receiver [2].

The noise budget calculations that considers RF loss in the beam splitter, input cryostat window and 110 K and 15 K infrared filters predicts the noise temperature at the input of the cold optics in the range 14-17 K for the IF 4-20 GHz. Fig. 5 displays the noise temperature of the receiver measured at the input of the beam-splitter and averaged over the IF bandwidth.

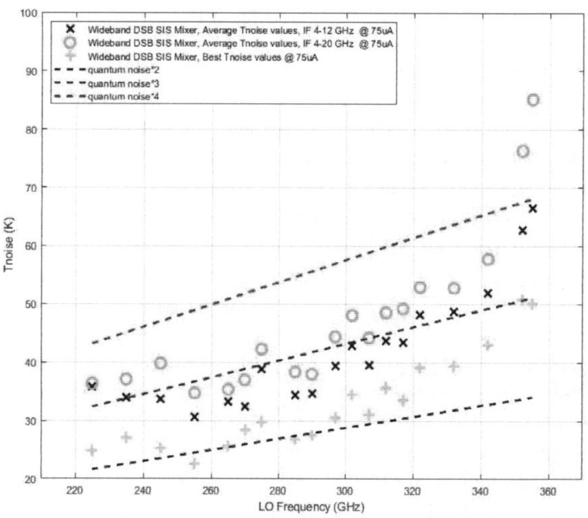

Fig. 5. Measured non-corrected DSB receiver noise temperature vs. LO frequency. Cyan crosses show the best noise temperature obtained for particular LO frequency. The black crosses show the receiver noise temperature averaged over the IF band 4 – 12 GHz and green circles indicate the receiver noise temperature averaged over the IF band 4 – 20 GHz. Red, blue and black dashed lines show the level of 4-, 3- and 2-times quantum noise respectively.

IV. Discussion

The measured noise temperature of the mixer increase follows in general the linear dependence vs. LO frequency. The best detected noise temperatures are close to two times the quantum noise, hf/k, while the averaged over the IF band noise temperatures are between times 3 and 4 the quantum noise. However, above 340 GHz LO, the performance of the mixer started to be influenced by slightly detuned SIS mixer (SIS junctions RnA=17 instead of the designed value RnA=14). At this frequency the 30 μm thick Mylar beam-splitter becomes close to 1/8 the wavelength, and introduces additional factor of the RF loss and contributes to the increase of the noise temperature.

Interestingly, we observed nearly constant difference between the ultimately best measured noise temperature (still averaged over 100 MHz band) and the averaged over the IF band noise temperature, Fig. 5. The best noise temperature points in the plot, Fig. 5, follow the same pattern as the noise temperature averaged over the IF band, and thus should be considered as part of real mixer performance. Suggesting that at the point of detecting the lowest noise temperature, the mixer exhibits high conversion gain, the difference between the best and the averaged over the IF 4-12 GHz band noise temperatures reveals the contribution of the IF part of the receiver, and is about ~7 - 8 K. This is consistent with the noise temperature of the IF amplifier ~4 K providing the averaged gain (of the mixer + optics loss + LO injection beam-splitter loss) in total is about -3 dB.

As mentioned above, the noise budget predicts 14 - 17 K noise temperature at the input of the cold optics. If we speculate that the 2SB mixer integrated in an ALMA cartridge will have ~2.8 times higher noise temperature, this based on our ALMA Band 5 experience, we end up with 39 – 48 K noise temperature that should fulfill ALMA specifications for both Band 6 and Band 7 [7]. Planned replacement of the room temperature IF amplifier that will perform over the entire IF band 4 - 20 GHz should improve the IF performance even above 18 GHz.

ACKNOWLEDGEMENTS

This work was supported by ESO under Collaboration Agreement No. 94468/20/103859/HNE and co-funded by Onsala Observatory. We would like to acknowledge colleagues from Yebes Observatory, Spain for lending the state-of-the-art 4-20 GHz cryogenic low-noise amplifier.

REFERENCES

[1] J. Carpenter, D. Iono, L. Testi, N. Whyborn, A. Wootten, N. Evans, "THE ALMA DEVELOPMENT ROADMAP", (AKA ALMA 2030), on-line https://www.almaobservatory.org/wp-content/uploads/2018/07/20180712-alma-development-roadmap.pdf

[2] D. Meledin, et. al., "SEPIA345: A 345 GHz dual polarization heterodyne receiver channel for SEPIA at the APEX telescope", A&A, A2, Volume 668, December 2022, https://doi.org/10.1051/0004-6361/202244211

[3] T. Kojima et al., "275–500-GHz Wideband Waveguide SIS Mixers," in IEEE Transactions on Terahertz Science and Technology, vol. 8, no. 6, pp. 638-646, Nov. 2018, doi: 10.1109/TTHZ.2018.2873487.

[4] A. Pavolotsky, et. al., "SIS technology development to serve Next Generation receivers for ALMA", in Proceedings of 32nd IEEE International Symposium on Space THz Technology (ISSTT 2022), Baeza, Spain, October 16-20, 2022.

[5] V. Belitsky, M.A. Tarasov, "SIS Junction Reactance Complete Compensation", IEEE Trans. on Magnetic, 1991, MAG- 27, v. 2, pt. 4, pp. 2638-2641.

[6] Isaac López-Fernández, Juan Daniel Gallego, Carmen Diez, Inmaculada Malo, "Development and Prototyping of a Cryogenic IF Low Noise Amplifier: Final Project Report", CDT Technical Report 2023-5, https://icts-yebes.oan.es/reports/doc/IT-CDT-2023-5.pdf

[7] Juande Santander-Vela, Giorgio Siringo, John Carpenter, Carla Crovari, Todd Hunter, Takafumi Kojima, Hiroshi Nagai, Neil Phillips, Wenlei Shan, Kamaljeet Saini, Gie Han Tan", Report of the ALMA Front-end & Digitizer Requirements Upgrade Working Group", v2, ALMA-05.00.00.00-3009-2-REP, 2022-09-09.

Production of ALMA Band 2 Cryogenic 1st Stage LNA

P. Pütz*[1], S. Türk[1], C. Leinz[1], F. Thome[2], A. Leuther[2], S. Heyminck[1], and G. Wieching[1]

***Abstract*— We report on the ongoing production and qualification of the cryogenic 1st stage low noise amplifiers for the ALMA Band 2 receiver. The RF performance results for the first production batches are presented and are evaluated against the technical specifications. The measurement results so far agree with the successful pre-production run in 2022. In particular, the state-of-the-art noise performance over the extended W-band (67 GHz – 116 GHz) is achieved.**

***Keywords*—Cryogenic, high-electron-mobility transistor (HEMT), low-noise amplifier (LNA), metamorphic HEMT (mHEMT), millimeter wave (mmW), monolithic microwave integrated circuit (MMIC), radio astronomy, W-band.**

I. Introduction

The European Southern Observatory (ESO) has awarded the MPIfR/Fraunhofer IAF consortium the production of the cryogenic 1st stage LNA units for the upcoming ALMA Band 2 receivers.

Decisive was the successful demonstration of repeatable state-of-the-art performance for a small series pre-production run of LNA units in a production-like environment [1].

Band 2 covers the entire atmospheric window in an extended W-Band from 67 GHz – 116 GHz. Complying to the specifications, in particular the noise performance, over this large frequency range sets challenges not only in the design and fabrication of the LNA, but also the measurement instrumentation required for qualification.

II. LNA Production

The cryogenic LNA has been adapted according to the envelope set by ESO's ICD with WR10 waveguide interface for the RF and an 9-pin socket female Omnetics Bi-lobe Nano-D type connector for DC. The LNA is of traditional E-plane split-block design with fused silica substrate waveguide probes at input and output. All components are populated into the lower half (Fig. 1). For the housing all precision machining is performed by the mechanical workshop at the MPIfR, recently upgraded with a Kern Micro HD featuring 5-axis as well as batch processing capabilities. The housing parts are likewise gold plated in-house to established recipe.

A production wafer with the ultra-low noise MMIC using 50-nm gate length metamorphic HEMT (mHEMT) technology is fabricated at Fraunhofer IAF. Here the active device layers are grown on GaAs substrates by means of a metamorphic buffer. This technology has matured over the recent years to provide state-of-the-art cryogenic noise performance, concluding in a benchmark W-band LNA design that is basis of our ALMA Band 2 LNA contribution [2].

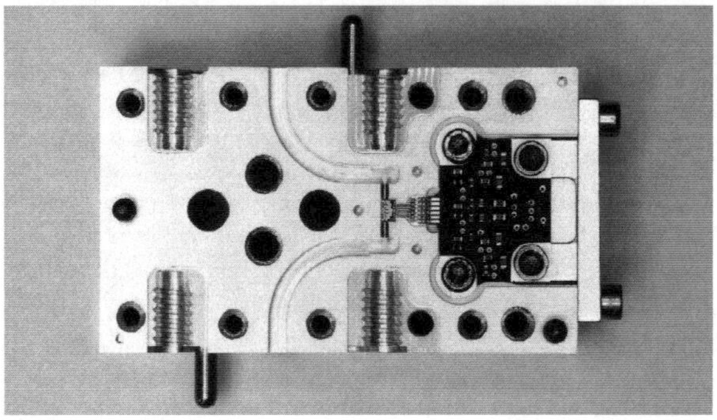

Fig. 1. View into bottom split-block half of one of the ALMA Band 2 production LNA. RF input into the WR10 waveguide is from the top and RF output to the bottom side. The LNA is biased through a Nano-D connector on the right side according to the ICD set by ESO. The depicted lateral dimensions of the split-block are 32.5 mm x 18.92 mm.

III. Qualification

We report on the current status of LNA production and qualification with measurement results. We will detail on the newly developed semi-automated cryogenic measurement setup that features dual measurement capability for S-parameter and noise power based measurements. A custom cryostat employs WR10 waveguide feedthroughs with sliding seals to allow for cryogenic S-parameter calibration. In order to minimize vibrations and ensure quick turnaround of the time-consuming cooling cycles a pulse tuber cooler is employed.

References

[1] S. Türk et al., "Results from ALMA Band 2 cryogenic LNA pre-production run", Proc. of the 32nd International Symposium on Space Terahertz Technology, Baeza, 2022, p. 85. [Online]. Available: https://www.isstt2022.com/

[2] F. Thome, et al., "A 67–116-GHz Cryogenic Low-Noise Amplifier in a 50-nm InGaAs Metamorphic HEMT Technology", IEEE Microwave and Wireless Components Letters, vol. 32, no. 5, pp. 430-433, May 2022, doi: 10.1109/LMWC.2021.3134462.

[1]Max Planck Institute for Radio Astronomy, Bonn, 53121, Germany; [2]Fraunhofer Institute for Applied Solid State Physics IAF, Freiburg im Breisgau, 79108, Germany. *Corresponding author (email: ppuetz@mpifr-bonn.mpg.de).

NOTES:

A New Technique for Measurement of the IF Output Impedance of SIS Mixers

P. Dindo[1], A. Navarrini[1], A.R. Kerr[1], S.-K. Pan[1], M. Pospieszalski[1], C. Jarufe[1], J. Lambert[1], J. Effland[1], and K. Saini[1]

***Abstract*—A new method is presented to measure the Intermediate Frequency (IF) output impedance (Z_{IF}) of superconducting tunnel junction (SIS) mixers at cryogenic temperatures. The setup uses a commercial Vector Network Analyzer (VNA) with high-sensitivity enhancements to increase the dynamic range of calibration and to perform low-power one-port measurements. The coupler inside the VNA is bypassed and replaced with an equivalent cold coupler inside the cryostat. A bias module connected to an isolator provides the bias to the mixer. A cryogenic low-noise amplifier (CLNA) in the return path to the VNA increases its dynamic range. One-port calibration standards (Short, Open, and Load) are connected to a cryogenic SP4T switch and VNA Software is used to de-embed fixtures. The new method allows direct measurement of the amplitude and phase of the reflection coefficient Γ_{IF} and hence Z_{IF}, from 2 to 16 GHz with very low power levels at the device under test.**

***Keywords*—VNA, SIS Mixer, CLNA, Γ_{IF}, Z_{IF}.**

I. Introduction

The ALMA Band 6 version 2 receiver upgrade specifies a wide IF band of 4–16 GHz and potentially 4–20 GHz [1]. The use of a cryogenic isolator between the SIS mixer and CLNA is currently under investigation. However, another approach is contemplated in which the SIS mixer is connected directly to the CLNA with the potential of obtaining lower noise across the IF band. To achieve high IF bandwidth it has been proposed that SIS mixers with higher critical current density J_C could lower the mixer's IF output impedance Z_{IF} to a range of values closer to 50 Ω where the available cryogenic IF amplifiers deliver optimum low-noise performance [2]. With the resulting lower junction capacitance it is possible to achieve wider RF and IF bandwidths as required by the receiver upgrade. This approach requires a careful mixer design as well as optimization of the IF network between the SIS junctions and IF amplifier.

From Tucker's quantum theory of mixing the IF output impedance at the junctions of an SIS mixer is known to be dependent on the intermediate frequency, but it has been shown [3] that for ratios $F_{IF}/F_{LO} < 0.1$ (our case) this variation is small. The IF output impedance, Z_{IF}, at the output port of the SIS mixer chip is calculated using Tucker's quantum theory of mixing and electromagnetic simulations of the embedding mixer circuit.

The goal of the present work is to describe the new SIS mixer IF output impedance measurement technique and compare with predicted Z_{IF} values over the full IF band.

II. The Measurement Setup

VNAs can easily measure one-port impedances at room temperature and medium power levels, but challenges arise when measuring devices inside a cold cryostat and where the incident signal power levels should be low enough to avoid saturating the SIS mixer. While a single port measurement normally uses the same cable for the incident and reflected signals, for accurate measurements at very low power levels it is necessary to separate the outgoing and returning signals between the VNA and the device under test as shown in Fig. 1.

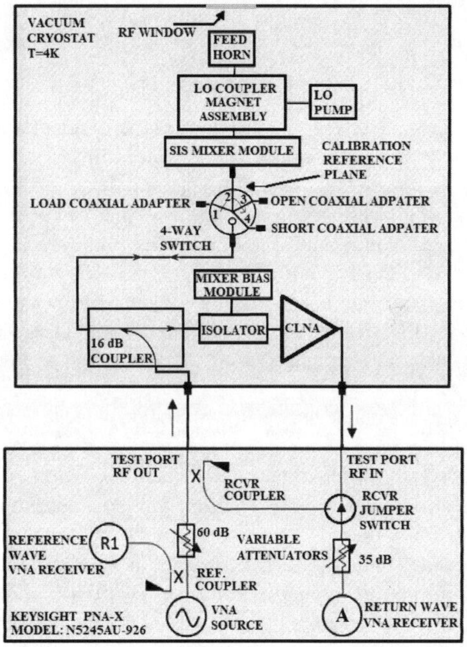

Fig. 1. The one-port Z_{IF} measurement setup. The lower block is a schematic of one of the four ports of the VNA. The upper block represents the schematic of the Band 6 receiver. The outgoing and returning signals are separated inside the VNA and inside the cryostat.

The lower block in Fig. 1 is a schematic of one of the four ports of the VNA. It shows the VNA source, the reference- and return-wave receivers, the reference and receiver couplers, two

[1] NRAO Central Development Laboratory, Charlottesville, VA 22903, USA. *Corresponding author (email: pdindo@nrao.edu).

The National Radio Astronomy Observatory is a facility of the National Science Foundation operated under cooperative agreement by Associated Universities, Inc

variable attenuators, the receiver jumper switch, and the RF input and output test ports. The base VNA model is upgraded to include a hardware kit (Keysight N5245AU-926) to decrease the source power from -30 to -90 dBm. The added variable source attenuator increases the range of the Automatic Source Levelling Circuit (ALC) and allows higher power during calibration than during measurement without losing calibration integrity. The VNA is equipped with front panel jumpers which can be used to bypass the receiver coupler and select an alternative return path for the RF. This allows the use of an external cold amplifier to increase the level of signals returning to the VNA and increase the dynamic range. The additional gain of the external amplifier, and removal of the 16 dB loss of the receiver coupler, compensates for the attenuation of the test signal.

The upper block in Fig. 1 shows the components of the SIS receiver: the feed-horn, LO coupler and SIS mixer module and LO source, a 4-way switch, 16 dB coupler, mixer bias module, IF isolator, and the cold IF amplifier. Cryogenic open, short, and load coaxial standards are connected to three positions of the computer controlled 4-way switch [4], while an 8-inch long semi-rigid coaxial cable is connected between the switch and the SIS mixer module.

III. The Calibration Technique

The location of the measurement plane is important. The previous technique referenced in [5] placed this plane deeper within the mixer chip, at the SIS junctions. Therefore, one has to transform the measurement to the IF output pad of the chip; and the amplitude and phase may become less accurate. Furthermore, the previous technique used the low resistance, high resistance, and linear resistance regions of the unpumped SIS mixer's I(V) curve as the calibration loads. These regions approximate short circuit, open circuit, and resistive calibration loads.

The new technique locates the calibration plane at the output port of the mixer chip, which is convenient for accurate determination of the complex reflection coefficient. we use coaxial short, open, and 50 Ω terminations so that the one-port 3-term errors of the VNA (directivity, port match, and tracking) are accurately determined and the amplitude and phase of the measurements are properly corrected. The VNA calibration reference plane located at the short-circuit termination on the 4-way switch must be moved electrically past an 8-inch semi-rigid coaxial cable to the mixer chip inside the SIS mixer module. The Automatic Fixture Removal (AFR) Software utility from Keysight Technologies [6], built into the VNA, is used to de-embed the coaxial cable and fixture circuitry between the calibration reference plane and the mixer chip output port. The utility requires a calibration step in which the fixture is terminated with an electrical short at the intended "de-embedding" reference plane.

Calibration and measurement accuracy hinges on using an appropriate incident stimulus power level at the mixer port. Too much power causes the SIS mixer to saturate, altering its characteristics and leading to unreliable data. We characterized the reflection coefficient of the SIS mixer biased at the gap voltage and observed that it undergoes compression at high power levels. We determined that to avoid saturation the RF power level at the mixer port should not exceed approximately -72 dBm. This allowed us to define a safe operating range below the saturation threshold, ensuring accurate measurements for our experiment.

IV. Calibration Technique Validation

Verification of the calibration quality in the previous technique [5] may be difficult. A key advantage of our technique is the ability to verify the calibration by measuring the output impedance of other cryogenic devices at room temperature, such as a low-noise IF amplifier, within the cryostat. This allows for a comparison with direct VNA measurements.

We validated the new calibration technique by measuring the output reflection coefficient S22 of an unbiased low-noise amplifier inside the cryostat at room temperature and compared it with direct measurements on the VNA. We found good agreement of amplitude and phase across the 2 to 16 GHz frequency range. The maximum phase error was 6°, and maximum magnitude error was 0.5 dB.

V. Z_{IF} Measurement Results

The SIS mixer chip used to demonstrate the output impedance measurements was a production chip from ALMA Band 6 which operates in sideband separating mode with 211–275 GHz RF and 4–12 GHz IF [7]. AWR Microwave Office [8] was used to model the SIS mixer using data from the complex five-frequency conversion admittance and noise-current correlation matrices of the pumped SIS junction based on Tucker's quantum theory of mixing [9] [10], and electromagnetic simulation of the RF and IF circuits.

The Γ_{IF} characteristics of the pumped SIS mixer at LO frequencies from 221 to 265 GHz, with IF swept from 2 to 16 GHz were measured showing good agreement with simulated results at lower IFs. A discrepancy between simulation and measurement observed at higher IF frequencies is being investigated.

VI. Conclusions

We presented a new technique for measuring the IF output impedance at the output IF bond pad of the SIS mixer using a commercially available VNA with dynamic range enhancement. Additionally, a simulation program was used to predict the IF output impedance and compare with measurements. While good agreement between measurements and simulations was observed at low IF, some discrepancies emerged at higher frequencies.

References

[1] A. Navarrini et al., "ALMA Band 6v2 receiver development status," 2023 XXXVth Gen. Assem. Sci. Symp. Int. Union Radio Sci. (URSI GASS), vol. 00, pp. 1–4, 2023, doi: 10.23919/ursigass57860.2023.10265563.

[2] T. Kojima, M. Kroug, K. Uemizu, Y. Niizeki, H. Takahashi, and Y. Uzawa, "Performance and Characterization of a Wide IF SIS-Mixer-Preamplifier Module Employing High-Jc SIS Junctions," IEEE Trans. Terahertz Science and Technology, v. 7, no. 6, pp. 694-703, Nov 2017.

[3] S.-K. Pan and A. R. Kerr, "SIS Mixer Analysis with Non-Zero Intermediate Frequencies," Proc. Symposium on Space THz Tech., Charlottesville, p. 195, March 1996.

[4] RF Switch SP6T, Model R573423600, Radiall USA, Tempe, AZ 85284

[5] P. Serres, A. Navarrini, Y. Bortolotti and O. Garnier, "The Output Impedance of SIS Mixers," IEEE Trans. Terahertz Science and Technology, v. 5, no. 1, pp. 27-36, Jan 2015.

[6] www.keysight.com, Keysight Technologies, CA, USA.

[7] A. R. Kerr, S.-K. Pan, S. Claude, P. Dindo, A. W. Lichtenberger, J. Effland, E. F. Lauria, "Development of the ALMA Band-3 and Band-6 Sideband Separating SIS Mixers," *IEEE Trans. Terahertz Sci. Technol.*, vol. 4, no. 2, March 2014.

[8] AWR Microwave Office, Cadence Design Systems, San Jose, CA 95134

[9] J. R. Tucker, "Quantum limited detection in tunnel junction mixers," IEEE J. Quantum Electron., vol. QE-6, no. 11, pp. 1234–1258, Nov. 1979.

[10] J. R. Tucker and M. J. Feldman, "Quantum detection at millimeter wavelengths," Rev. Mod. Phys., vol. 57, no. 4, pp. 1055-1113, Oct. 1985.

Embedding Impedance Recovery in a Twin- Junction SIS Mixer

Ghassan Yassin[1], Jakob Wenninger[1,†] and Boon-Kok Tan[1]

***Abstract*—We describe a method for calculating the embedding impedance of an SIS mixer employing twin junction tuning. Calculation of the embedding impedance is done using equivalent transmission line circuits of the mixer, the measured pumped and unpumped IV curves of the mixer chip and the Tucker theory expressions of the tunneling currents. Our simulation method does not require the individual IV curves of the two junctions.**

***Keywords*— Belitsky tuning, SIS mixer, Twin junction mixer.**

An important step in the design of SIS mixers is the matching of the RF feeding circuit to the SIS junction which is made difficult as a result of the large capacitance of the junction. The mixer circuit therefore includes a mechanism for tuning out of the junction capacitance. An elegant design of this tuning mechanism is the twin-junction device which was proposed by Belitsky et al [1] and was subsequently employed in SIS mixers at several frequency bands. It consists of two parallel SIS tunnel junctions connected by a transmission line, typically a microstrip line. The principle of operation is based on the fact that when a transmission line with carefully selected length and characteristic impedance separating the two SIS junctions is used, the reactance of the two junctions can be made to cancel each other and the input impedance of the device becomes real, hence can easily be matched to the feeding RF circuit.

I. Method of Impedance Recovery

Recovery of the embedding impedance in SIS mixers is a powerful tool that allows assessment of the design of the mixer circuits, hence gives additional tools to improve the next version of the mixer chip design. Recovery of the embedding impedance of a single mixer has already been reported [2], [3] and several computational techniques were employed to implement the calculations, based on the knowledge of the device pumped and unpumped curves. Applying the same method for the twin- junction mixer is not straightforward since the transmission line connecting the two junctions causes them to be pumped at different levels and the currents flowing through the two junctions will have different phases. Nevertheless, we were able to express the twin junction arrangement in a simple Thevenin Circuit albeit the current through the device became more complex. The Thevenin equivalent circuit of a twin-junction mixer can be represented by the equation

$$V_{\mathrm{Source}} = V_{\mathrm{D}} + Z_{\mathrm{emb}} I_{\mathrm{D}} \,,$$

where V_{Source} is the voltage of the local oscillator, V_{D} is the complex voltage across the device and Z_{emb} is the embedding impedance. The current across the device can be written as

$$I_{\mathrm{D}} = I_1(V_0, |V_1|) + I_2(V_0, |V_2|)\cos(\beta l) + i\frac{V_2(V_0)\sin(\beta l)}{Z_0} \,,$$

where l, β and Z_0 are respectively the length, propagation constant and characteristics impedance of the microstrip line. The current I_k through each junction may be taken from Tucker theory but a phase multiple must be included:

$$I_k(V_0, V_k) = \sum_{n=-\infty}^{\infty} J_n(\alpha_k)[J_{n-1}(\alpha_k) + J_{n+1}(\alpha_k)]$$

$$[\, I_{\mathrm{DC},k}(V_0 - nV_{\mathrm{Ph}}) + i I_{\mathrm{KK},k}(V_0 - nV_{\mathrm{Ph}})] e^{i\phi_k} \,,$$

where V_0 is the bias voltage, V_k and α_k are respectively the magnitude of the voltage and the pump factor across either junction and ϕ_k is a phase factor. The other symbols have their usual meaning. Recovery of the embedding impedance can then proceed as described in the references but two transcendental equations are now needed to recover the two voltages across the two junctions. This can make calculation of the error surfaces lengthy but we developed software techniques that allows reasonably fast iterations.

We have demonstrated our method by recovering the embedding impedance of two twin junction SIS mixer, one operating at 230 GHz and the other at THz mixer frequencies. We compared the performance of several computational techniques and also discuss the influence of the IF circuit on the integrity of the computations.

References

[1] V. Y. Belitsky, S. W. Jacobson, L. V. Filippenko, and E. L. Kolberg, "Broadband twin-junction tuning circuit for submillimetre SIS mixers," *Microwave and Optical Technology Letters,* vol. 10, no. 2, pp. 74-78, 1995.

[2] A. Skalare, "Determining embedding circuit parameters from DC measurements on Quasiparticle mixers," *International Journal of Infrared and Millimeter waves,* vol. 10, no. 11, pp.1339-1353, 1989.

[3] S. Withington, K. G. Isaak, S. A. Kovtonyuk, R. A. Panhuyzen, and T. M. Klapwijk, "Direct detection at submillimetre wavelengths using superconducting tunnel junctions," *Infrared Phys. Technol.*, p. 17, 1995, doi: 10.1016/1350-4495(95)00058-5.

[1]University of Oxford, Oxford, OX1 3RH, United Kingdom; [†]Now at Rosenberger Hochfrequenztechnik GmbH & Co. KG, 83413 Fridolfing, Germany

NOTES:

SESSION 7

Overview of ESO ALMA development studies

Carlos De Breuck[1*]

***Abstract*—The development program is an integral part of the ALMA project and aims to keep ALMA at the forefront of technology. While the development studies are run per executive, the development studies are independent, following the local specifics. I will present the ESO development studies, which are issued on a 3-year cycle to allow for more substantial studies to be performed. A detailed overview of the studies is given on https://www.eso.org/sci/facilities/alma/development-studies.html . Since the start of ALMA operations, ESO has issued five calls for proposals for development studies, covering a wide range of topics, from receiver components and prototyping, digitizers, software upgrades, observing modes to archive use enhancements.**

***Keywords*—Receiver technologies, miscellaneous.**

I. Introduction

Roughly half of the approved studies are related to receiver development. These have been key preparation for the construction of ALMA Bands 5 and 2. With the completion of all originally planned ALMA receiver bands, the studies are now concentrating on the upgrades of existing bands to cover the ambitious requirements of the ALMA Wideband Sensitivity Upgrade (WSU) to expand the instantaneous IF bandwidth by up to four times the current one. For example, several studies are looking into the upgrades of Band 7 and 9 using SIS technologies, while another study aims to use an integrated system on chip approach for Band 4+5.

Further down the signal chain, several ESO development studies have performed a market study to select new digitizers allowing to sample an IF range from 2 to 20 GHz with a substantially increased overall system efficiency. This has led to a development project to install these new digitizers and upgrade the digital signal transport, which is a core part of the ALMA WSU.

Some of the studies are not specific to receivers but intend to cover multiple bands. One example is development of cryogenic low noise amplifiers on Monolithic Microwave Integrated Circuits (MMICs), which aims to cover IF bandwidths from 4 to 20 GHz, a key requirement of the WSU. Another ongoing study is looking into the development of advanced waveguide technologies to reduce the losses using microfabrication rather than milling. This is particularly important for OrthoMode Transducers (OMT) at frequencies of 300-720 GHz. Similar synergies between Bands 7 and 9 are sought in the development of small new area junctions allowing to extend the IF bandwidth.

The ESO studies aim to closely involve the member state institutes in ALMA development. The studies allow to take more risks than the projects, which is inherent in the process to go well beyond the current capabilities. ESO hopes to issue the next call for development studies in 2025 and is looking forward to a continued great interest from our member state institutes.

[1]European Southern Observatory, Karl Schwarzschild Straße 2, 85748 Garching, Germany. [*]Corresponding author (email: cdebreuc@eso.org).

NOTES:

4.7-THz Schottky Diode Harmonic Mixer: Design, Fabrication, and Performance Optimization

Divya Jayasankar[1,2], Nick Rothbart[3], Heinz-Wilhelm Hübers[3,4], and Jan Stake[1]

***Abstract*—This paper focuses on the ongoing development of supra-THz harmonic mixers at Chalmers University of Technology. The planar, single-ended, ×8-harmonic mixers based on Schottky-barrier diodes were realized on a 2-μm GaAs substrate with integrated pyramidal horn. In contrast to our previous design with a diagonal horn, the new design addresses sensitivity to E-plane misalignment, which previously compromised mixer performance. The availability of THz harmonic frequency converters operating at ambient temperature is pivotal in realizing high-resolution THz heterodyne receivers.**

***Index Terms*—Harmonic mixers, Schottky diodes, Terahertz electronics**

I. Introduction

Terahertz (THz) heterodyne spectroscopy is a valuable tool for understanding the physics, distribution profile, and concentration of molecular and atomic gas in space. Identifying gas species like atomic oxygen (OI) at 4.7 THz [1] can enhance climate and weather prediction models. Quantum-cascade Lasers (QCLs) offer a few mW of output power while operating in continuous-wave (CW) mode, an optimal choice for THz heterodyne receivers. However, frequency instability arises from temperature and bias current fluctuations. Hence, frequency stabilization of QCLs is critical [2].

An efficient solution is to phase-lock the QCLs to a stable microwave source using a harmonic mixer. Danylov *et al.* [3] demonstrated phase locking of a 2.32-THz QCL using a balanced-Schottky diode ×21-harmonic mixer, which exhibited a conversion loss of about 110 dB. Later, Bulcha *et al.* [4] designed single-ended Schottky diode harmonic mixers yielding a conversion loss of 30 dB for fourth-harmonic mixing. Subsequently, Jayasankar *et al.* demonstrated single-ended planar Schottky diode ×6-harmonic mixers realized on a 2-μm GaAs substrate with 59-dB conversion loss. More recently, Reck *et al.* presented a 2.5-THz,×4-harmonic mixer with anti-parallel diodes with 26-dB conversion loss [5]. This work presents the design and fabrication of 2nd-generation of THz harmonic mixer with integrated pyramidal horn.

II. Design and Fabrication

The incoming RF signal from the QCL is coupled to the diode using a pyramidal horn integrated into the RF rectangular waveguide WM-48. The mixer is pumped by a Schottky varactor ×64-LO multiplier source. The radial LO probe was optimized to provide wide-band LO matching to the diode around 600 GHz. Fig. 1 shows the integrated circuit assembled on an E-plane split-block [6].

[1]Chalmers University of Technology, SE-412 96 Gothenburg, Sweden; [2]Research Institutes of Sweden, SE-504 62, Borås, Sweden; [3]German Aerospace Center, Institute of Optical Sensor Systems, 12489 Berlin, Germany. [4]Humboldt-Universität zu Berlin, Department of Physics, 12489 Berlin, Germany.

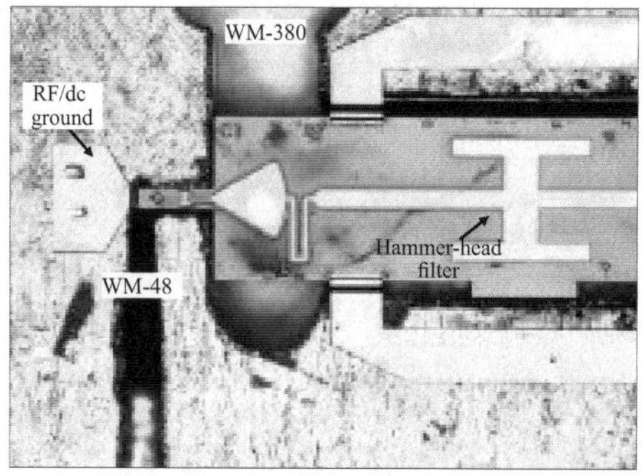

Fig. 1. Micrograph of integrated planar, single-ended 4.7-THz, ×8-harmonic Schottky diode mixer circuit assembled on an E-plane split block.

The realization of terahertz integrated circuits demands the alignment of patterns with high accuracy and precision in sub-micron order. Hence, we have developed a fabrication process entirely based on electron-beam lithography; the wafer structure can be found in [6].

III. RF Characterisation

The 4.7-THz QCL is placed in a cryocooler, and a TPX lens is used to focus the incoming THz signal on the harmonic mixer. The harmonic mixers were pumped using a ×64 Schottky varactor-based multiplier source. The 200-MHz IF signal is amplified and detected using a spectrum analyzer. RF characterization is ongoing, characterization results and new measurement technique [7] will be presented at the conference.

References

[1] H. Richter *et al.*, "Direct measurements of atomic oxygen in the mesosphere and lower thermosphere using terahertz heterodyne spectroscopy," *Commun. Earth Environment*, vol. 2, no. 1, Jan. 2021.
[2] H. Richter, N. Rothbart *et al.*, "Phase-Locking of Quantum-Cascade Lasers operating around 3.5 THz and 4.7 THz with a Schottky-Diode Harmonic Mixer," *submitted to IEEE Trans. THz Sci. Technol.* , Jan. 2024.
[3] A. Danylov *et al.*, "Phase locking of 2.324 and 2.959 terahertz quantum cascade lasers using a Schottky diode harmonic mixer," *Opt. Lett.*, 2015.
[4] B. T. Bulcha *et al.*, "Design and characterization of 1.8–3.2 THz Schottky-based harmonic mixers," *IEEE Trans. THz Sci. Technol.*, vol. 6, no. 5, pp. 737–746, Sep. 2016.
[5] T. J. Reck *et al.*, "Design of a 2.5 THz Schottky-Diode Fourth-Harmonic Mixer," *IEEE Trans. THz Sci. Technol.*, vol. 13, no. 6, pp. 580–586, 2023.
[6] D. Jayasankar *et al.*, "A 3.5-THz, x6-Harmonic, Single-Ended Schottky Diode Mixer for Frequency Stabilization of Quantum-Cascade Lasers," *IEEE Trans. on Terahertz Science and Technology*, 2021.
[7] D. Jayasankar, T. Reck *et al.*, "A Broadband Conversion Loss Measurement Technique for Terahertz Harmonic Mixers," *submitted to IEEE Trans. THz Sci. Technol.*, 2024.

1.90THz-2.06 THz Schottky Receiver with 4000-6000K DSB Noise Temperature at Room Temperature

A. Maestrini[1*], J. Siles[1], C. Lee[1], R. Lin[1] and I. Mehdi[1]

***Abstract*— We report on recent progress made toward highly sensitive room temperature heterodyne receivers at 2THz. Schottky mixer and frequency multiplier devices, designed and fabricated at NASA Jet Propulsion Laboratory, have enabled the demonstration of a room temperature heterodyne receiver sensitive enough to measure the wind velocities, temperature and density in the Earth's thermosphere by observing the emission of the atomic oxygen at 2.06THz.**

***Keywords*—heterodyne, Schottky mixer, THz.**

I. Introduction

Observing the atomic oxygen line at 2.06 THz in the Earth's 100-200km altitude region with a high spectral-resolution limb-sounder onboard a low Earth orbit satellite can provide global measurements of the neutral winds, temperature and density of the lower thermosphere, which are critical to better understand the underlying mechanisms of the upper atmospheric composition / dynamics/ temperature variability and the role of neutral dynamics on the ionospheric variability. Currently, only heterodyne receivers can provide these measurements with complete local time coverage and desired spatial resolution, precision and accuracy. Before our first reported results [1], only cryogenically cooled Hot-Electron-Bolometer mixers had the necessary sensitivity at 2 THz for such receivers. JPL Schottky technology, however, enables for the first time the construction of a *room-temperature* receiver with sufficient sensitivity to perform these critically needed measurements.

II. Receiver Design and New RF results

The configuration for the 2.06 THz receiver has been presented in [1]. The current version relies on the same Local Oscillator chain which is described in [2], but the sub-harmonic mixer uses a newly fabricated Schottky device that features a bias-able anti-parallel pair of diodes monolithically integrated on a thin GaAs membrane. The design of this mixer (see Fig.1) is rooted on the state-of-the-art 1.2THz sub-harmonic mixer designed for the Submillimeter-Wave-Instrument of the JUpiter ICy moon Explorer [3]. At 2.06THz, a DSB receiver equivalent noise temperature of 6,080K was recorded at room temperature, which is believed to be the best performance of a heterodyne receiver ever reported at this frequency and operating temperature. The mixer is optimally pumped with slightly less than 2mW of LO power and does not require biasing at this pump level. The DSB mixer noise temperature and gain were measured respectively at 5440K and -12.9dB. When operating at 1.92THz the receiver DSB noise temperature reaches a minimum of 4135K with a DSB mixer noise temperature and gain of respectively 3675K and -11.5dB. The measurements were performed in air with a 6cm optical path and 33% humidity at 23°C and with a 5.8-6.5GHz IF filter. Data were corrected for atmospheric attenuation based on version 11.0 of *am*, a program for radiative transfer computations at microwave to submillimeter wavelengths [4]. Raw data for the DSB receiver noise temperature at 2.06THz and 1.92THz are respectively 6322K and 4653K.

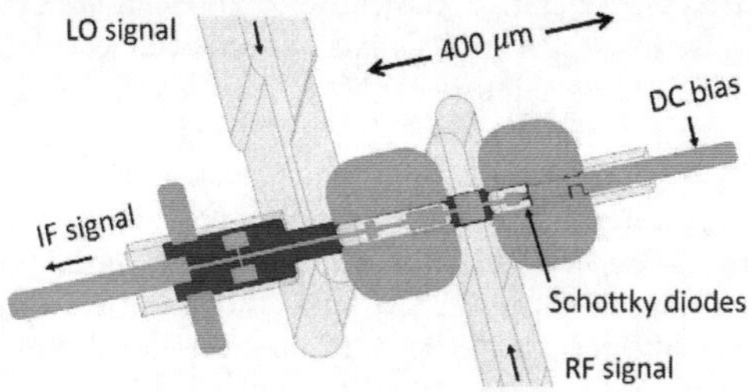

Fig. 1. Schematics of JPL 2.06THz bias-able antiparallel-diode Schottky mixer.

References

[1] A. Maestrini et al. "2 THz Receiver for Thermospheric Science with 7000K DSB Noise Temperature at Room Temperature," 2023 48th International Conference on Infrared, Millimeter, and Terahertz Waves (IRMMW-THz), Montreal, QC, Canada, 2023, pp. 1-2, doi: 10.1109/IRMMW-THz57677.2023.10299161

[2] J. Siles, et. al., "A New Generation of Room-Temperature Frequency Multiplied Sources with up to 10x Higher Output Power in the 160 GHz-1.6 THz Range", Proc. of the IEEE Transactions of Space Terahertz Communications, Nov. 2018

[3] A. Maestrini *et al.*, "The 1200GHz Receiver Frontend of the Submillimetre Wave Instrument of ESA JUpiter ICy moons Explorer," *2018 43rd International Conference on Infrared, Millimeter, and Terahertz Waves (IRMMW-THz)*, 2018, pp. 1-2, doi: 10.1109/IRMMW-THz.2018.8509935.

[4] https://doi.org/10.5281/zenodo.640645

[1]Jet Propulsion Laboratory, California Institute of Technology, Pasadena, CA 91109, USA. *Alain Maestrini (email: alain.e.maestrini@jpl.nasa.gov).

NOTES:

Development Status the ALMA B6v2 SIS Mixers

Joseph G. Lambert[1*], Alessandro Navarrini[1], Anthony R. Kerr[1], John Effland[1], Philip Dindo[1], Alec Handy[1,2], Arthur W. Lichtenberger[2], and Michael Cyberey[2]

***Abstract*—We report on the development of SIS mixers for the upgraded ALMA Band 6 sideband separating receiver currently in design, called "Band 6v2". The new Band 6v2 receiver has several challenging specifications, requiring exceptionally low noise and high image rejection ratio over wider Intermediate Frequency (IF) and Radio Frequency (RF) bandwidths. To meet these requirements, several IF architectures are being explored, with the baseline configuration consisting of a wide-band low-loss isolator between the mixer and low-noise amplifier. Alternative IF configurations are also being investigated, and different mixer designs are in development with the goal of optimizing the receiver performance for these configurations. Our report includes some recent experimental results, and new mixer designs in progress.**

***Keywords*—ALMA, mixer, SIS, silicon-on-insulator, sideband separation.**

I. INTRODUCTION

THE first-generation ALMA Band 6 (Band 6v1) sideband separating receiver has mixers directly connected to the IF low noise amplifiers (LNA), covering 4.5-10 GHz IF over 211-275 GHz RF [1]. The IF band for Band 6v2 is at least 4-16 GHz with a goal of as much as 4-20 GHz [2,3]. Achieving greater IF bandwidth with lower noise is challenging with directly connected LNAs, largely due to the typically high IF output impedance of the SIS mixers. For this reason, the baseline architecture for the Band 6v2 receiver includes a wide-band low-loss isolator [4] between the mixer and LNA (Fig. 1).

Two iterations of mixer designs have been designed, fabricated and tested. In the following, we briefly review the design parameters of each and show preliminary double-

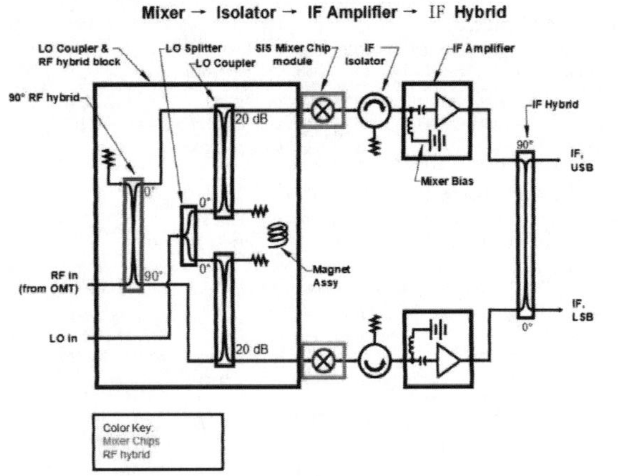

Fig. 1. The baseline architecture of the ALMA Band 6v2 receiver frontend

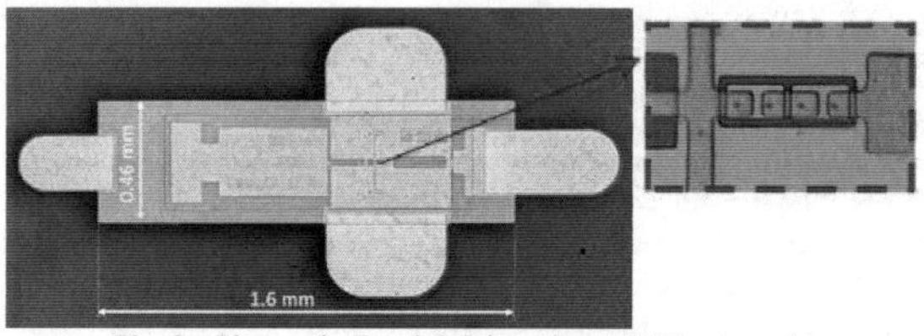

Fig. 2. Photo of a Band 6v2 iteration 1 (M1) mixer chip.

sideband results of the first iteration. We briefly describe a third iteration of B6v2 mixer design currently in progress.

II. MIXER DESIGN PARAMETERS

All the Band 6v2 mixers employ silicon-on-insulator (SOI) technology with thin (5-μm) gold beamleads for electrical connections and mechanical support. The first two iterations (denoted M1 and M2) employ Nb/Al-AlO_x/Nb junctions, with a target critical current density of J_c = 6.9 kA/cm^2. A comparison of the target design parameters for B6v1 and both B6v2 mixer designs is shown in TABLE I.

The primary goals for the M1 mixer designs compared to the existing Band-6 mixers [1] were to modestly increase the RF bandwidth, flatten the RF response, and utilize SOI technology with gold beamleads for easier assembly [5,6]. A micrograph of an M1 mixer chip is shown in- Fig. 2Fig. 2. The larger top and bottom beamleads provide DC and RF ground connection to the mechanical block and the right beamlead provides the DC and IF connection. The beamlead on the left provides support to the waveguide RF probe and serves as part of the RF probe design to match to the 50-Ω suspended strip line. The RF probe was designed to couple to perpendicularly oriented WR3.7 full height waveguide. The oval waveguide backshort is machined into the mixer chip DSB module (see e.g. Fig. 4).

The primary goals of M2 mixer design were to modestly decrease the conversion loss, modestly decrease the IF output capacitance for flatter IF response, include narrower chip designs with an improved RF probe and increased density of chips on a wafer, and to test 2-junction series array designs. Two wafers have been fabricated and testing is in progress. An example of a narrower chip (300 μm) is shown in Fig. 4.

TABLE I. TARGET DESIGN PARAMETERS FOR BAND-6 MIXERS

Parameter	B6v1 [1]	B6v2 M1	B6v2 M2
Jc (kA/cm^2)	5.2	6.9	6.9
Number of junctions	4	4	4 & 2
Junction diameter (μm)	1.7	1.5	1.5 & 1.1
Array R_N (Ω)	61	59	59
Substrate	Fused quartz	5-μm SOI	5-μm SOI

[1]National Radio Astronomy Observatory, Central Development Laboratory, Charlottesville, VA, 22903, USA; [2]Innovation in Fabrication (IFAB) Facility, University of Virginia, Charlottesville, VA, 22904, USA; [*]Corresponding author (email: jlambert@nrao.edu). The National Radio Astronomy Observatory is a facility of the National Science Foundation operated under cooperative agreement by Associated Universities, Inc.

The design of the 3rd iteration mixers (M3) is in progress with primary goals of lower output impedance, and lower conversion loss. Lower output impedance may facilitate alternate receiver architectures, e.g. with the IF hybrid directly following the mixers, but does not preclude the use of an isolator. Based on preliminary results of the M2 mixer tests, we plan to use the narrower chip design with 2 junctions in series. At present we are targeting a critical current density of 10 kA/cm^2 and an array normal resistance of 15 Ω

III. Double Sideband Measurements

The parasitic on-chip capacitance becomes increasingly problematic at higher IF. An off-chip IF compensation network was designed and tested to counter the on-chip capacitance and achieve a broad IF of at least 4-20 GHz. The compensation network is a microstrip circuit on a 10-mil thick Rogers Duroid 6002 substrate. The circuit is inserted between the IF isolator and the mixer IF output port, and is designed to maintain a 50-Ω real impedance presented to the mixer junctions. It consists of parallel connected open and short circuit stubs to cancel the on-chip capacitance and flatten the IF response. The short circuit stub is terminated by a 10 pF capacitor to provide a DC blocking RF ground (the mixer is DC biased from the LNA through the IF port). Between the stubs and the isolator is a 3-section impedance transformer.

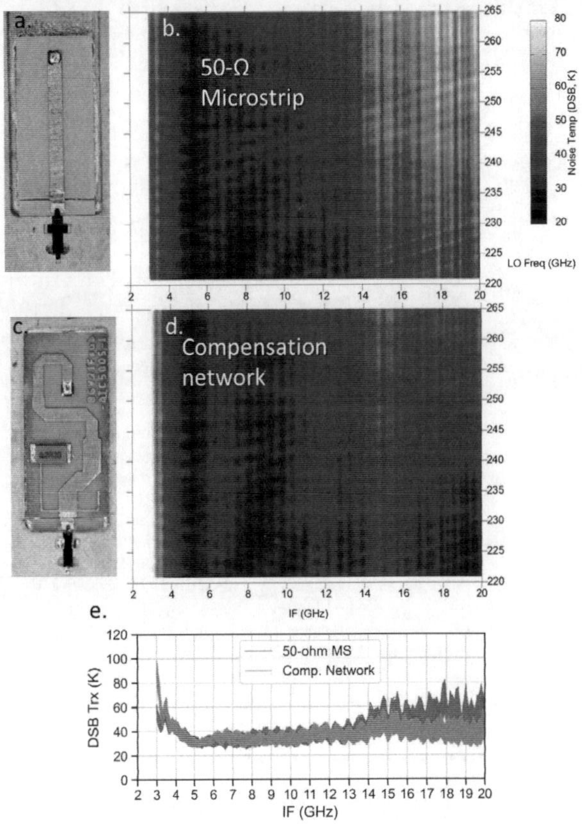

Fig. 3. Double-sideband (DSB) noise temperature from a M1 B6v2 mixer chip. The receiver chain consists of the DSB mixer module, wideband edge mode isolator [4], and Low Noise Factor LNA model LNF-LNC6_20D [7]. The two false-color plots show the DSB noise as a function of IF and LO, with 50-Ω line (a. and b.) and IF compensation network (c. and d.). The same data is shown in e. versus IF for both cases.

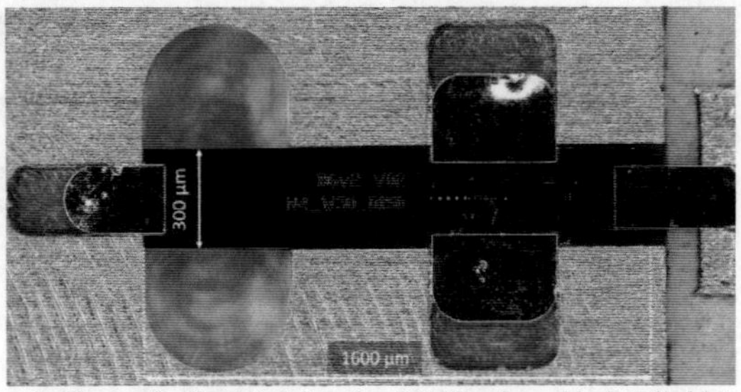

Fig. 4. Photo of a Band 6v2 iteration 2 (M2) mixer chip installed into a DSB module.

The performance of the compensation network is demonstrated in Fig. 3. The top panels (a. and b.) show the results with a simple 50-Ω microstrip line and the bottom panels (c. and d.) show the results with the compensation network. The same M1 mixer chip was used in both cases; the Duroid substrate was cut into two sections so that the top portions could be replaced without damaging the mixer chip. The measurements cover LO 221-265 GHz in 100 MHz steps. The plot in Fig. 3 e. shows the same noise data versus IF only, for all LO frequencies.

The same IF network is in use for the M2 mixers and shows flatter IF noise temperatures versus RF, indicating that the parasitic capacitance is indeed lower for M2 compared to M1 mixers. We will likely redesign the off-chip IF compensation network for the M3 mixer designs.

IV. Conclusion

We presented the design summary of the first 2 iterations of the ALMA B6v2 mixer designs, and the design plans of the 3rd iteration currently in development. We also reported on the design and demonstration of an off-chip microstrip network to compensate for the parasitic on-chip capacitance of the SIS mixers.

This work is a collaborative effort between the National Radio Astronomy Observatory (NRAO) and the Innovations in Fabrication (IFAB) at the University of Virginia. The project is funded by the North American ALMA Cycle 9.

References

[1] A. R. Kerr et al., "Development of the ALMA Band-3 and Band-6 Sideband-Separating SIS Mixers," IEEE Transactions on Terahertz Science and Technology, vol. 4, no. 2, pp. 201–212, 2014.

[2] J. Carpenter, C. Brogan, D. Iono, T. Mroczkowski, "The ALMA2030 Wideband Sensitivity Upgrade", ALMA Memo 621, Nov. 2022.

[3] A. Navarrini et al., "ALMA Band 6v2 receiver development status," 2023 XXXVth Gen. Assem. Sci. Symp. Int. Union Radio Sci. (URSI GASS), pp. 1–4, 2023, doi: 10.23919/ursigass57860.2023.10265563.

[4] L. Zeng, C.-Y. E. Tong, and S. N. Paine, "A Low-Insertion Loss Cryogenic Edge-Mode Isolator With 18 GHz Bandwidth," *IEEE J. Microw.*, vol. PP, no. 99, pp. 1–9, 2023.

[5] R. B. Bass, J. C. Schultz, A. W. Lichtenberger, R. M. Weikle, S.-K. Pan, E. Bryerton, C. K. Walker, J. Kooi. "Ultra-Thin Silicon Chips for Submillimeter-wave Applications," Fifteenth International Symposium on Space THz Technology (pp. 392-399), 2004

[6] N. S. Barker, M. Bauwens, A. W. Lichtenberger, R. Weikle II, "Silicon-on-Insulator Substrates as a Micromachining Platform for Advanced Terahertz Circuits," Proceedings of the IEEE, Vol 105, No 6, June 2017.

[7] https://lownoisefactory.com/product/lnf-lnc6_20d/

Lumped-element Model Analysis for THz HEB Mixer Based on Sputtered MgB_2 Thin Films

Changyun Yoo[1], Changsub Kim[1], Daniel C. Cunnane[1], Chris A. Curwen[1], and Boris S. Karasik[1*]

***Abstract*—We experimentally and theoretically investigate the mixer properties of THz hot-electron bolometer (HEB) mixer based on sputtered MgB_2 thin (7 nm) films. By simultaneously measuring the U-factor and the Y-factor of the device, we measure the voltage bias dependence of the conversion gain, output noise temperature, and the mixer input noise temperature. The great uniformity of the sputtered MgB_2 film allows us to apply the lumped-element bolometric model that describes the mixer characteristics surprisingly well. The measured double-sideband (DSB) receiver noise temperature after correcting for optical loss is ~2,000 K at 2.52 THz at 5 K, with a -3dB intermediate frequency (IF) bandwidth of ~3 GHz and a required local-oscillator (LO) power of ~ 2 μW.**

***Keywords*—THz heterodyne receiver, MgB_2 THz hot-electron bolometer**

I. Introduction

While the superconducting Hot-Electron Bolometer (HEB) mixer has been around for more than three decades, there is still a poor understanding why the noise temperature in practical mixers is still far from the quantum limit, partially due to the lack of the adequate model applicable to non-uniform HEB devices. If the device is uniform, however, a relatively simple lumped-element bolometric model [1, 2] can be applied to describe the mixer characteristics, which is important for understanding the mixer performance and giving a guidance on how to optimize the material and device parameters for better performance.

THz HEB mixers based on magnesium diboride (MgB_2) films have recently become an attractive option due to their higher critical temperature and short electron-phonon interaction time, useful for high temperature operation above 20 K and a wide -3dB intermediate frequency (IF) bandwidth ~ 10 GHz. Recently, we began employing sputtered MgB_2 thin films for the fabrication of HEB mixers. The great uniformity of these films allows us to fabricate small (0.75 μm (L) × 2 μm (W)) devices with reproducible characteristics and to apply the lumped-element model [1,2] that describes the mixer parameters well. Below, we present a set of experimental data and an analysis showing what can be learned about the HEB mixer when a detailed model is available.

II. Results

By simultaneously measuring the U-factor and the Y-factor, we investigated the voltage bias dependence of the conversion gain, output noise temperature, and the mixer input noise temperature of the MgB_2 HEB device (Fig. 1). The lumped-element model analysis using the experimental current-voltage (IV) characteristics and the temperature dependence of the device resistance (Fig. 1(a)) reproduces the bias dependence of the conversion gain, output noise temperature, and mixer input noise temperature reasonably well (Fig. 1(b-d)).

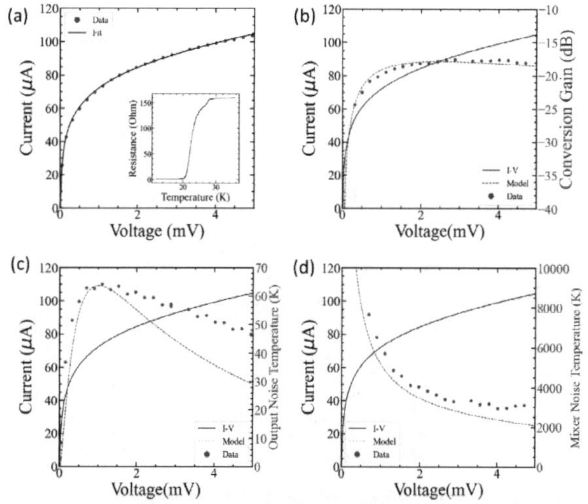

Fig. 1. Lumped-element model analysis results for THz HEB mixer based on sputtered MgB_2 thin films. (a) current-voltage (IV) characteristics when the device is pumped with a local-oscillator (LO) power at 2.52 THz at 5 K. The inset shows the temperature dependence of the device resistance. (b-d) bias dependence of conversion gain (b), output noise temperature (c), and mixer input noise temperature (d). Red dots indicate the experimental data, and the red dashed lines show the modelling results.

References

[1] B. S. Karasik and A. I. Elantev, "Analysis of the Noise Performance of a Hot-Electron Superconducting Bolometer Mixer," in Sixth International Symposium on Space Terahertz Technology, March 01, 1995, pp. 229-246.

[2] B. S. Karasik and A. I. Elantiev, "Noise temperature limit of a superconducting hot-electron bolometer mixer," Applied Physics Letters, vol. 68, no. 6, pp. 853-855, 1996, doi: 10.1063/1.116555.

[1] Jet Propulsion Laboratory, California Institute of Technology, Pasadena, CA, 91109, USA *Corresponding author (email: boris.s.karasik@jpl.nasa.gov).

NOTES:

SESSION 8

Dual Band 1.3mm/3mm Receivers for the NOEMA Observatory

A.-L. Fontana[1], P. Serres[1], J. Reverdy[1], Y. Bortolotti[1], D. Maier [1], S. Mahieu[1], M. Berton[1], M. Parioleau[1], J. Ballot[1], B. Pissard[1], Q. Moutote[1], C. Risacher[1*]

***Abstract*—The NOEMA observatory was completed in 2022 with its last 12th antenna delivered and commissioned successfully together with the track length extensions. Next step for the observatory upgrades is the implementation of a dual-frequency observation mode where the 1.3 mm and 3 mm bands (based on SIS mixers) can be observed simultaneously using a dichroic mirror, all in dual polarization, full 4-12 GHz IF bands. We will present the current status of this development which is due to be completed by end of 2024.**

***Keywords*—heterodyne, SIS mixers, dichroic, mm bands**

I. Introduction

THE NOEMA observatory (Northern Extended Millimeter Array) is the successor of the Plateau de Bure Interferometer (PdBi), which consisted of 6x15m antennas located at the Plateau de Bure, in the French Alps at an altitude of ~2500 m. The interferometer had two tracks: a North-South of ~370m length and an East-West track of about 760m. The NOEMA project goals were to provide 6 additional 15 m antennas and extend substantially the East-West track from 760m to 1700m, while operating a state-of-the-art heterodyne receiver, observing in 4 bands (3mm, 2mm, 1.3mm and 0.85 mm), using sideband separation SIS-based receivers, in a dual polarization mode, with 4-12 GHz IF bandwidth. The NOEMA observatory was successfully complete in 2022.

The next step in the observatory upgrade is to provide a dual-band capability, as we are now limited to observing one band at the time. The goal therefore is to be able to observe the 1.3 mm and 3 mm bands simultaneously, co-aligned on sky while maintaining the state-of-the-art performance of the system.

II. Dual Band receiver specifications and design

The Plateau de Bure Interferometer already had dual-band capabilities, but that was done selecting the bands via polarization. Therefore, the system sensitivity was decreased during dual-band operations. Now, the choice for combining the bands relies on dichroic filters, where the selection is done by frequency, with minimum losses, while keeping the dual-polarization capability for each band. IRAM is developing this dichroics using several layers of metallization on quartz substrates.

As of March 2024, 7 antennas out of the 12 have being equipped with the new mechanical and optical required components. In summer and fall 2024, the remaining antennas will be equipped, together with the installation of the new warm IF components, and second correlator unit that will allow operating in dual band.

Element	3mm Band	1.3mm Band
Frequency range (RF)	70-119 GHz	196-276 GHz
Mixer topology	2SB – SIS mixers	2SB – SIS mixers
IF bandwidth	4-12 GHz	4-12 GHz
Configuration	2SB mode - dual polarization	2SB mode – dual polarization
Dichroic losses	Max 5% losses	Max 5% losses

The full design of the system will be presented, together with the required hardware changes and the first preliminary performance assessment.

This upgrade using warm frequency separation dichroic will possibly be followed in the future with a receiver implementing the dichroic at cold (inside the cryostat), to minimize the loss contributions to the system temperature.

[1]Institut de Radioastronomie Millimétrique, Saint Martin d'Hères, 38330, France; [*]Corresponding author (email: risacher@iram.fr).

NOTES:

Findings for the OSAS-B 4.7-THz heterodyne spectrometer for atomic oxygen in the mesosphere and lower thermosphere

Martin Wienold[1*], Alexey Semenov[1], E. Dietz[1], S. Frohmann[1], P. Dern[1], B. Klein[2] and Heinz-Wilhelm Hübers[1,3]

***Abstract*— We present results and findings from the first flight of the OSAS-B instrument on a stratospheric balloon. OSAS-B is a heterodyne receiver for the 4.7-THz emission of atomic oxygen which is based on a hot-electron bolometer mixer and a quantum-cascade laser as local oscillator.**

I. Introduction

The Oxygen Spectrometer for Atmospheric Science on a Balloon (OSAS-B) has been developed for probing neutral atomic oxygen (OI) in Earth's mesosphere and lower thermosphere (MLT) from a stratospheric balloon. Atomic oxygen is the dominant species in the MLT region and thus plays an important role for the chemistry and energy balance [1,2]. OSAS-B is a heterodyne receiver for the ground state fine-structure line of OI at 4.745 THz, which cannot be observed from ground due to the absorption by water vapor in the troposphere.

II. Instrument Design and First Flight in 2022

Figure 1(a) shows a photography of the instrument. The cryostat comprises the receiver front end. It contains a liquid helium stage for the hot-electron bolometer (HEB) as the heterodyne mixer and a solid nitrogen stage for the quantum-cascade laser (QCL), which acts as the 4.7-THz local oscillator. A rotatable 35-mm mirror in the optics compartment collects the atomic-oxygen emission for different elevation angles or alternatively the blackbody radiation from one of the two calibration loads at ambient temperature and 400 K. The backend of the receiver comprises a digital Fourier transform spectrometer for data acquisition.

The first flight of the instrument took place in September 2022 from Esrange, Sweden. Atmospheric data were recorded for several hours during one day. Figure 1(b) depicts measured spectra for 0° (horizontal) and 60° elevation along with synthetic spectra as expected from atmospheric models [3, 4]. The spectrum for 0° elevation exhibits a pronounced wing structure, which is due to an interplay of emission and absorption from different altitudes, and reflects the radiation transport in the MLT region. Results show that OSAS-B allows for observing the subtle line-shape details as predicted for different atomic-oxygen distribution models. In certain spectra, we found signatures of shear winds in the MLT region, which manifest in a relative Doppler shift of components from different altitudes. For the radiometric calibration, we found that incorporating the direct-detection effect plays a crucial role for obtaining a reasonable agreement between the experimental and the expected brightness temperatures of the OI line. We will present further results of the balloon flight and discuss the in-flight performance of OSAS-B.

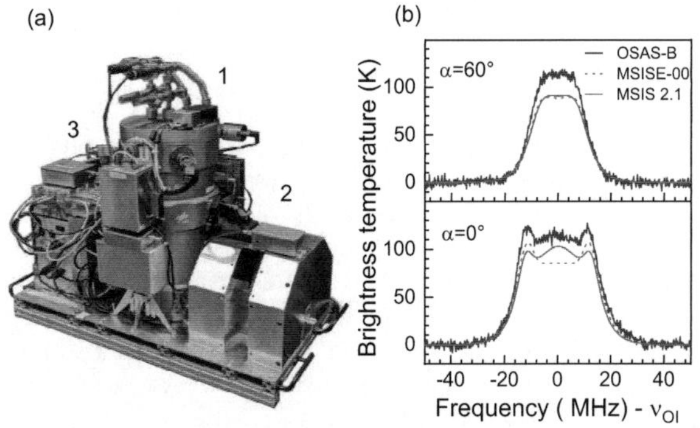

Fig. 1: (a) Photography of the instrument. 1: cryostat, 2: optics compartment, 3: backend electronics. (b) Measured and simulated spectra for 60° and 0° elevation Observation: 2022/09/07 9:14 UTC, 68°19′N,19°40′E, height: 32.6 km , azimuth direction: 23° NNE. Simulated spectra based on the NRL MSISE-00 [3] and MSIS 2.1 model [4].

References

[1] M. G. Mlynczak and S. Solomon, "A detailed evaluation of the heating efficiency in the middle atmosphere," *J. Geophys. Res.*, vol 98, 10517–10541, 1993.

[2] H. Richter et al., "Direct measurements of atomic oxygen in the mesosphere and lower thermosphere using terahertz heterodyne spectroscopy," Commun. Earth Environ., vol. 2, 19, 2021.

[3] Picone et al., "NRLMSISE-00 empirical model of the atmosphere: Statistical comparisons and scientific issues," J. Geophys. Res., vol. 107, no. A12, p. 1468, 2002.

[4] J. T. Emmert et al., "NRLMSIS 2.1: An empirical model of nitric oxide incorporated into MSIS," J. Geophys. Res., vol. 127, no. 10, p. e2022JA030896, 2022.

[1]DLR Institute of Optical Sensor Systems, Rutherfordstr. 2, Berlin, 12489 Berlin, Germany; [2]Max-Planck-Institut für Radioastronomie, Auf dem Hügel 69, 53121 Bonn, Germany [3]Humboldt-Universität zu Berlin, Department of Physics, Newtonstr. 15, 12489 Berlin, Germany (email: martin.wienold@dlr.de).

NOTES:

Cryogenic Receiver System for the Black Hole Explorer

C. Edward Tong[1*], Paul Grimes[1], Janice Houston[1], Michael D. Johnson[1], Dan Marrone[2], and Hannah Rana[1]

***Abstract*—The Black Hole Explorer (BHEX) is a space Very Long Baseline Interferometer mission under development, aimed at advanced Black Hole studies. This paper focuses on the receiver system, outlining the instrument concept, cryogenic and receiver designs.**

***Keywords*—Black Hole, multi-band receiver, SIS receiver, space cryogenics, VLBI.**

I. Introduction

THE Black Hole Explorer (BHEX) mission is a NASA Small Explorer (SMEX) mission under planning, that aims to produce the sharpest images in the history of astronomy by extending submillimeter Very Long Baseline Interferometry (VLBI) to space. The BHEX will discover and measure the bright and narrow "photon ring" that is predicted to exist in images of black holes. To achieve this goal, several key technologies are indispensable. In this paper, we will focus on the receiver system under design.

II. Instrument Concept

The science instrument for the BHEX is a coherent receiver system designed to operate within the 80-350 GHz frequency range. It is coupled to a 3 – 4 meters diameter antenna, which is addressed by another paper within this Symposium [1]. The BHEX receiver front end will observe 2 bands simultaneously in dual linear polarizations, around 90 GHz and either 230 or 345 GHz. The 90 GHz channel will be based on HEMT amplifiers, operating at around 20 K, providing input for the Frequency-to-Phase Transfer Techniques, which will improve the coherence time of high-frequency observations. For the higher frequency channels (230/345 GHz), SIS receivers will be used due to their quantum-limited sensitivity. As a result, the 230/345 GHz front end must be cooled to about 4 K.

III. Cryogenics and Receiver Designs

The use of liquid helium is excluded as a cooling option for the 2-year long BHEX mission because of weight considerations for the small satellite. A 2-stage Stirling cooler, employed for the JEM/SMILES mission [2], is being considered instead. Similar mechanical coolers are currently deployed on board the Japanese satellite, X-Ray Imaging and Spectroscopy Mission (XRISM). This cryocooler provides cooling power at 100 K, 20 K and 4.5 K. The required thermal budget of the BHEX front end is projected to be 30 mW at 4.5 K and 200 mW at 20 K.

A pair of optical diplexers [3] will be used to multiplex the beam from the telescope to feed the 3 receivers. The 230 and 345 GHz receivers will feature a state-of-the-art dual polarized sideband separating (2SB) architecture, operating with a 4-12 GHz IF. The receivers are expected to produce a data stream of nominally 192 Gbits/sec.

Owing to the limited DC power available on a SMEX spacecraft and the cryogenics is a major consumer of power, the receiver design must be carefully worked out to reduce the power draw. An alternate plan B, based on a dual band dual-polarized receiver set, is also under study. Instead of 2 SIS bands, a single high-band receiver will cover a wide frequency range, from 220 to 300 GHz, with the possibility of an even wider RF tuning range, extending the upper frequency coverage to 345 GHz. In addition, the use of Double-Side-Band (DSB) mixers is being considered, which offer simpler architecture and halving the data output for the SIS receivers. A trade-off study of the 2 receiver plans is being conducted, balancing the quest for higher sensitivities and the resource constraints, as well as complexities of design.

The high data rate necessitates the use of a state-of-the-art laser communication system to transport the data. That is why DSB mixers, with half the data output, may prove to be important.

IV. Conclusion

The BHEX mission, with its innovative cryogenic receiver system, emerges as a pioneering endeavor in Black Hole research. An international team is being formed to promote the BHEX mission.

References

[1] Sridharan *et al*, "The Black Hole Explorer: Mission Overview and Antenna Concept," to be presented in this Symposium.

[2] Ochiai *et al*, "Performance of JEM/SMILES in orbit," in *Proc. 21st ISSTT*, pp.172-184, Oxford, UK, March 2010.

[3] Carter, Tong and Zeng, "A low-loss optical duplexing scheme for Millimeter to Terahertz Waves," to be presented in this Symposium.

[1]Center for Astrophysics | Harvard & Smithsonian, Cambridge, MA 02138, USA; [2]University of Arizona, Steward Observatory, Tucson, AZ 85719, USA; [*]C. Edward Tong (email: etong@cfa.harvard.edu).

NOTES:

The Terahertz Intensity Mapper: Design, Modeling, and Characterization of the Cryogenic Receiver

Jianyang Fu[1*], Shubh Agrawal[2], James E. Aguirre[2], Hrushi Athreya[3], Justin S. Bracks[2,4], C. Matt Bradford[4,5], Brockton S. Brendal[3], Bruce Bumble[1], Nick Emerson[6], Jeffrey P. Filippini[3], Christopher E. Groppi[7], Steve Hailey-Dunsheath[4], Reinier M.J. Janssen[4,5], Dylan Joralmon[8], Garret K. Keating[9], Victor Gasho[6], Ryan P. Keenan[10], Mikolaj Kowalik[11], Lun-Jun Liu[4], Ian N. Lowe[12], Alex Manduca[2], Daniel P. Marrone[12], Philip Mauskopf[7], Evan C. Mayer[12], Rong Nie[3], Vesal Razavimaleki[3], Talia Saeid[7], Isaac Trumper[13], Jessica A. Zebrowski[14], Joaquin D. Vieira[1,11]

***Abstract*— The Terahertz Intensity Mapper (TIM) [1,2], a balloon-borne telescope funded by NASA, is set to measure [CII] emission from star forming galaxies over a vast cosmic volume. We will present an overview of TIM's cryogenic receiver design, its thermal modeling, and the ongoing characterization efforts in preparation for the Antarctic flight scheduled for 2026.**

***Keywords*— Cryogenic Receiver, Balloon, Far-Infrared, Intensity Mapping, Kinetic Inductance Detector, Spectrometer, Terahertz**

I. Design

TIM's cryogenic receiver is based on the proven design of the BLAST-TNG's receiver [3]. To accommodate our two grating spectrometers, we've expanded the cold plate and increased the capacity of the liquid helium tank to 270 liters. The receiver features three distinct cryogenic stages (see Fig. 1): a 4 K inner stage that houses the liquid helium tank and two concentric vapor-cooled shields (VCSs), VCS1 and VCS2, constructed from Aluminum 1100. These shields minimize thermal loading and provide heat sinking for various IR filters, cables, and plumbing. Each of the two VCSs is cooled by a multi-channel copper heat exchanger [4], featuring 66 parallel 3.2 mm channels. This highly efficient, low-impedance design allows the venting of cold helium gas through them even during high flow occasions like initial filling, effectively minimizing liquid helium consumption and reducing system cooldown time. The instrument space is positioned below the helium tank, with the bottom of the 4 K tank acting as a direct heat sink for the cold optics and the helium sorption refrigerator. This closed-cycle, 3-stage 3He/4He sorption refrigerator, developed by Chase Research Cryogenic, is designed to achieve a base temperature of 250 mK for our Microwave Kinetic Inductance Detectors (MKIDs) to ensure photon noise-limited performance. The cryostat assembly weighs approximately 430 kg, excluding liquid helium, cold optics, and cables.

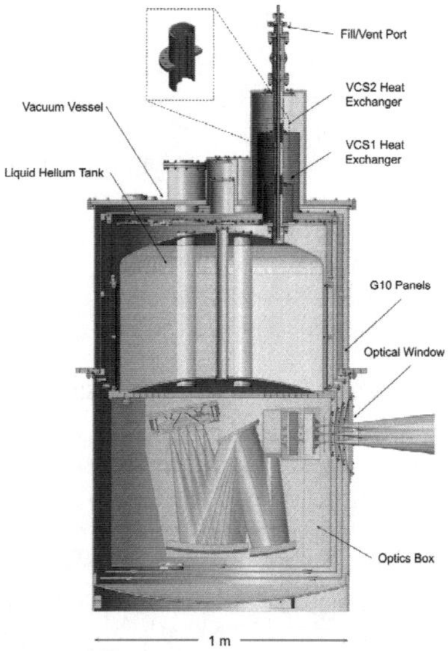

Fig. 1. Design of the TIM cryogenic receiver, highlighting the cryogenic stages and the instrument space, with a close-up view of the heat exchanger.

[1]Department of Astronomy, University of Illinois Urbana-Champaign, Urbana, IL 61801, USA; [2]Department of Physics and Astronomy, University of Pennsylvania, Philadelphia, PA 19104, USA; [3]Department of Physics, University of Illinois Urbana-Champaign, Urbana, IL 61801, USA; [4]Department of Astronomy, California Institute of Technology, Pasadena, CA 91125, USA; [5]Jet Propulsion Laboratory, California Institute of Technology, Pasadena, CA 91109, USA; [6]Steward Observatory, University of Arizona, Tucson, AZ 85721, USA; [7]School of Earth and Space Exploration, Arizona State University, Tempe, AZ 85287, USA; [8]School for Engineering of Matter, Transport and Energy, Arizona State University, Tempe, AZ 85287, USA; [9]Center for Astrophysics, Harvard & Smithsonian, Cambridge MA 02067, USA; [10]Max-Planck-Institut fur Astronomie, Heidelberg, D-69117, Germany; [11]NCSA, University of Illinois Urbana-Champaign, Urbana, IL 61801, USA; [12]Department of Astronomy, University of Arizona, Tucson, AZ 85721, USA; [13]ELE Optics, Tucson, AZ 85705, USA; [14]Kavli Institute for Cosmological Physics, University of Chicago, Chicago, IL 60637, USA. *Corresponding author (email: fu32@illinois.edu).

NOTES:

II. MODELING AND CHARACTERIZATION:

Our team has developed a Python-based cryogenic thermal model to simulate the loading and optimize the operations of the cryostat. This model accounts for various thermal transfer modes and iteratively solves for both stable state and transient responses. The equilibrium temperatures of the two VCSs are estimated to be at 60 K and 160 K, respectively. Our model predicts a thermal loading of 400 mW at the 4 K stage, with a ground-based hold time of 20 days, extending to 30 days under flight conditions.

Since its delivery to the University of Illinois in July 2024, we have assembled the cryostat and successfully cooled it to 4.2 K. The observed loading and equilibrium temperatures closely match our predictions from the model. Early tests also indicate a heat exchange efficiency exceeding 90% for both exchangers. Integration of the refrigerator and cold optics is on track for completion by summer 2024.

REFERENCES

[1] J. Vieira, et al., "The Terahertz Intensity Mapper (TIM): an Imaging Spectrometer for Galaxy Evolution Studies at High Redshift" *Proceedings of the 30th International Symposium on Space Terahertz Technology*, Apr 2019 [Online]. Available: https://arxiv.org/pdf/2009.14340.pdf [Accessed: Jan. 27, 2024].

[2] D.P. Marrone, et al., "The Terahertz Intensity Mapper: a balloon-borne imaging spectrometer for galaxy evolution", *Millimeter, Submillimeter, and Far-Infrared Detectors and Instrumentation for Astronomy XI, vol. 12190 of Society of Photo-Optical Instrumentation Engineers (SPIE) Conference Series*, 1219008, 2022.

[3] I. N. Lowe, et al., "Characterization, deployment, and in-flight performance of the BLAST-TNG cryogenic receive", *Millimeter, Submillimeter, and Far-Infrared Detectors and Instrumentation for Astronomy X*, 2020.

[4] H. Okamoto, D. Chen, "A low-loss, ultrahigh vacuum compatible helium cryostat without liquid nitrogen shield" *Review of Scientific Instruments*, 72, 2001.

NOTES:

Highly-Compact Terahertz Planetary/Cometary Instruments

Goutam Chattopadhyay, Adrian Tang, Subash Khanal, Sven van Berkel, Jacob Kooi, Cecile Jung-Kubiak, Alain Maestrini, Choonsup Lee, and Imran Mehdi

Jet Propulsion Laboratory, California Institute of Technology, Pasadena, CA 91109, USA

There has been only a handful of terahertz planetary science instruments that were baselined for space missions in recent years. The primary reason is that the terahertz heterodyne spectrometer instruments are relatively more power hungry and require more mass and volume compared to instruments at other frequencies.

Over the last several years, groups around the globe have made tremendous strides in reducing mass, volume, and power requirements for terahertz heterodyne instruments by focusing on innovative component developments and packaging solutions. Usage of commercial silicon foundries to develop low-power CMOS based components as well as low-profile antenna technology developments helped in this regard. As a result, it is possible to field highly-compact and low-power high spectral resolution heterodyne instruments for various planetary applications for space missions.

These developments have led to the consideration of using other platforms such as balloon-borne, SmallSats, and CubeSats. Traditionally, SmallSats and CubeSats and their related instrumentation development was primarily done by university undergraduate student teams. However, in recent years, the international space agencies have been actively looking into CubeSats and SmallSats as useable platforms to supplement main missions as well as use them for standalone scientific missions. NASA has flown such missions even to Mars for providing communication infrastructure during entry-descent-landing [1].

Development of compact and low-power terahertz scientific instruments for planetary applications, particularly on SmallSat and CubeSat platforms, poses a host of challenges: the lack of available space, severe restrictions on DC power availability, and antenna size limitations are some of them. For example, one has to be innovative in the design of the antennas as traditional high gain reflector antennas (for scientific payload as well as for data communication) are not practical for these platforms. Deployable antennas are preferred; however, even though some efforts are being made in this area, such antennas at terahertz frequencies are either not available yet, or the performance is not satisfactory. In recent years, metasurface and lens based low-profile antennas have been getting a lot of attention to address some of these challenges.

At JPL, we have developed a fully functional submillimeter-wave standalone scientific instrument for planetary science applications. The modular instrument architecture we adopted allows the instrument to be used for standard planetary science missions to the outer-planets, asteroids, or comets, and on CubeSat/SmallSat platforms.

Specifically, we have developed a low-mass (~2 kg) and low-power (~5W) 500-600 GHz high-resolution spectrometer instrument capable of remotely measuring water isotopes and deuterium/hydrogen (D/H) ratios on comets. The broadband instrument can also do trace-gas spectroscopy at this band. The instrument's 18 cm diameter aperture (which can be easily changed to larger diameter aperture, if required) consists of a novel low-profile leaky-wave lens-based antenna with a waveguide feed. We also have developed a low-profile reflector-based antenna that can be implemented instead of the lens antenna if so desired. A microelectromechanical system (MEMS) based calibration switch is integrated along with a variable temperature waveguide-based calibration load, all of which are integrated with the receiver front-end. A low-power (~300 mW) CMOS-SoC-based synthesizer at W-band with Schottky diode-based frequency multipliers generates the LO signal to pump a Schottky-diode based subharmonic mixer with state-of-the-art mixer noise temperature (~1000K). We have also developed a low-power (~1.5 W) CMOS-based backend spectrometer with 3 GHz available bandwidth and 4096 channels. This allows the overall instrument mass and power to be in the range ideal for implementation on planetary/cometary missions to outer planets. We have flown this instrument on a balloon-borne platform for technology demonstrations and TRL advancement.

In this presentation, we will provide the design and implementation of the 500-600 GHz spectrometer instrument with details of innovative packaging solutions, antenna technology, and low-power backend solutions suitable for future planetary science missions.

References

[1] S. Asmar et al., "Mars Cube One (MarCO) Shifting the Paradigm in Relay Deep Space Operations," in the Proceedings of the 14th International Conference on Space Operations (SpaceOps 2016), Daejeon, Korea, May 2016.

Enhancing the IRAM30m telescope for the next 15 years

Carlos A. Durán[1*], Miguel Sanchez[1], Manuel Castillo[1], Salvador Sanchez[1], Joaquín Santiago[1] and Antonio J. Vázquez [1]

***Abstract*—The IRAM30m millimeter-wave radio telescope started its operations in 1985. Since then, it has undergone multiple upgrades, including those related to instrumentation, thermal control, as well as telescope hardware and software. To ensure the radio telescope remains competitive over the coming 15 years, it requires more substantial renovations and updates. The first major step in that direction involved replacing the entire servo control system of the radio telescope. This affects both the main axes (AZ/EL) and the sureflector hexapod, as well as the subreflector wobbling system. This modernization is currently undergoing and being funded by Regional European Funds. Another significant improvement involves enhancing the main reflector surface by restoring the characteristic 30m white paint to maximize the radio telescope optimal performance across the entire elevation range. Along with these refurbishment taks, essential pieces of software for the telescoope control and science operations are being updated or replaced. After this upgrade is completed, science operations will resume. Meanwhile discusssions about further improvements that would include new instruments, surface degradation tracking and external spourious signal tracking and monitoring are taking place.**

***Keywords*— Instrumentation, IRAM, Telescope, upgrade.**

I. Introduction

IRAM 30m telescope, built in 1981, has been successfully operating since 1985. The contribution of the IRAM30m to science during these almost 40 years is undeniable. Moreover, the 30m has been source of inspiration and reference for many other telescopes/observatories [1] around the globe, not only from the purely scientific point of view, but also for the technical (hardware, software, techniques, etc.) and operational aspects.

The telescope has reached the point where more substantial and disruptive changes are required to keep it at the forefront. These changes include the upgrade of critical hardware like the servo system (see Fig 1.) and interface software as first steps. Still in this long-term view of 15 years, the telescope -observatory- would require also major upgrades on instrumentation, control system software, data reduction software and critical KPI monitoring tools. This talk presents the extent of some of the most relevant tasks carried out so far, and provides an overview of what we are aiming for the future of the 30m.

Fig. 1. IRAM30m telescope during the hexapod replacement, as part of the servo control system upgrade. July 2023.

References

[1] R. Guesten et al, "*The Atacama Pathfinder EXperiment (APEX) – a new submillimeter facility for southern skies–*" A&A 454 (2) L13-L16 (2006) [Online]. Available. https://doi.org/10.1051/0004-6361:20065420 [Accessed: Jan. 26, 2023].

[1]Instituto de Radioastronomía Milimétrica IRAM, Granada, 18012, Spain.
[*]Carlos A. Durán / cduran@iram.es

NOTES:

POSTER SESSION

Progress towards a focal plane unit for CHAI based on superconducting planar circuits

I. Barrueto[1], U. U. Graf[1], C.E. Honingh[1], M. Justen[1], K. Jacobs[1], F. Münning[1], M. Schultz[1], K. Vynokurova[1], S. Wulff[1], J. Stutzki[1] and D. Riechers[1]

***Abstract*— The LO distribution network is a crucial part of any heterodyne array. For the waveguide-based LO distribution for CHAI, we address problems previously found with a 3-dB quadrature planar power divider and present a new candidate for a 3-db power division based on superconducting planar circuits.**

***Keywords*— Heterodyne mixer, superconducting circuits, 90° hybrids, power division.**

I. Introduction

The CCAT Heterodyne Array Instrument (CHAI) is a 2x64 dichromatic heterodyne receiver, that is currently under development for the CCAT-prime observatory [1]. The bands intended for observation are the LFA (455-495 GHz) and the HFA (780-820 GHz), with each band consisting of a square array of 64 pixels composed of 16 sub-arrays of 1x4 pixels each. These 4-pixel blocks constitute the basic unit of the array where the LO signal is received and distributed to the four pixels. The 1-to-4 power division is achieved by a cascade network of three 3 dB power dividers.

II. Power dividers and 2-pixel block

The testing platform for the dividers is a 2-pixel block that allows for the testing of a single power divider in a simple network (Fig.1,2). This test block is fabricated in CuTe split block waveguide technology, receiving the LO power at its back through a rectangular 460x230 μm^2 waveguide interface. The power dividers are based on superconducting Nb planar circuits on a 9 μm silicon membrane. The mixers chosen to populate the test units are all based on the design presented in [2]. These mixers are used as direct power detectors by measuring the current LO-induced in the SIS junctions at a fixed voltage bias.

A first generation of power dividers based on 3 dB quadrature hybrids with on-chip loads was presented by the authors in [3], where it was observed that the power transmission to the mixers suffered from standing waves and a reduction of the LO coupling to the junctions in a limited frequency range (Fig.3a).

The reason behind the observed variation in coupling as a function of frequency was that the reflections caused by the load

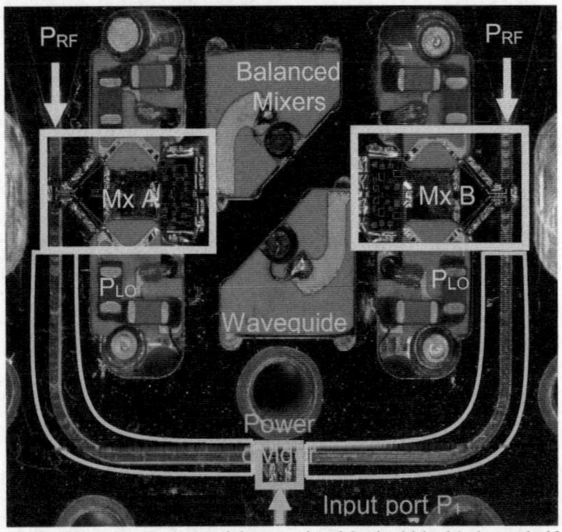

Figure 1: Microscope photo of the populated 2-pixel block's lower half.

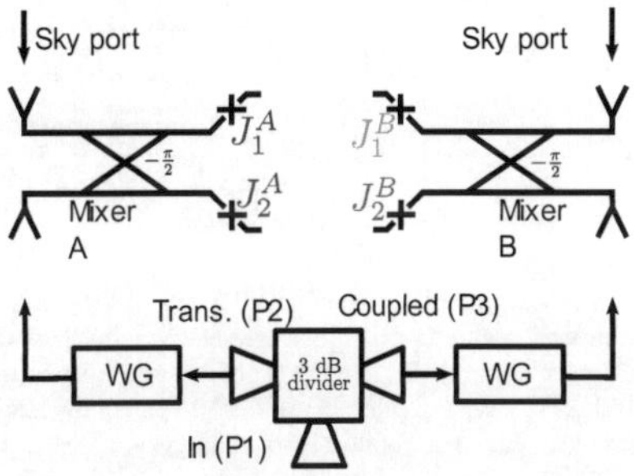

Fig. 2. Schematic of the 2-pixel block with the two mixers and power divider connected by the long waveguide.

at P_4 and the $\Delta\varphi$ introduced by the hybrid caused standing waves cavities of different electrical lengths in the LO-paths to the mixer A and B. This results in the out-of-phase behaviour observed in the power coupled to the junctions in the mixers A and B (Fig.3a). This effect is seen clearer if the ratio in power coupled to mixers A and B is studied, where the small. imbalance of the 90° hybrid is greatly enhanced by the out-of-

[1]I. Physikalisches Institut, Universität zu Köln, Köln, 50937, Germany.
*barrueto@ph1.uni-koeln.de

NOTES:

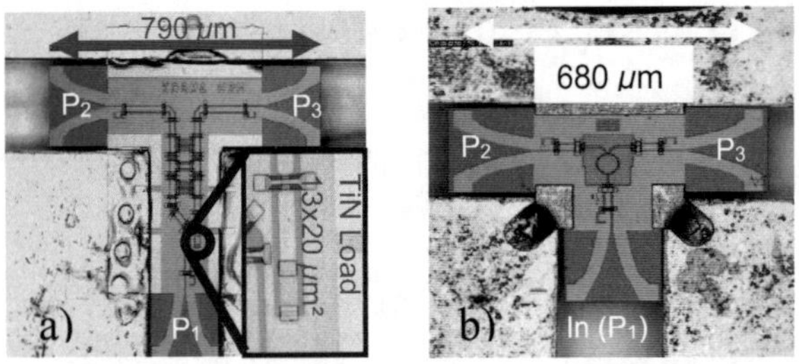

Figure 3: Microscope photos of the 2-pixel block and both power dividers. a) Bottom half of the populated w-pixel block. b) 90° three branch line CPW power divider, terminated with a resistive load c) Wilkinson power divider

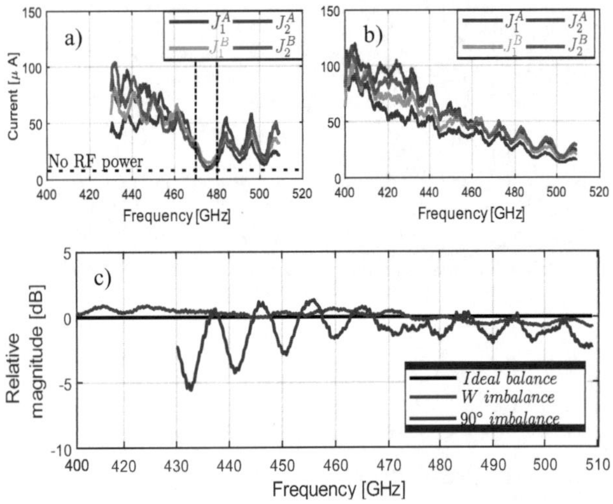

Figure 4: Measured current at constant V_{bias}=1.8 mV of the 4 SIS junctions of the 2 balanced mixers in the 2-pixel block while sweeping the local oscillator over the frequency band and deducted power division balance of the power dividers (a) 90° hybrid power divider. (b) Wilkinson power divider. (c) Balance of the power dividers, here J_1 of both mixers A and B is compared, effectively yielding the amplitude imbalance of the power divider.

phase behaviour of the standing waves (Fig.4c). In a scenario where the existence of standing waves cannot be avoided, the imbalance between mixers can be minimized by using an in-phase power divider, since with an in-phase power divider the standing waves in the LO-paths towards the two mixers are the same, resulting in a balanced power distribution, even if the total transmission is affected by standing waves. Fig.3b shows a fabricated microstrip Wilkinson power divider. It can be seen how the induced currents in the SIS junctions in the mixers A and B have the same behaviour as a function of frequency (Fig.4b) and how the imbalance is considerably better than with the CPW 90° hybrid (Fig.4c).

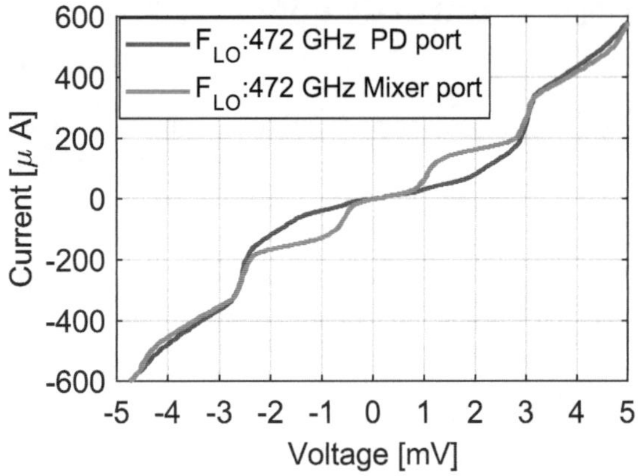

Figure 5: I-V characteristics of the SIS junctions with RF power coupled through the power divider (violet curve) and the horn (blue curve) for the same junction.

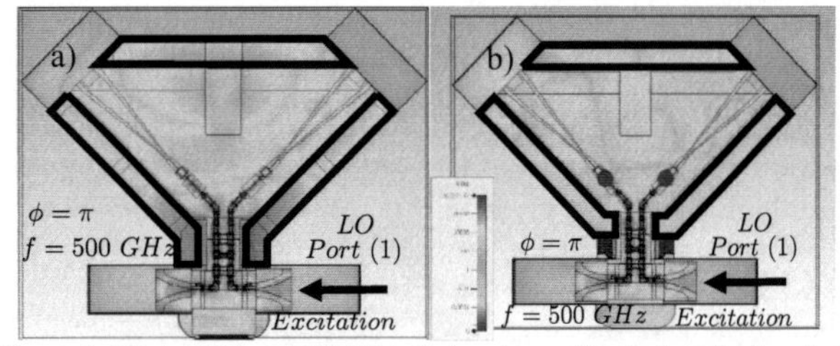

Figure 6: 3D simulation of IBAMI in its housing cavity, the beamlead pocket footprints is highlighted in the black outline. (a) Field simulation at f = 500 GHz in the original cavity. (b) Field simulation at f = 500 in the modified cavity.

The second feature observed with the CPW 90° hybrid as a LO divider was the range of low induced current between 470-480 GHz. In addition to this range of low induced current, anomalous IV curves were observed (Fig.5) when injecting the LO through the power divider. These features were only observed when the LO was coupled through the power divider and not when coupling the LO through the RF ports. The reason behind this was cavity resonances in the mixer cavity that coupled to beamlead pockets present in the mixer cavity, shown in Fig.6a delineated with the black lines. These beamlead pockets, shown in more detail in Fig.7, were included to prevent the silicon chip's mechanical degradation from several open/close cycles of the 2-pixel block. Since the addition of beamlead pockets substantially prolongs the lifetime of the mixer chips, they cannot be completely removed. Fig.6b shows a modified beamleads footprint that prevents the coupling to the resonance. This modified beamlead footprint was implemented in the Wilkinson 2-pixel block, where no range of low induced current was observed (Fig.4b).

NOTES:

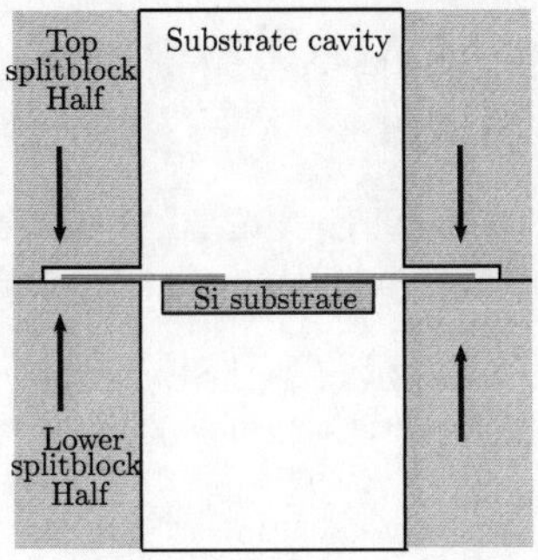

Figure 7: Cross-section of the housing cavity of the IBAMI with beamlead pockets depicted. These pockets are milled over the beamleads to prevent mechanical degradation from several open/close cycles of the block.

III. Summary

The two issues initially observed in the 2-pixel block have been solved. The enhanced imbalance measured with the CPW 90° hybrid was addressed by replacing the hybrid with a microstrip Wilkinson power divider, whose phase characteristics allow for a more balanced power distribution, even in the presence of standing waves. The range of low induced current was discovered to be caused by resonances coupling to the beamlead pockets in the mixer cavity. A modified beamlead footprint avoids coupling to the cavity resonance and solves the issue.

References

[1] http://www.ccatobservatory.org
[2] M. P. Westig, K. Jacobs, J. Stutzki, M. Justen, and C. E. Honingh, Supercond. Sci. Technol. 24, 085012 (2011)
[3] I. Barrueto, K. Jacobs, S.Wulff, M.Schultz, M. Justen, C.E. Honingh and J. Stutzki, "LO power division circuits for the CCAT-prime Heterodyne Array Instrument (CHAI)", In Proc. 32nd ISSTT, 2022, pp. 174-177.

NOTES:

Waveguide Circuitry for the Prototype ALMA Band 6v2 Sideband Separating SIS Mixer

P. Dindo[1*], D. Monasterio[1], J. Lambert[1], A. Navarrini[1], A. R. Kerr[1], B. Casto[1], G. Morris[1], P. Astudillo[1], C. Jarufe[1], J. Effland[1], and K. Saini[1]

***Abstract*—We describe the design and analysis of the waveguide circuitry proposed for the sideband separating (2SB) mixer of an upgraded ALMA Band 6 (Band 6v2) receiver. This unit, optimized for the 209–281 GHz Band 6v2, incorporates a waveguide quadrature RF hybrid, a cold termination for image frequencies, and an LO power splitter. It connects two Band 6v2 double sideband (DSB) SIS (Superconductor-Insulator-Superconductor) mixer modules to the input LO waveguide and the RF orthomode transducer. The output IF band is 4–16 GHz, with a goal of 4–20 GHz.**

The Band 6v2 2SB receiver must have a sideband rejection ratio (SRR) ≥ 15 dB over 90% of the bandwidth. Multiple reflections between the RF hybrid, LO splitter, image termination, and the DSB SIS mixer modules contribute to the SRR degradation. We use an analytical procedure to predict the contribution of the RF hybrid and LO couplers to the SRR.

The component waveguide branch-line RF hybrid was independently fabricated as a test module and characterized using a vector network analyzer (VNA). Simulated and measured results of the coupling, phase and amplitude balance, and return loss of the quadrature RF hybrid are presented along with their predicted contribution to the SRR.

***Keywords*—2SB, RF hybrid, SRR, 4K image termination.**

I. Introduction

Sideband separating SIS mixers, used in most ALMA receiver front ends, help to minimize the single sideband (SSB) system noise by reducing the atmospheric noise contribution from the unwanted image sideband. The Band 6v2 prototype receiver is based on a compact design in which two DSB SIS mixer chip modules are connected to the sideband separating waveguide module (2SB module assembly) along with the superconducting electromagnet used to suppress Josephson currents in the SIS junctions. The 2SB module assembly is an E-plane split block containing a 3-dB quadrature RF hybrid, 4-K image termination, and an in-phase LO power splitter. The ALMA Band 6 Wide Sensitivity Upgrade (WSU) receiver upgrade specifications require RF coverage to be expanded from the current 211-275 GHz to 209-281 GHz, and the SRR to be increased from ≥ 10 dB to ≥ 15 dB over 90% of the band [1].

To meet the new bandwidth and SRR specifications, the 2SB assembly has been designed with a layout which minimizes its size and insertion loss. This paper describes the design of the RF hybrid and the LO splitter. The full 2SB mixer module has a non-standard interface, therefore, it does not lend itself to direct testing with a VNA. We therefore used a test module with RF hybrid identical to that in the 2SB module but with accessible waveguide ports for connection to a VNA. The waveguide size WR-3.7 (0.0185" x 0.037") supports TE_{10} single mode propagation throughout the 209–281 GHz operating band. The SRR of the 2 SB assembly was calculated from measured RF hybrid data and using the analytical model described in [2].

II. Design of 2SB Module Assembly

The new 2SB module is shown in Fig. 1. The quadrature RF hybrid and the LO splitter waveguide structure is shown in Fig. 2. They were designed with electromagnetic field simulation in CST Studio Suite [3].

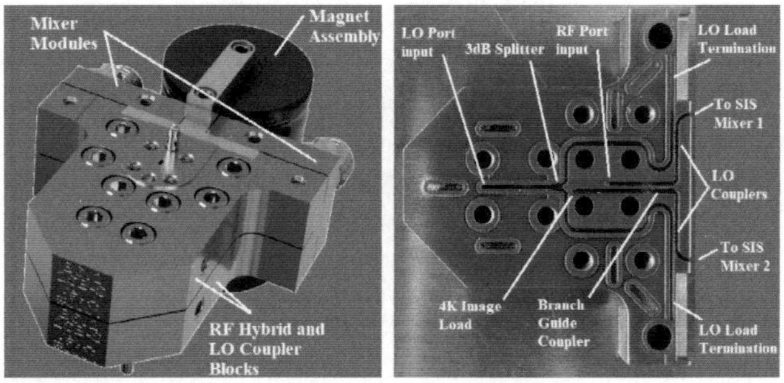

Fig. 1: Left: 2SB assembly with two mixer modules and a superconducting solenoid attached. Right: View of one half of the split-block showing the waveguide circuitry, 4K image termination and waveguide ports.

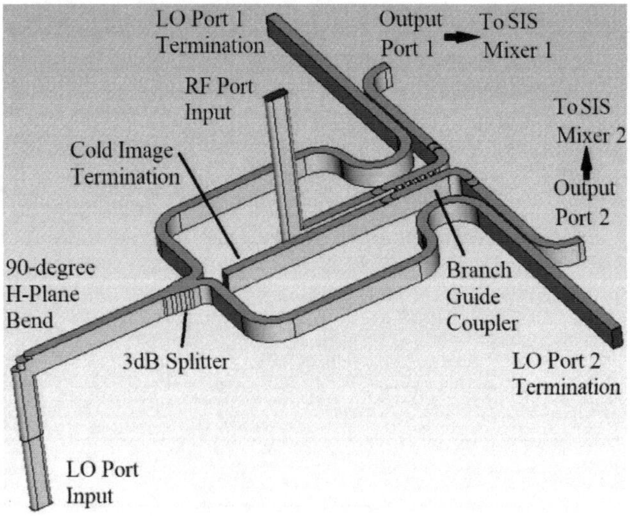

Fig. 2: The structure of the waveguide circuit of the 2SB module assembly simulated in CST Studio. The circuit includes the LO splitter, quadrature RF hybrid, and the 4K image termination.

[1] NRAO Central Development Laboratory, Charlottesville, VA 22903, USA*Corresponding author (email: pdindo@nrao.edu). The National Radio Astronomy Observatory is a facility of the National Science Foundation operated under cooperative agreement by Associated Universities, Inc.

The E-plane branch-line coupler is designed following the approach in [4] so that power incident on any port is divided equally between two coupled ports with a 90-phase difference, while the fourth port is isolated. There is a plane of symmetry around the plane perpendicular to the broad walls of all the waveguides and intersecting them in the middle. As no currents flow across this plane of symmetry for the TE_{10} mode, imperfect contact between the two halves should not impact the performance of the circuit.

Details of the branch guide coupler are shown in Fig. 3. It consists of two WR-3.7 waveguides that start at full height (b = a/2) and then increase in height in two steps. There are six quarter-wave branch guides whose dimensions and separations are optimized to obtain 3 dB coupling over the entire RF band with an amplitude imbalance ≤ 1 dB and phase balance within 1.0°. The input of the branch guide coupler is connected to the WR-3.7 RF waveguide port via an H-plane 90-degree bend, while the two output waveguides are connected to the output ports with 90° E-plane bends. A wedge-shaped load of Eccosorb MF116 material, which has a minimum return loss of 35 dB across the RF band [5] is installed as the image termination for the 2SB mixer.

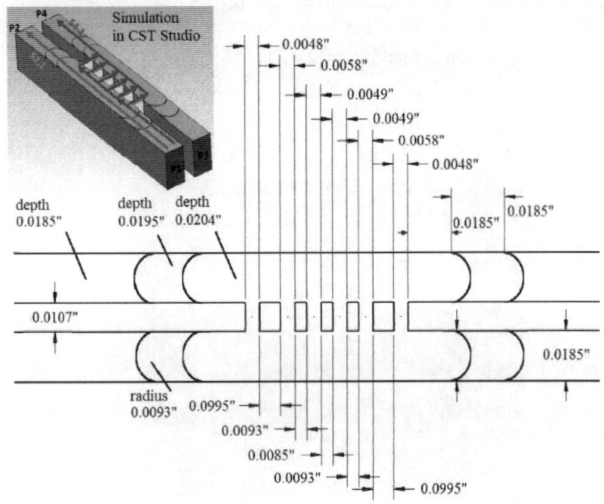

Fig. 3: The WR-3.7 quadrature hybrid consists of two parallel waveguides coupled through a series of six branch waveguides and fabricated as a split block structure. The lengths of the branch guides and their spacings are approximately a quarter of a guide wavelength at the center frequency of operation.

The electromagnetic field simulation of the LO power splitter is described in [6]. For the present application in the 2SB assembly, we have added an H-plane 90-degree bend followed by a two-stage transition from WR-3.7 to the WR-3.4 (0.034" x 0.017") LO input port.

The LO power is coupled into the RF signal waveguides through directional couplers as shown in Fig. 4. Each coupler has six gold plated coupling probes between the two waveguides similar to the one described in [7]. The probes are 0.004" wide x 0.010" long x 0.001" thick and are supported on fused quartz standoffs 0.004" x 0.004" x 0.003" high in channels 0.004" long x 0.010" wide x 0.008" high.

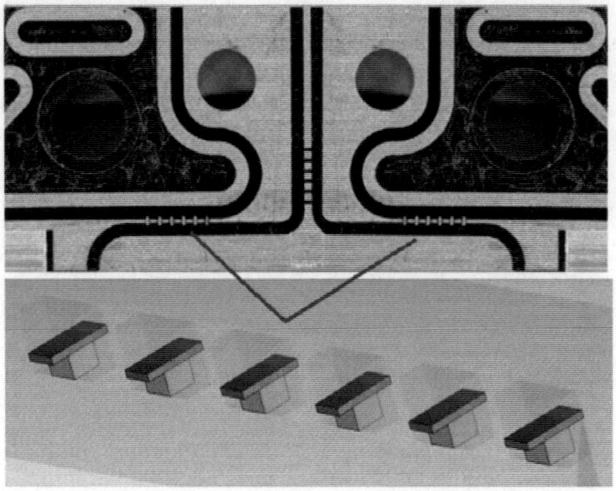

Fig. 4: Photograph shows a section of the branch guide coupler and the two sets of LO coupling waveguide channels. The bottom picture shows the CST simulation of the coupling of six gold probes on quartz standoffs inside each set of the LO coupling waveguide channels.

The coupling of the LO to the two output ports was simulated using CST Microwave Studio [3] and found to be 22.9 ± 1.0 dB across the RF bandwidth. Table I shows the simulated performance characteristics of the complete 2SB module assembly.

TABLE I.

Frequency Range	209-281 GHz
Amplitude Imbalance, max	0.60 dB
Phase Imbalance, max	+/-0.59°
Port Return Losses	≥ 19 dB
Isolation	≥ 22.5 dB
Sideband rejection Ratio	≥ 27.5 dB
LO to RF Coupling	-22.9 ± 1.0 dB
Excess Insertion Loss, Cold	≤ -0.21 dB
Excess Insertion Loss, T=298K	≤ -0.45 dB

The sideband rejection ratio of the 2SB module is simulated using the approach taken in [2], which includes the effect of multiple reflections inside the module. The SRR is not only affected by the phase and amplitude imbalance of the individual components, but also by two reflection mechanisms interfering inside the 2SB module: (1) reflections from the output ports pass back and interfere at the imperfect 4K load termination inside the isolated port as illustrated in Fig. 5(a), and, (2) reflections at the two output ports due to input return loss of the mixers and the finite isolation of the quadrature RF hybrid as illustrated in Fig. 5(b).

We implemented in AWR Microwave Office [8] the SRR analytical model [2], and simulated these two reflection mechanisms to determine their impact on amplitude imbalance, phase imbalance and SRR. We simulated a 4K image noise termination with return loss = 35 dB. We also simulated two cases of mixer reflection coefficients: -3 dB and -9 dB.

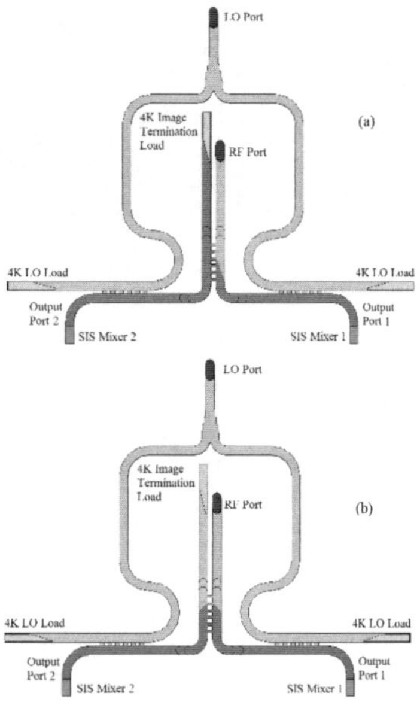

Fig. 5: Schematic illustration of the two reflection paths shown in orange color in the RF waveguide structure: (a) shows the reflections between the mixer ports and the 4K image termination and (b) corresponds to the reflections between the mixer ports due to reverse coupling through the branch-guide coupler.

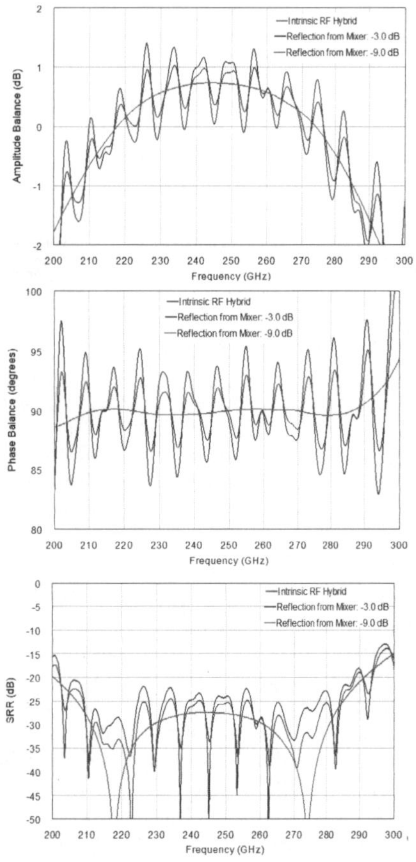

Fig. 6: The effects of reflections from the SIS mixer ports and the 4K image termination on the amplitude balance, phase balance, and sideband rejection ratio.

Fig. 6 shows results of these simulations where we superimpose the characteristics of the intrinsic 2SB module assembly with no reflections from the 4K image and output ports. Results show that amplitude imbalance can degrade by about 1.0 dB and phase balance can degrade by up to 10 degrees. It is seen that with these degradations in phase and amplitude imbalance, the sideband rejection ratio remains ≥ 20 dB which would meet the ALMA Band 6v2 specifications.

III. Measurement of RF Hybrid Test Modules

We fabricated a test module incorporating the quadrature RF hybrid waveguide structure identical to that of the 2SB module, but with WR-3.7 waveguide access ports for measurement with a VNA and WR-3.4 (0.017" x 0.034") 200-300 GHz millimeter wave extenders from Virginia Diodes Inc. [9]. The RF hybrid test module as well as the 2SB mixer module were machined from C360 brass at NRAO Central Development Laboratory with a Matsuura LX-160 5-Axis NC Mill and plated with 1.5 μm of gold [10]. The RF hybrid test module is shown in Fig. 7, and was tested on the VNA system shown in Fig. 8. The effects of the abrupt but small discontinuities at the interfaces between the WR-3.4 waveguides of the VNA extender and the WR-3.7 waveguides of the RF hybrid had negligible impact on results.

Fig. 7: Photograph showing a test RF hybrid module.

Fig. 8: Photograph showing RF hybrid modules under test using Keysight PNA-X 5245B network analyzer with VDI WR-3.4 200-300 GHz millimeter wave extenders.

The simulated and measured amplitude and phase balance are shown in Fig. 9. The simulated and measured SRR are shown in Fig. 10, while simulated and measured RF hybrid characteristics including coupler amplitudes, return losses and sideband rejection ratios are shown in Fig. 11. In all cases, measured results agree well with simulations.

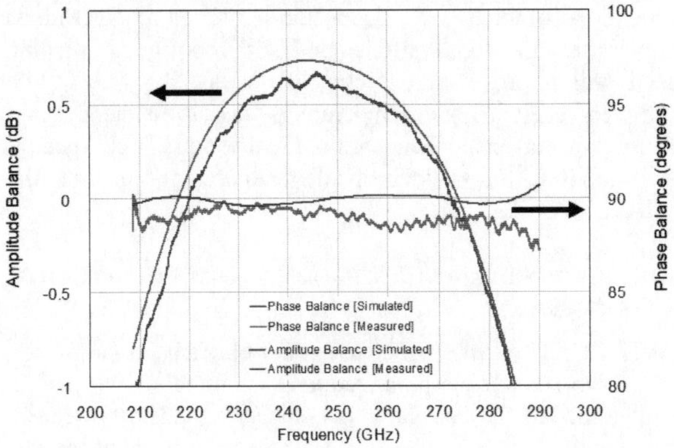

Fig. 9: Simulated and measured amplitude and phase balance of the test RF hybrid.

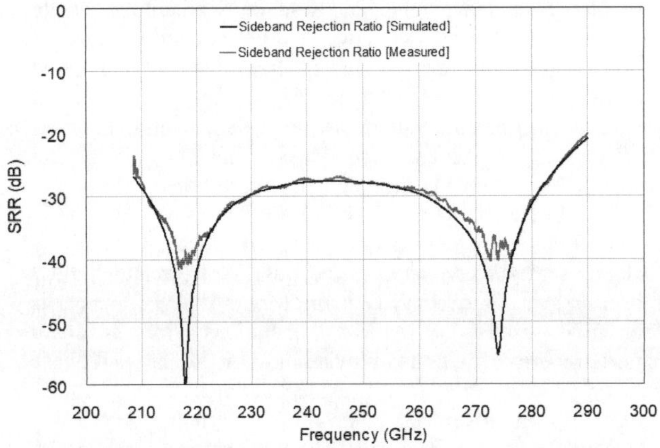

Fig. 10: Simulated and measured sideband rejection ratio of the test RF hybrid. The SRR of the 2 SB assembly was calculated from simulated and measured RF hybrid data and using the analytical model described in [2].

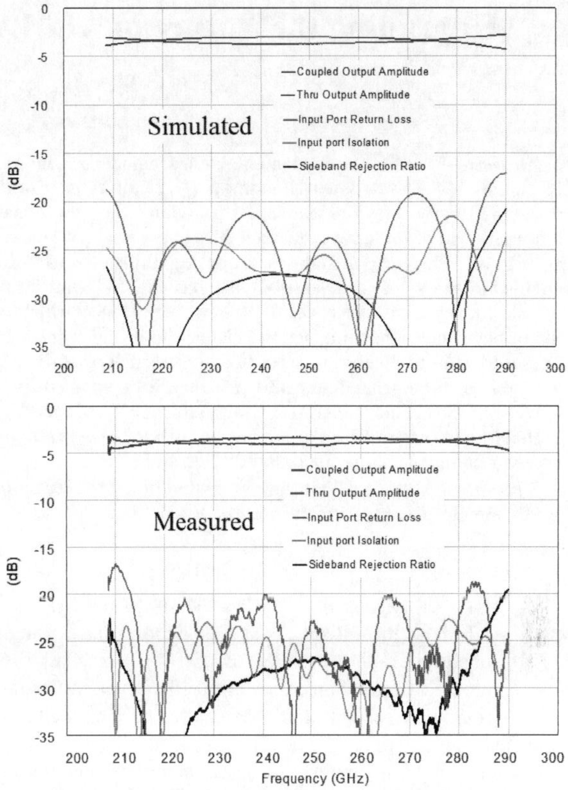

Fig. 11: Simulated and measured RF hybrid characteristics. They include output amplitudes of the thru (red color) and coupled (blue color) ports, input return loss (green color), isolation (Orange color), and sideband rejection ratio (black color).

IV. CONCLUSIONS

The waveguide assembly of the prototype sideband separating SIS mixer for the ALMA Band 6v2 receiver is analyzed to determine its contribution to the sideband rejection. It is shown that the design meets the new receiver specifications.

REFERENCES

[1] A. Navarrini et al., "ALMA Band 6v2 receiver development status," 2023 XXXVth Gen. Assem. Sci. Symp. Int. Union Radio Sci. (URSI GASS), vol. 00, pp. 1–4, 2023, doi: 10.23919/ursigass57860.2023.10265563.

[2] A. Khudchenko at al., "Design and Analysis of a Waveguide Structure for 211–275 GHz 2SB SIS Mixer," IEEE Transactions on Terahertz Science and Technology, vol. 13, no. 6, pp. 645-653, Nov. 2023.

[3] CST Studio Suite, https://www.3ds.com/products/simulia.

[4] S. Srikanth and A. R. Kerr, "Waveguide Quadrature Hybrids for ALMA Receivers," ALMA Memo 343, 11 Jan. 2001, http://www.alma.nrao.edu/memos/.

[5] D. Monasterio, 2024, Private communication.

[6] A. R. Kerr, "Elements for E-Plane Split Block Waveguide Circuits," ALMA Memo 381, 5 July 2001, http://www.alma.nrao.edu/memos/.

[7] A. R. Kerr and N. Horner, "A Split-Block Waveguide Directional Coupler," ALMA Memo 432, 26 Augut 2002, http://www.alma.nrao.edu/memos/.

[8] AWR Microwave Office, Cadence Design Systems, San Jose, CA 95134

[9] VNA Extender Modules (VNAX), Virginia Diodes, Inc, https://vadiodes.com.

[10] Plating specification: ASTM B 488-01, https://library.nrao.edu/public/memos/edtn/EDTN_229.pdf

Upgrading the Future of ALMA: the Wideband Sensitivity Upgrade

J. Donovan Meyer[1*], R.A. Loomis[1], C. Brogan[1], and T. Hunter[1]

Abstract—**Meeting the fundamental science goals put forward in the ALMA Development Roadmap [1] requires a series of dramatic upgrades to the telescope throughout the signal chain, and users will begin to see these upgrades in the next few years. In this Proceeding, we demonstrate the capabilities specifically enabled by the Wideband Sensitivity Upgrade (WSU) with science use cases, as well as aspects of these new technologies which will likely have the highest impact on science observing from a PI's perspective. In particular, we introduce the spectral resolution and available instantaneous bandwidth which will be enabled by the Advanced Technology ALMA Correlator (ATAC), and we highlight the improvements coming to two of the three first receiver bands to be upgraded (Bands 2, 6, and 8).**

Keywords—**Millimeter astronomy, Submillimeter astronomy, Radio interferometry, Instrumentation.**

I. Introduction

The Wideband Sensitivity Upgrade (WSU) is an ALMA-wide initiative to at least double (and eventually quadruple) the system bandwidth with improved receiver performance and digital efficiency. Much of the ALMA hardware (except the antennas), as well as the Online and Offline software, is being improved. New wideband hardware being implemented includes: new receivers, with the new Band 2 receiver (ESO, NAOJ, NRAO) and upgraded Bands 6v2 (NRAO, NAOJ) and 8v2 (NAOJ) coming first; digitizers and digital signal processing (ESO); Data Transmission System (DTS: NAOJ, NRAO); a correlator to handle both 7-m and 12-m Array observations (ATAC: NRC, NRAO); and an upgraded Total Power GPU Spectrometer (TPGS: KASI, NAOJ). The goal of the ALMA Project is to start WSU Science Operations by 2030, though some receiver bands will come later (since the legacy receivers can be used with the new digital system).

II. Enhanced Capabilities with the WSU

The scientific improvements expected with the 2x bandwidth WSU system, yielding 16 GHz of instantaneous observing bandwidth per polarization, are numerous. At low spectral resolution, the system will provide a relatively modest 2x *correlated* bandwidth increase, but at high spectral resolution (0.1 km/s), the increase in *correlated* bandwidth depends on the frequency of the receiver (up to 4x at the highest frequency Band 10 and up to 68x at the lowest frequency Band 1).

Similarly, the system will provide a spectral scan speed increase of 2x for low spectral resolution, but at high spectral resolution (0.1 km/s), the increase in spectral scan speed will be up to 4x at the highest frequency (Band 10) and up to 68x at the lowest frequency (Band 1). The spectral line imaging speed will improve by a factor of 2.2 due to the improved receiver noise temperatures and digital efficiency, and the continuum imaging speed will improve by a factor of at least 4.8 due to the improved receiver noise temperatures, digital efficiency, and increase in correlated bandwidth. Finally, ultra-high spectral resolution (0.01 km/s) will be available to users for the first time at all ALMA frequencies.

III. The Advanced Technology ALMA Correlator (ATAC)

ATAC will provide an increase in correlated bandwidth from 7.5 GHz to 16 GHz per polarization in the initial delivery of 2x bandwidth, but the correlator is being designed to be readily expandable to 4x bandwidth. In fact, the correlator ingest will already handle 4x bandwidth with 2x bandwidth selected for fine channelization and correlation. Up to 1.2 million channels across 80 x 200 MHz frequency slices will be produced, with flexible channel averaging and stitching available to create wider science spectral windows at full polarization. Spectral resolutions as fine as 0.1 km/s across the full correlated bandwidth will be available at any ALMA frequency.

ATAC will be able to correlate up to 68 antennas, either all together or in separate subarrays. Flexible subarraying capabilities will enable independent observations with the 12-m (Main) Array and 7-m Array (ACA).

The improvement in digital sensitivity to be provided by ATAC is 13.4% compared to the Baseline Correlator (BLC) observing in Frequency Domain Mode (FDM), which is currently used in ALMA operations for standard interferometry. It will also provide all-digital delay and phase tracking with no delay rate-dependent anomalies.

Finally, improvements relevant to current Very Long Baseline Interferometry (VLBI) and pulsar observing modes include simultaneous coarse spectral resolution visibilities and a single antenna VLBI mode (A1-VLBI) without disturbance to interferometric observing.

[1]National Radio Astronomy Observatory, Charlottesville, VA, USA.
[*]Corresponding author (email: jmeyer@nrao.edu).

NOTES:

IV. Band 2 (67-116 GHz)

The wide RF frequency coverage of Band 2, coupled with the ATAC, yields the availability of at least one critical molecular gas tracer (carbon monoxide, or CO, transitions from J=1-0 to J=10-9) at all redshifts and [CI], or atomic carbon, beyond z=3. These lines will enable characterizations of the gas content in untargeted galaxy searches out to z=10. The most efficient way to set up a spectral scan for the 2-18 GHz IF of Band 2 is shown in Fig.1 for a galaxy at z=6.

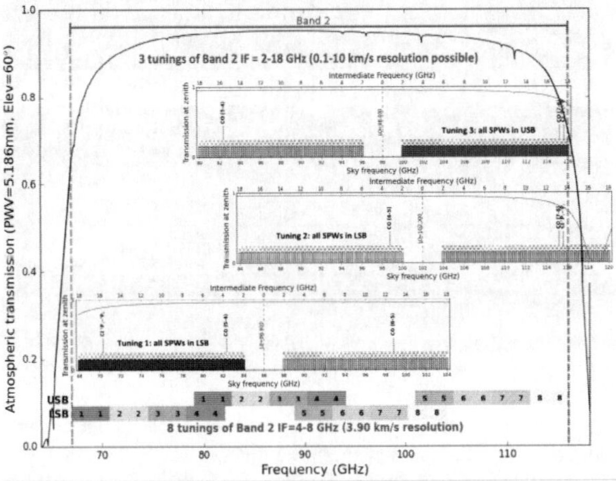

Fig. 1. Only three tunings (in blue, yellow, and green) are needed to cover the full RF range of Band 2 with ATAC (at 2x bandwidth), compared to eight tunings currently required with the BLC (shown in the numbered colored boxes along the bottom of the Figure).

V. Band 6v2 (209-281 GHz)

The extended frequency range of Band 6v2 is demonstrated with a high spectral resolution protoplanetary disk astrochemical survey, modeled on a current ALMA Large Program (Fig. 2). Flux ratios of different chemical tracers allow for characterization of the carbon-to-oxygen (C/O) ratio in disks. Improved performance of the IF edges results in better observing efficiency for setups that include all CO isotopologues.

VI. Summary

The dramatic upgrades coming to the entire ALMA system as part of the WSU will begin to appear for users in the next few years. In this Proceeding, we summarize the improved capabilities coming to the entire system from a user's perspective in terms of the instantaneous and correlated bandwidth, imaging speeds, and ultra-high spectral resolution. The improved correlation capabilities to be performed by ATAC are described in terms of the correlated bandwidth, number of channels, increased digital efficiency, subarraying capabilities, and VLBI modes. Finally, we show two example science setups using ATAC with Band 2 and Band 6v2 (two of the first three receiver bands to be upgraded) to show the improvement in observing capabilities.

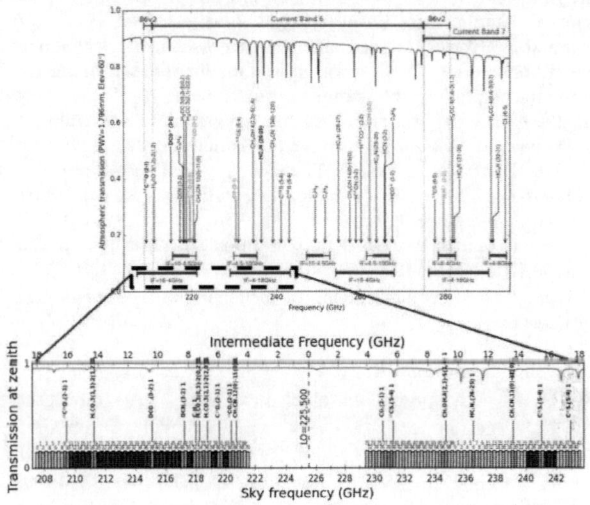

Fig. 2. (Top) Three spectral setups with the BLC (red) and two scientifically equivalent setups with ATAC (blue) are shown. Lines in orange are required to characterize the C/O ratio, but many other potentially interesting species can be simultaneously covered. *A factor of five* less observing time is needed with the WSU. (Bottom) Example tuning for Band 6v2. Colored regions show individual spectral windows, demonstrating that they can be different widths (in increments of 200 MHz) and unequally distributed across sidebands. The orange and blue colors represent two different choices of spectral resolution: orange is higher spectral resolution for targeted lines, while blue is for lower spectral resolution "continuum" in this example.

References

[1] Carpenter, Brogan, Iono, and Mroczkowski. "*The ALMA 2030 Wideband Sensitivity Upgrade.*" ALMA Memo Series 621 (2022).

NOTES:

Broadband Microfabricated Waveguide Terminations for Low-power Applications at Terahertz Frequencies

Karl Birkir Flosason, Cristian López*, Denis Meledin, Leif Helldner, Sven-Erik -Ferm, Victor Belitsky and Vincent Desmaris

***Abstract*— We present the design of broadband microfabricated waveguide terminations, for prospective use as drop-in components in waveguide blocks. Two types of terminations were designed and characterized using frequency extension modules. The first type of termination is based on quartz substrate, employing an E-probe to couple to a waveguide and on-substrate Ti-N alloy resistive absorber integrated with broadband tuning circuitry, exhibiting a return loss better than 20 dB over the 260-370 GHz frequency range. The second type of load features a finline to slot-line transition made of a 30 µm thick Si membrane covered with high resistivity Ti-N alloy. This termination exhibits a measured return loss better than 20 dB at all frequencies in the 200-380 GHz band.**

***Keywords*—Microfabrication, termination, waveguides, Terahertz.**

I. Introduction

Waveguide terminations are an important component used in, e.g., power combining waveguide circuits with 3-dB hybrids, directional couplers for LO injection, etc. The terminations load the idle port of such hybrids or directional couplers and absorb unwanted signals [1]. In receiver systems based on 2SB mixers, having high-quality terminations has been shown to be crucial to the mixers' sideband rejection performance [2]. The traditionally used of absorbing material, e.g. Eccosorb™ [3], becomes difficult to machine at high frequencies because o very small dimensions and the material performance at cryogenic temperature is not accurately known. Furthermore, distributed loads [4] require a bigger space for longer waveguides that make the mixer block bulkier and additionally complicates the fabrication of the mixer block by requiring special RF absorbing coating, e.g., high resistivity Titanium film on the milled structures, along with conventional gold plating. The matched loads in the instrumentation applications are exposed to quite low power and thus could be engineered without special considerations for heatsinks. In addition, with a specific design of the loads, the performance of the waveguide terminations becomes predictable and reliable even at cryogenic temperatures. In this paper, we present 2 novel wideband waveguide terminations, Fig. 1, based on chips with E-probe and finline designs. These drop-in chips can be easily integrated into existing systems, achieving return loss levels of better than 20dB over the designed frequency range.

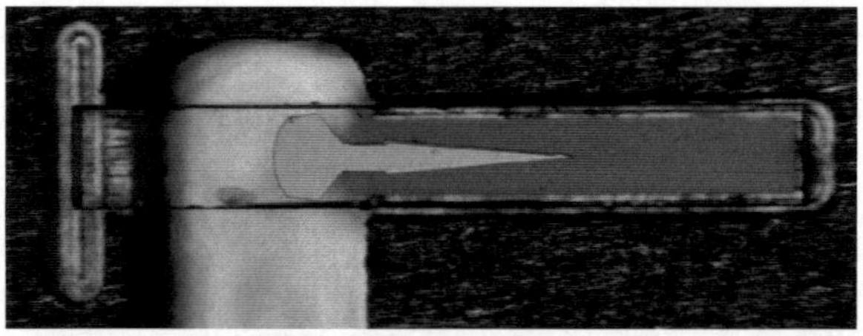

(a)

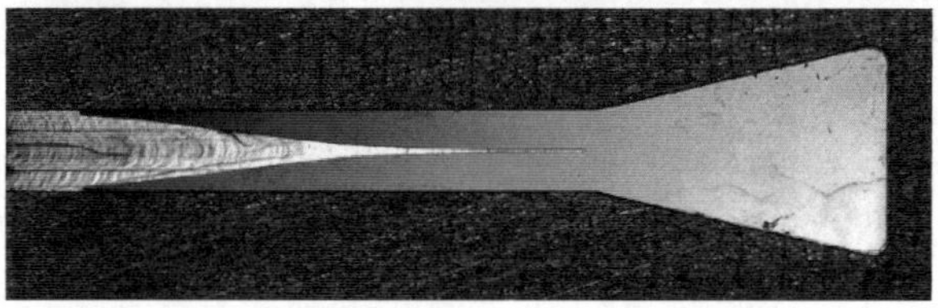

(b)

Fig.1 Fabricated Waveguide terminations. (a) E-probe chip. (b) Finline based load. The waveguide dimension are 380 µm x760 µm.

II. Termination Design and Simulation

The proposed waveguide terminations were designed and optimized in Ansys HFSS employing realistic parameters for the materials.

The first waveguide load is designed to cover the frequency range 260-370 GHz. The design employs a quartz substrate and resembles a mixer chip, comprising an E-probe to couple the power from the waveguide. Moreover, it includes a simple matching microstrip circuitry made of Gold, and a thin film absorber based on Ti-N alloy, with resistivity 30 Ω/□.

In order to achieve a broader bandwidth, 210-380 GHz, the second type of termination employs a waveguide to slotline transition using a finline profile described in [5]. The load is made off the 30 µm thick Si membrane covered with Ti-N alloy with a resistivity of 30 Ω/□. In Fig. 1 both types of loads are depicted.

All authors were with the Group for Advanced Receiver Development, Department of Space, Earth and Environmental Sciences, Chalmers University of Technology, Gothenburg, SE-41296, Sweden. *Corresponding author *Cristian Lopez, cristian.lopez@chalmers.se.*

NOTES:

III. Results and Discussion

The designs were characterized using a Keysight PNA-X VNA and frequency extension modules. To cover the designed frequency range, two extension modules were employed: WR 3.4 (200-330 GHz) and WR2.2 (325-500 GHz). In both cases, a 1-port TRL calibration was performed. Additionally, waveguide adapters were used to accommodate the waveguide dimensions for each extension module. These adapters were de-embedded through post-processing.

Fig. 2 displays the simulated and measured performance for each termination. Since the measured return loss levels are below 20 dB for most of the band, small discrepancies in positioning could have a significant impact on the measured performance, particularly for the design based on E-probes, which are more sensitive to their position relative to the backshort. However, finlines are less sensitive to their position within the waveguide. In this case, the misalignment between the input waveguide and the extension modul also plays a role in performance, as demonstrated in Fig.2b, where a 30 µm misalignment is introduced in the simulation. The simulations shows a reasonable agreement with the measurement results.

From the graphs, it is clear that the E-probe waveguide load shows a return loss of better than 20 dB over the entire frequency band of 260-375 GHz. Meanwhile, the finline-based waveguide load outperforms the former with a return loss of better than 25 dB over a broader frequency range of 210-380 GHz.

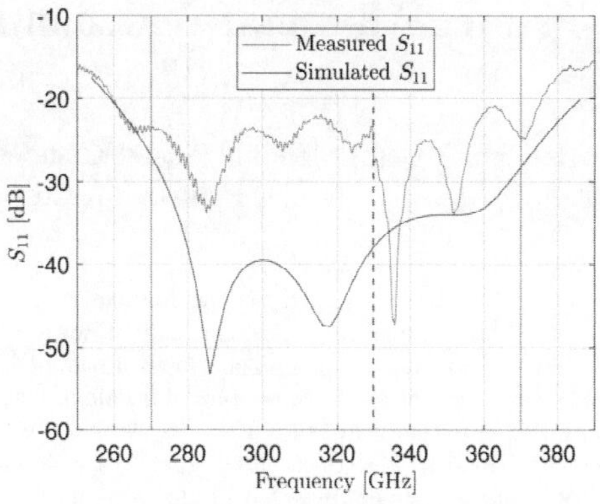

(a)

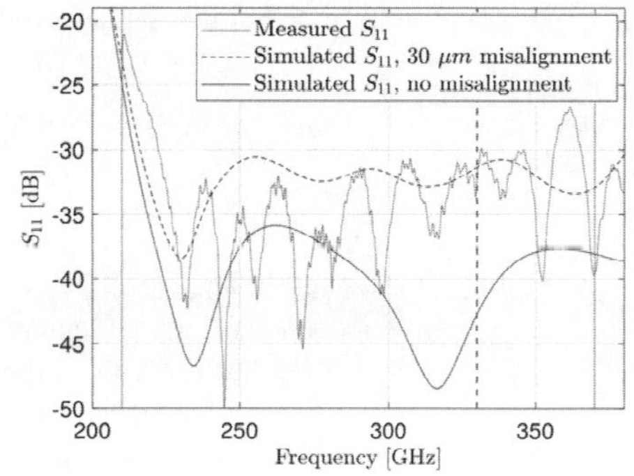

(b)

Fig.2 Simulated and measured performance of the waveguide terminations. (a) Simulated and measured performance for E-probe termination (b) Simulated and measured performance for finline based termination. A simulation with 30 µm misalignment beteween the upper and lower parts of the split-block is included.

IV. Conclusion

In this paper, we have presented the design, fabrication and characterization of two types of broadband microfabricated waveguide terminations for low-power applications at terahertz frequencies. The first design, based on an E-probe and Ti-N alloy resistive absorber, achieved a return loss better than 20 dB over 260-375 GHz. The second design, featuring a finline to slot-line transition on a Si membrane covered with Ti-N alloy, demonstrated superior performance with a return loss of 25 dB over a broader frequency range of 210-380 GHz. These results indicate that the proposed waveguide terminations are suitable for integration into existing systems, offering reliable and predictable performance even at cryogenic temperatures and with appropriate scaling to even higher frequencies.

References

[1] A. Gouda et al. "Millimeter-wave wideband waveguide power divider with improved isolation between output ports", IEEE TST Science and Technology vol. 11, no. 4, pp. 686-693, Jul. 2021, doi:. 10.1109/TTHZ.2021.3078876

[2] A Khudchenko et al., "Efficiency of the Image Band Suppression in Sideband Separating SIS Receivers for Radio Astronomy". In: *2020 7th All- Russian Microwave Conference (RMC)*. Nov. 2020, pp. 19–20. doi: 10.1109/RMC50626.2020.9312246.

[3] G.A. Ediss et al., "FTS measurements of Eccosorb MF112 at room temperature and 5 K from 300 GHz to 2.4 THz". In: *ALMA memos* #273 (Sept. 1999).

[4] R. Hesper, et al., "A High-Performance 650-GHz Sideband-Separating Mixer—Design and Results," in IEEE TST and Technology, vol. 7, no. 6, pp. 686-693, Nov. 2017, doi: 10.1109/TTHZ.2017.2758270.

[5] C. López et al., "Waveguide-to-Substrate Transition Based on Unilateral Substrateless Finline Structure: Design, Fabrication, and Characterization". In: IEEE *TST* (Nov. 2020), pp. 668–676. issn: 2156-3446. doi: 10.1109/TTHZ.2020.3020683

NOTES:

3D-Printed All-metal Wideband Dual-Polarization Cryogenic Dichroic Filters

Leif Helldner[1*], Denis Meledin[1], Saeed Khademzadeh[2], Yu Cao[2], Lars Nyborg[2], Igor Lapkin[1], Sven-Erik Ferm[1], Victor Belitsky[1], Alexey Pavolotsky[1], and Vincent Desmaris[1]

***Abstract*—In this work we report the performance of a 3D-printed single metallic layer high pass dichroic filter with an improved spectral response at non-normal beam incidence. The dichroic takes advantage of the extra degrees of freedom provided by metal additive manufacturing to enable the slanted design of the perforations of the dichroic metal plate. The dichroic demonstrates a good agreement between simulated and measured performance allowing transmission better than 90% and cross-polarization levels below -40 dB over a fractional bandwidth of about 28-30% in the Q-Band and could be scaled to higher THz frequency advantageously.**

***Keywords*—dichroic filters, mm-wave, additive manufacturing, metamaterial, 3D-printing.**

I. Introduction

DURING the last decade multiband receivers have gained significant attention in radio astronomy and environmental science [1-3]. Such receivers allow an improvement in the phase calibration for the higher frequency band, using the lower frequency band phase shift information due to the impact of atmospheric changes in the optical path. Moreover, faster mapping speed and enhanced UV-plane coverage make multiband receivers even more attractive, for example, to use in very large baseline interferometry (VLBI) observations. Frequency-selective surfaces (FSSs) commonly employed in multi-band receivers spatially divide a signal beam into two ones based on their difference in frequency. A dichroic filter, as a type of FSS, often employs a sequence of several patterned metallic and dielectric layers [4]. The geometrical properties such as the patterns' shapes and sizes, their periodicity, and the thickness of the metal and dielectric layers determine the frequency response of such dichroic.

The use of an all-metal design [5] offers the advantages of employing the dichroic filter at cryogenic temperatures, thereby minimizing system noise contributions through the improved metal conductivity and the reduced RF insertion loss, as well as reducing issues with outgassing and thermal contraction causing delamination.

In this work, we present a 3D-printed single-layer metallic dichroic filter for prospective reliable use at cryogenic temperatures. The filter allows the transmission of both polarization components of the input signal at a selected part of the Q-band (37-50 GHz), while reflecting the signal at the K-band (18-27 GHz), as it could be further used in a VLBI Tri-band receiver planned for Onsala Space Observatory (OSO) 20 m antenna. Yet, the design is perfectly scalable with an approach to explore additive manufacturing of a dichroic allowing transmission in other frequency bands.

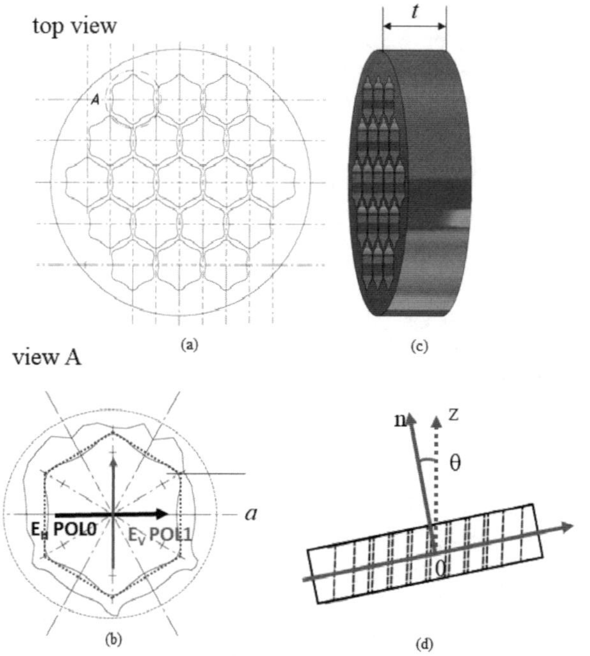

Fig. 1. The geometry of the designed perforated plate: (a) top view from *z* direction; (b) detailed top view A of the dichroic aperture from *z* direction; (c) and (d) - side view of the dichroic plate. The incident wave illuminates the dichroic plate with the angular offset of the apertures θ from the direction of z=∞. The side length of the initial hexagonal aperture is marked with a red color dot line.

II. Dichroic Design

The single-layer dichroic metal plate features perforations, as shown in Fig. 1, at a slant angle, θ, to minimize the angular degradation. The shape of the perforation has been first approximated using the equations in [5] and further optimized using HFSS simulations to reduce the insertion loss, eliminate in-band resonance spikes, and allow the transmission of both polarization of the input signal within the Q-band. A value of θ=13 degrees was considered acceptable for the proposed optical layout of the OSO Tri-band receiver, ensuring a fractional bandwidth of over 30%. Fig.2 shows the simulated transmission (TR) and cross-polarization characteristics (XPOL) as a function of frequency at 13 degrees of the waveguide aperture tilt, representing POL0 and POL1 polarizations, respectively. The incident radiation wave vector is always coaxial to the waveguide apertures. The transmission

[1]Group for Advanced Receiver Development, Department of Space, Earth and Environmental Sciences, Chalmers University of Technology, Gothenburg, SE-41296, Sweden; [2]Materials and manufacture, Department of industrial and Materials Science, Chalmers University of Technology, Gothenburg, SE-41296, Sweden. *Corresponding author (email: leif.helldner@chalmers.se).

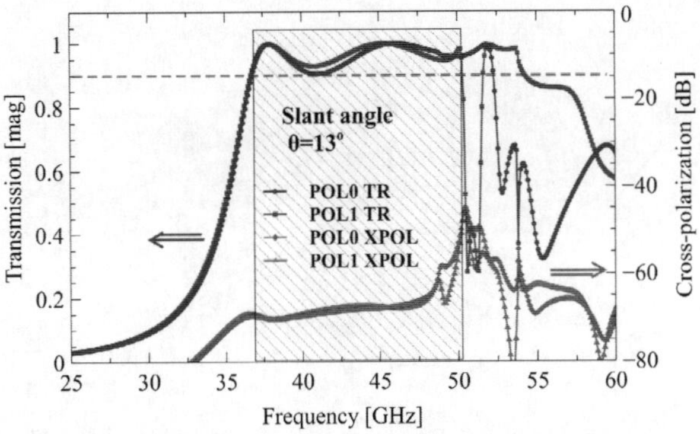

Fig. 2. Simulated transmission (TR, left axis) for horizontally (E_H) oriented E–field (POL0) and vertically (E_V) oriented E–field (POL1), and cross-polarization characteristics (XPOL, right axis) between the polarizations of the designed dichroic filter as a function of the frequency at the designed value of slant angle θ=13 degrees. The frequency ranges with transmission exceeding 90% are indicated by the hatched areas. This transmission level is indicated by a gray dashed line.

for frequencies lower than 27 GHz is expected to be below 5%. The cross-polarization simulations show a negligible effect on the performances, consistently staying well below the threshold value of -40 dB.

III. DICHROIC FABRICATION

The dichroic has been fabricated using additive manufacturing to enable the electromagnetic design of the dichroic with slanted perforations. For this study, gas-atomized $Al_{10}SiMg$ powder was selected as the starting material. The prototype was fabricated using the Powder Bed Fusion Laser Beam (PBF-LB) additive manufacturing process in an EOS M290 machine with a 100 μm spot size and a 400 W (370 W nominal power) Yb-fiber laser. Optimum precision process parameters were used to achieve the maximum density and minimum deviation from the predesigned model.

Fig. 3. Manufactured dichroic filter. The insert provides a close-up view of a single perforation cell, highlighting the details of its structure. The view is taken from the *n*-direction (see Fig. 1).

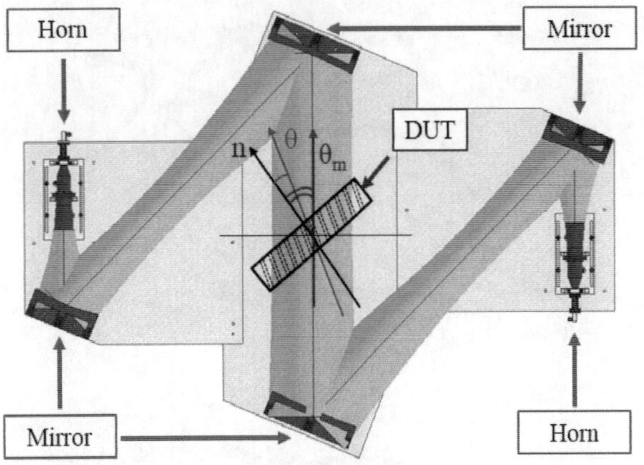

Fig. 4. The schematic layout of the optical assembly for the characterizing the dichroic filter, the grey-shaded volume represents the calculated quasi-optical beam indicating the path of the beam within the system; A photograph of the measurement setup can be seen in [5]

Prior to cutting the printed part from the build plate, stress relief heat treatment was conducted at 300°C for 3 hours, followed by furnace cooling. The printed dichroic plate was then cut from the platform using a cold saw. To inhibit the deformation of thin walls in honeycomb structures during cutting, the voids were filled with hard wax. To achieve further dimensional accuracy, the bottom of the dichroic plate was then surface-finished in a precision milling machine, followed by wax removal.

The manufactured dichroic filter prototype is 7.11 mm thick and its features are shown in Fig. 3.

IV. DICHROIC CHARACTERISATION AND PERFORMANCE

The electromagnetic performance of the 3D-printed dichroic was obtained using a Vector network analyzer coupled to a quasi-optical system that provides a plane-wave-like beam at the position where the tested dichroic is installed. The setup consists of four ellipsoidal mirrors with horn antennas at the transmitting and receiving sides. The layout of the measurement setup, visualizing the propagated beams, is illustrated in Fig. 4. To prevent any parasitic reflections, the area outside of 4 times the beam at the dichroic plate is shielded with an absorber. Moreover, polarization grids have been incorporated into the optical system to minimize cross-polarization levels. The waveguide aperture tilt θ=13degrees is indicated in Fig. 4 with a blue arrow, the incident angle θ_m represents the apparent angle between the beam axis, indicated by a red arrow, and the normal direction to the metal plate *n*, marked with a dark grey arrow.

The comparison between simulated and measurement transmission as well as cross-polarization characteristics was carried out for POL1 as depicted in Fig.5a. It is evident from the figures that the measurement results closely align with the simulated transmission data exhibiting a fractional bandwidth of about 30%. The measured low edges of the filter (defined as 90% transmission) passband match well with simulations with

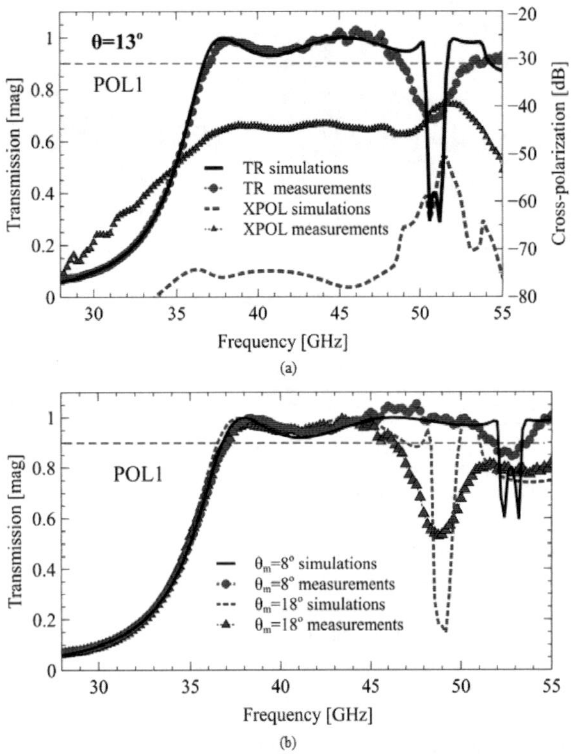

Fig. 5. The comparison between simulated and measurement transmission as well as cross-polarization characteristics for POL1: a.) transmission (TR, left axis), and cross-polarization characteristics (XPOL, right axis) between polarizations of the designed dichroic filter as a function of the frequency at the designed value of slant angle θ=13 degrees. b.) transmission at $\Delta\theta=\pm5$ degrees: A transmission level of 90% is indicated by gray dashed lines.

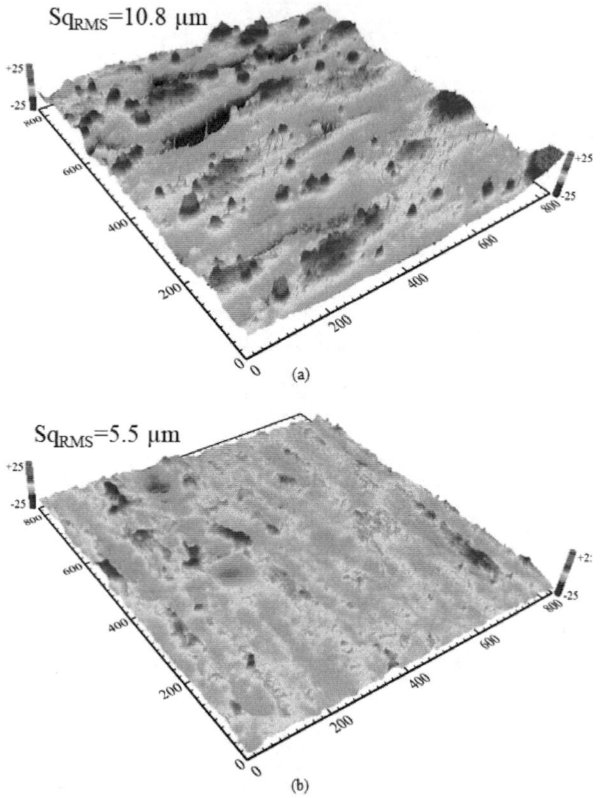

Fig. 6. The 3D surface scan of a). the 3D-printed dichroic filter presented in this work; b), Laser milled fabricated dichroic filter presented in [5]

an accuracy of about 0.2 GHz. The measured cross-polarization leakage remains better than 40 dB across the entire passband for both polarizations, fully satisfying the practical design requirements. For the cross-polarization plots in Fig.5a, the only case represented is when POL1 was excited and POL0 was detected.

In order to verify the accuracy of our model, we performed the measurements of the fabricated dichroic with a manufactured aperture tilt of θ =13 degrees at different values of $\theta_m=\theta\pm\Delta\theta$, where $\Delta\theta$ was set to ±5 degrees (see Fig.4). In the HFSS simulations, the angle θ was kept always at 13 degrees and $\Delta\theta$ values were an angle between the waveguide walls axis and the incident wave axis. In Fig.5b we depict a comparison between the measured transmission and the corresponding simulated values for POL1 at $\Delta\theta = \pm5$ degrees, respectively. The simulated transmission closely matched the measured results at both values of θ_m varied in the experiments. The measurement results prove the feasibility of the dichroic filter design with desirable frequency and polarization performance.

It is important to acknowledge that our simulations did not take into account the roughness of the metal plate surface and the perforation walls. However, we compared surface quality between the 3D printed dichroic presented in this work and the laser-cut dichroic filter with tilted perforations for the Q-band reported in [5]. Measured RMS surface roughness (Sq_{RMS}) values presented in Fig.6 for both devices demonstrate surface roughness of the same order of magnitude (relative to the skin depth on such material at mm-wave frequencies) for the 3D printed dichroic and the other filter fabricated employing laser cutting technology. Therefore, despite the difference in surface roughness values, both dichroic filters demonstrate very similar performances allowing fractional bandwidth of about 30% with transmission better than 90% and exceptional cross-polarization levels below -32 dB.

V. CONCLUSION

We have proposed a design for the dichroic filter based on a perforated metal plate fabricated using 3D printing with additive manufacturing. The proposed design can reach 90% of the transmission about 30% of the RF bandwidth around 43 GHz central frequency. The measured performance showed a very good agreement with the simulations. The presented design is scaleable to higher frequencies and could be employed as a cold dichroic filter for providing simultaneous operation at higher frequencies, for instance, e.g. 230 and 345 GHz channels of the Event Horizont Telescope. Despite a difference in surface roughness between the 3D-printed dichroic fabricated by additive manufacturing and the one manufactured using water-guided laser cutting, both dichroic filters show very similar performances in the Q-band frequency range. Thereby shows that 3D metal printing is very suitable for producing advanced mm-wave dichroic and opens up for novel and more advanced designs of components in the mm-wave frequency range.

REFERENCES

[1] M. Carter, et al., "The EMIR multi-band mm-wave receiver for the IRAM 30-m telescope," *Astron. Astrophys.*, vol. 538, Art. no. A89, February 2012, Available: DOI: 10.1051/0004-6361/201118452. [Accessed: June. 24, 2024].

[2] S.-T. Han, et al., "Korean VLBI network receiver optics for simultaneous multifrequency observation: Evaluation," *Pub. Astron. Soc. Pac.*, vol. 125, no. 927, p. 539, May 2013. Available: DOI: 10.1086/671125. [Accessed: June. 24, 2024].

[3] N. Okada, et al., "Development of the multi-band simultaneous observation system of the Nobeyama 45-m Telescope in HINOTORI (Hybrid Installation project in NObeyama, Triple-band ORIented)", *Proceedings of the SPIE*, vol.11453, art. id. 1145349, December 2020. Available: DOI: 10.1117/12.2562137. [Accessed: June. 27, 2024].

[4] X. You, C. Fumeaux and W. Withayachumnankul, "Tutorial on broadband transmissive metasurfaces for wavefront and polarization control of terahertz waves", *J. Appl. Phys.*, vol. 131, art. id 061101, February 2022. Available: DOI: doi.org/10.1063/5.0077652. [Accessed: June. 27, 2024].

[5] D. Montofré, et al., "A Broad-Band Dual-Polarization All-Metal Dichroic Filter for Cryogenic Applications in Sub-THz Range," *IEEE Transactions on Terahertz Science and Technology*, vol. 14, no.2, p. 199, March 2024. Available: DOI: 10.1109/TTHZ.2023.3338472 [Accessed: June. 17, 2024].

A Turnstile OMT using Magic-Tees and Integrated Noise-Injection Couplers

Doug Henke[1*], Dave Atherton[2], Ivan Wevers[1], and Adam Densmore[1]

***Abstract*—We present an orthomode transducer (OMT) design based on the turnstile junction that provides balanced signal outputs, suitable for either linear or circular polarization. Hole couplers have been integrated for receiver calibration using noise-injection. The OMT uses magic-tees for recombination which has the advantage of terminating higher-order modes that could arise from fabrication uncertainty. Measurements will be shown for Q-band and applicable to ngVLA band 5.**

***Keywords*—Turnstile OMT, magic-tee, hole coupler, noise-injection.**

I. Introduction

An orthomode transducer (OMT) is used within the front-end of a receiver to separate out the received signal into two linear polarisations. Four common implementations of OMTs for millimetre-wave frequencies include: planar, Bøifot, turnstile junction, and dual-junction (some recent examples are [1]–[7]).

When using a turnstile as the polarisation discriminator, the input signal is split into 2 branch pairs (4 outputs total) where each pair is recombined into the respective polarisation. T-junction or Y-junction power combiners are typically used for recombination, and their effectiveness depends on phase and amplitude balance (180° difference with equal amplitudes). Under ideal symmetric conditions, each polarisation is wholly contained within one branch pair of signals and is fully recombined.

In practice, however, measured performance of OMTs often show sharp signal drops in the gain or "spikes" in the isolation response. There are several contributing factors for this degraded performance, as discussed in [8], and the most common include: branch path length imbalances and layer misalignment. The performance of an OMT is dependent on symmetry of the structure, and any deviation through misalignment or fabrication tolerance will degrade the signal throughput and isolation responses. In turnstile junction OMTs, magic-tees may be used in place of T- or Y-junctions, as shown in Fig. 1, to help mitigate asymmetries.

The advantage of using magic-tees as the branch combining element is that any signal imbalance can be terminated within the sum port of the magic-tee, instead of reflected internally. Fig. 1 also illustrates the inclusion of hole couplers that may be used for noise-injection and receiver calibration.

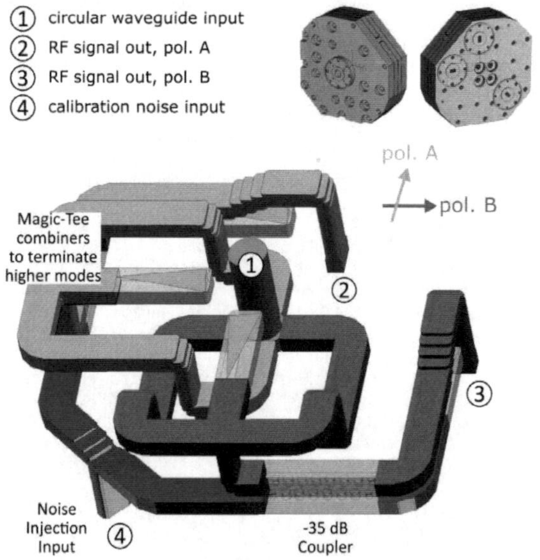

Fig. 1. Illustration of turnstile OMT using integrated noise-injection couplers and magic-tees to recombine turnstile outputs within each branch. Encircled numbers indicate ports as: (1) circular waveguide input; (2) and (3) signal outputs; and (4) noise-injection. The magic-T sum ports and hole couplers each have an internal load that is epoxied within the channel.

Fig. 2 shows the machining approach, whereby the block is sectioned into 4 layers. Waveguide channels are machined into both sides of each platelet and careful attention was given to flatness. While the design is suitable across millimetre-wavelengths, we have demonstrated the approach at Q-band towards the ngVLA Band 5 receiver (30.5–50.5 GHz).

Features of this particular design include:

- balanced amplitude and phase for signal and noise injected paths;
- integrated noise-injection couplers (~-35 dB);
- use of circular waveguide (direct connection to feed horn, no transition);
- integrated magic-T couplers to reduce signal drops from higher-order modes; and
- downward-facing waveguide outputs for symmetric cartridge assembly.

[1]National Research Council Canada, Herzberg Astronomy and Astrophysics (NRC-HAA), Victoria, BC, V9E 2E7. Canada; [2]National Research Council Canada, Design and Fabrication Services (NRC-DFS), Victoria, BC, V9E 2E7, Canada. *Corresponding author (email: Doug.Henke@nrc-cnrc.gc.ca).

NOTES:

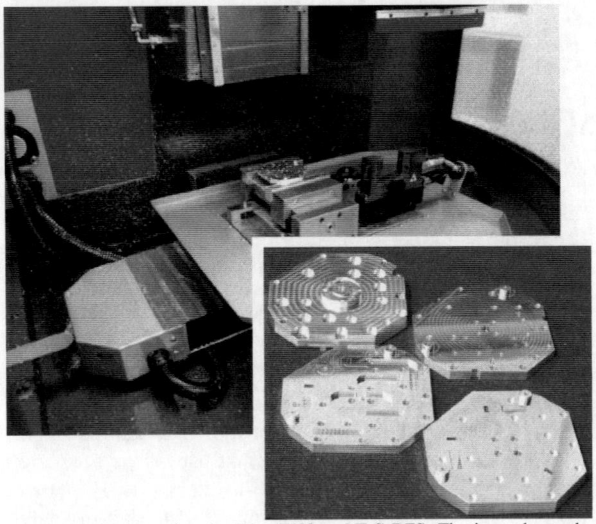

Fig. 2. Machining using the Kern EVO at NRC DFS. The inset shows the resulting 4 layers machined from aluminum.

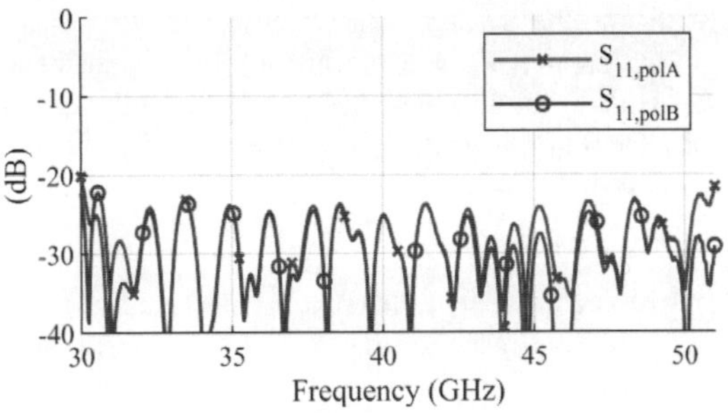

(a)

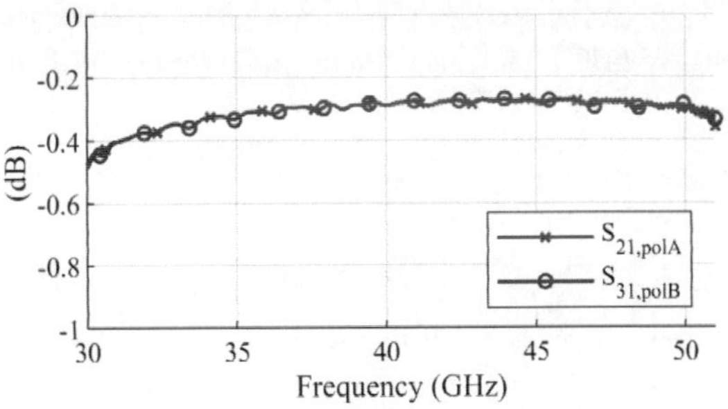

(b)

Fig. 3. Simulated reflected power and gain responses, (a) and (b) respectively, assuming a metal conductivity of 2×10^7 S/m to represent room temperature.

Fig. 3 shows the simulated gain and reflected power responses of the OMT.

REFERENCES

[1] J. Hubmayr, J. Austermann, J. Beall, J. Connors, S. Duff, J. McMahon, "Tolerance analysis of octave bandwidth millimeter-wave planar orthomode transducer," in *Proc. SPIE 12190, Millimeter, Submillimeter, and Far-Infrared Detectors and Instrumentation for Astronomy XI*, 121901K (31 August 2022)

[2] Shohei Ezaki and Wenlei Shan, "Development of Through-Substrate via Process for Silicon-Based Monolithic Microwave Integrated Circuits SIS Mixer", *IEEE Trans. Appl. Superconductivity*, vol.33, no.5, pp.1-5, 2023.

[3] J Wenninger, F. Boussaha, C. Chaumont, B.-K. Tan and G. Yassin, "Design of a 240 GHz on-chip dual-polarization receiver for SIS mixer arrays," *Supercond. Sci. Techn*ol., vol. 36, Apr. 13, 2023.

[4] A. Gonzalez and K. Kaneko, "Practical Aspects of the Design and Fabrication of High-Performance (sub)mm-Wave Dual-Ridged Waveguide Orthomode Transducers, and Application to a 205–280 GHz Design," *IEEE Trans. Terahertz Sci. Techn.*, vol. 13, no. 6, pp. 587-593, Nov. 2023, doi: 10.1109/TTHZ.2023.3308057.

[5] C. Stoumpos, J. Duran-Venegas, T. Pierré and M. García-Vigueras, "Orthomode Transducers in Additive Manufacturing for Broadband and High-Power Applications," in *17th European Conference on Antennas and Propagation (EuCAP)*, Florence, Italy, Mar. 23–26, 2023, pp. 1-5, doi: 10.23919/EuCAP57121.2023.10132846.

[6] D. Henke, N. Kelly, K. Marshall, I. Wevers and L. B. G. Knee, "A Turnstile Quad-Ridge Orthomode Transducer (OMT) for Octave-Bandwidth Receiver Front-Ends (24–51 GHz)," *IEEE Trans. Microw. Theory Tech.*, vol. 71, no. 11, pp. 4906-4921, Nov. 2023, doi: 10.1109/TMTT.2023.3267544.

[7] N. Fonseca, "Very Compact Waveguide Orthomode Transducer in the K-Band for Broadband Communication Satellite Array Antennas" *Sensors*, vol. 23, no. 2, Jan. 9, 2023. https://doi.org/10.3390/s23020735

[8] D. Henke and S. Claude, "Minimizing RF Performance Spikes in a Cryogenic Orthomode Transducer (OMT)," *IEEE Trans. Microw. Theory Tech.*, vol. 62, no. 4, pp. 840–850, Apr. 2014. http://dx.doi.org/10.1109/TMTT.2014.2309551

NOTES:

Development of the High-resolution Spectrometer of the Millimetron Space Observatory

I. V. Tretyakov, A. V. Khudchenko, K. I. Rudakov, I. V. Ivashentseva, N. S. Kaurova, B. M. Voronov, M. S. Kirsanova, T. I. Larchenkova, G. N. Goltsman, A. M. Baryshev, R. Hesper, F. V. Khan, E. S. Zhukova, A. V. Melentev, K. V. Zhivetev, A. V. Terentiev, V. P. Koshelets, and S. F. Likhachev

***Abstract*—The study of the origin and transport of water in the universe is an important part of the scientific program of the Millimetron space observatory. This will be made possible by observations conducted in single-dish mode using an onboard instrument - the High Resolution Spectrometer (HRS). This instrument incorporates heterodyne array receivers operating within the range $0.5-2.7$ THz, comprising 3-pixel arrays of superconductor-insulator-superconductor (SIS) mixers operating at frequencies below 1.3 THz and 7-pixel matrix receivers based on NbN HEB mixers observing above 1.3 THz. This paper presents the current status of development for all the mixers used in the HRS instrument of the Millimetron space observatory.**

***Index Terms*—Heterodyne receiver, THz range, matrix of detectors, NbN thin film.**

I. INTROLDUCTION

THE study of cosmic objects and processes in the Universe carried out from the surface of the Earth is limited by the atmosphere of the planet. In the submillimeter range $0.02-1$ mm absorption is due to molecules of water vapor H_2O, carbon dioxide CO_2 and oxygen O_2. The H_2O molecule is the most important molecule for studying the evolution of life in the Universe. Atmospheric water vapor makes it difficult to ground-based observations of the H_2O emission from space objects. The use of space observatories with broadband and sensitive receivers makes it possible to increase the number of directly observed water lines tens of times, which will ultimately make it possible to study the entire chain of its transformations in the interstellar medium from dense clouds to protoplanetary disks. The study of the origin and transport of water in the Universe is an important part of the scientific program of the Millimetron space observatory [1]–[3] - space telescope with a cooled 10-meters mirror designed to study various objects in the Universe in the millimeter and sub-mm bands. The goal of the "water" program is to determine the water content in various space objects from galaxies to comets, to study the mechanisms of water transfer between space objects of different types and the connection of interstellar water molecules with the emergence of life on Earth, as well as to study the ways of water formation in the gas and solid phases of interstellar matter [4].

Manuscript received [—]; revised [—]; accepted [—]. Date of publication [—]; date of current version [—]. NbN HEB mixers fabrication processes and experimental characterization was supported by a grant from the Russian Science Foundation (project No. 23-12-00187). SIS mixers development is supported by Ministry of Science and Higher Education of the Russian Federation (agreement No. 075-15-2024-538).

I. V. Tretyakov, A. V. Khudchenko, K. I. Rudakov, T. I. Larchenkova, and S. F. Likhachev are with the Astro Space Center of P.N. Lebedev Physical Institute of Russian Academy of Science, 119991, Moscow, Russia (e-mail: tretyakov@asc.rssi.ru).

M. S. Kirsanova is with the Astro Space Center of P.N. Lebedev Physical Institute of Russian Academy of Science, 119991, Moscow, Russia and the Institute of Astronomy of the Russian Academy of Sciences, 119017, Moscow, Russia.

I. V. Ivashentseva, N. S. Kaurova, B. M. Voronov, and G. N. Goltsman are with the Moscow State University of Education, 119435, Moscow, Russia.

E. S. Zhukova, K. V. Zhivetev, A. V. Terentiev are with the Moscow Institute of Physics and Technology (National Research University), 141700 Dolgoprudny, Russia

Valery P. Koshelets and F. Khan are with the Kotel'nikov Institute of Radioengineering and Electronics RAS, 125009, Moscow, Russia (e-mail:valery@hitech.cplire.ru)

A. M. Baryshev and R. Hesper are with the Kapteyn Astronomical Institute, University of Groningen, Groningen, 9747 AD, The Netherlands (e-mail: A.M.Baryshev@astro.rug.nl)

Digital Object Identifier [—]

II. HIGH-RESOLUTION SPECTROMETER GENERAL DESCRIPTION

As part of Millimetron "water" program, observations will be made in single-dish mode in the following frequency ranges: 500–600 GHz, 740–900 GHz, 1080–1230 GHz, $1300-1400$ GHz, $1890-1910$ GHz, $2390-2410$ GHz and $2660-2680$ GHz using heterodyne instrument "High-resolution spectrometer". The spectral lines located within these frequency bands, along with the scientific rationale for their observation, are described in details in [4]. The main technical parameters of the spectrometer are presented in TableI. For the first three bands it is planed to use 3 pixel receivers based on superconductor-insulator-superconductor (SIS) mixers. The intermediate frequency band is assumed to be $4-12$ GHz. At the same time, it can be reduced to $4-8$ GHz to save onboard resources; such a reduction will not have a critical impact on scientific tasks. At frequencies above 1300 GHz, it is proposed to use a 7 pixels DSB receivers with H/V polarisation based on Hot-electron bolometers (HEB) [5]. The expected single-sideband noise temperature is presented in the penultimate column. Further in this paper we will discuss in depth technical details regarding the mixers, with a view to elucidating the rationale behind the selected sensitivity.

The preliminary design of the High Resolution Spectrometer for Millimetron space observatory [6] is a logical continuation of the HiFi instrument [7]. The difference is in focus on facilitating mapping capabilities through the configuration of compact arrays and the modification of the instrument's pick-up mirror M3 from a fixed mode to a scanning regime.

III. SIS MIXER-BASED HETERODYNE RECEIVERS

Terahertz receivers based on SIS mixers are known to be the most sensitive heterodyne instruments for frequency range roughly 200–1300 GHz [8]. To operate the SIS mixer, the strong nonlinearity of the tunnel current in the current-voltage characteristic (IVC) of the SIS junction is used; under the influence of a local oscillator, quasiparticle current steps appear on the current-voltage characteristics of the SIS [9]; this process is called photon assisted tunneling. By their nature, SIS mixers can provide conversion gain. Important advantages of SIS mixers are low local oscillator power requirements and extremely low intrinsic noise. The noise temperature of the mixer in double sideband (DSB) mode is limited by the quantum value $hf/2k_B$ [10], where h and k_B are the Planck and Boltzmann constants, respectively. This is why SIS mixers are already successfully used both for ground-based radio telescopes [11]–[15], and for space missions [7], [16].

A. 500-600 GHz channel

This frequency range is well developed; SIS receivers have been implemented in space [7], [17] and at largest ground based multi-element interferometer ALMA [18]. The new SEPIA Instrument was installed and commissioned at the Atacama Pathfinder Experiment telescope (APEX) [19]. This range typical waveguide or quasioptical receivers are based on SIS junction incorporated into Nb thin film structures been manufactured on silicon or quartz substrates. The used Nb films are good superconductors at cryogenic temperatures with the energy gap above the frequency range. This Nb wiring delivers high and intermediate frequencies to and from the tunnel junction on the chips. Implementation of the SIS junctions with a high current density makes it possible to increase the operating frequency of SIS receivers and expand the bandwidth. However, there is a limit to increasing barrier transparency for SIS junctions with Al_2O_3 barrier. This limit is on the order of 10-15 kA/cm^2; with a further increase in current density, a sharp deterioration in the quality of SIS junctions occurs. In order to overcome this limitation, a technology was developed for the fabrication of Nb-Al/AlN-Nb tunnel SIS junctions with extremely high tunnel barrier transparency, made by nitridization of the Al surface in an RF plasma discharge. Implementation of the SIS junctions with AlN barrier allows not only realize DSB noise temperature well below 100 K at frequencies up to 650 GHz, but also provide a very flat Intermediate Frequency (IF) response from

TABLE I
FREQUENCY RANGES AND MAIN PARAMETERS OF HRS INSTRUMENT RECEIVERS.

Chanal	*Band* (GHz)	*Pixel*	*Technology*	T_n (K)	*Res.* (")
$M1$	500–600	3	*SIS*	200	14
$M2$	740–900	3	*SIS*	400	9
$M3$	1080–1230	3	*SIS*	1000	7
$M4$	1300 – 1400	7	*HEB*	1000	6
$M5$	1890 – 1910	7	*HEB*	1200	4
$M6$	2390 – 2410	7	*HEB*	1400	3
$M7$	2660 – 2680	7	*HEB*	1400	3

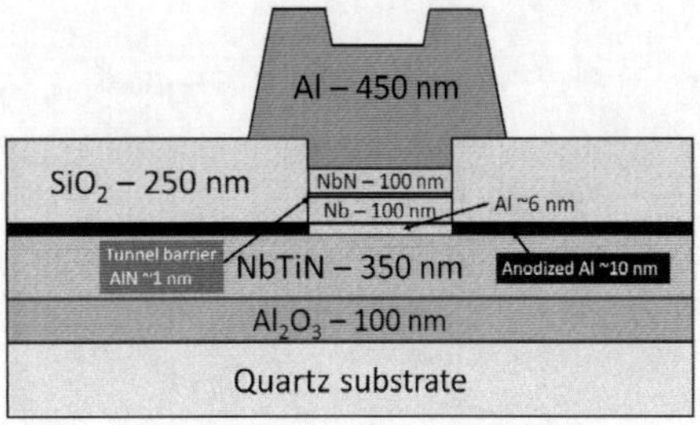

Fig. 1. Structure of the technological layers near $Nb/Al - AlN/NbN$ junction embedded into NbTiN/Al microstrip line in the SIS-receiver.

4 to 12 GHz with heterodyne waveguide receiver [18]. Based on the achieved sensitivities we put in the TableI expected single-sideband noise temperature of 200 K. Furthermore, we propose the utilisation of a sideband-separating SIS mixer [20], [21] to facilitate in-flight sideband calibration, with the objective of populating at least one pixel.

B. 740-900 GHz channel

Further improvement in the parameters of SIS mixers was achieved by development of Nb-AlN-NbN junctions, where niobium nitride is used as the top electrode instead of niobium. The utilisation of such structures enables not only an increase in the tunneling current density, but also a substantial elevation the total gap voltage of the junction V_g from 2.8 to 3.7 mV. This significantly enhances the potential for operation of such structures at high frequencies (above 700 GHz), when the size of the photonic step is hf/e exceeds V_g. In addition, at high current densities for Nb-AlN-NbN junctions, the quality parameter, determined by the ratio of the sub-gap resistance to the normal one, R_j/R_n, turns out to be noticeably higher than for purely niobium junctions. The R_j/R_n value reaches 20 at a tunnel current density of 70 kA/cm^2, which indicates the high quality of the tunnel barrier [22]. However, to ensure good matching between such high current density junctions and the antenna, submicron SIS junctions are required.

The operating frequency of SIS receivers based on niobium films is limited by the frequency of the Nb energy gap (approximately 700 GHz). A solution to this problem was found in the development of devices with microstrip lines based on Nb compounds with higher energy gap frequencies, in particular NbTiN. The top electrode of the line is usually made of a normal metal at these temperatures (typically Al), to avoid overheating of the SIS junction [23], [24]. We developed an SIS mixer based on high critical current density Nb-AlN-NbN tunnel junctions embedded in a microstrip line consisting of a 320 nm thick NbTiN bottom electrode (ground plane) and a 500 nm thick Al top electrode [25], [26]. Microstrip electrodes are separated by an insulating layer of SiO_2 250 nm thick. The SIS junction is located on the NbTiN film, and the top NbN layer is in contact with the top Al electrode (see cross-section in Fig.1).

For the frequency range 740-900 GHz we plan to build our instrument based on heritage gained during developed of the

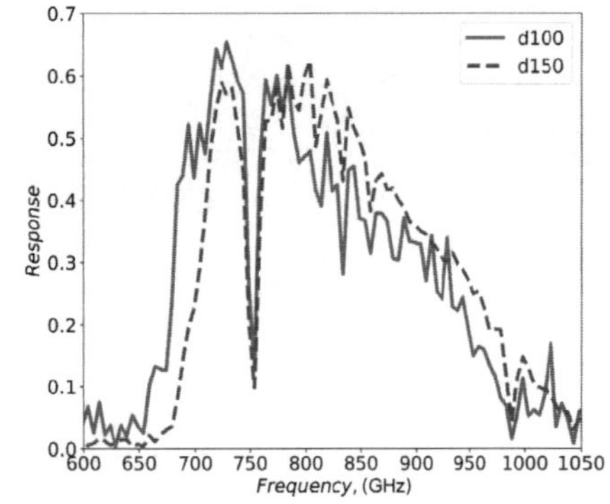

Fig. 2. The frequency response of SIS mixers for design "d100" and "d150" measured using a Fourier transform spectrometer.

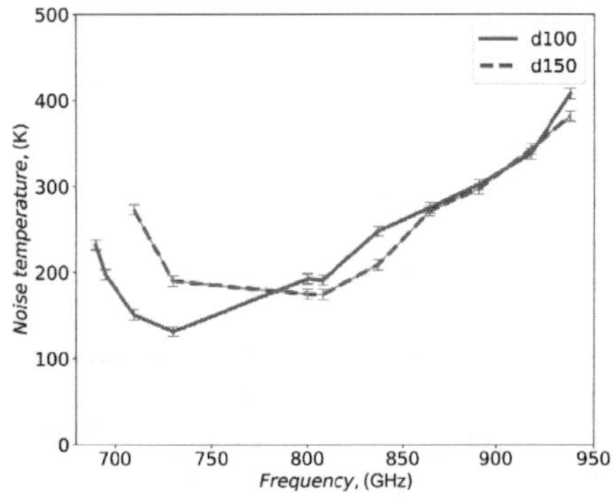

Fig. 3. The noise temperature of SIS mixers for design "d100" and "d150" measured by standard two-load (77K/300K) technique. The data is corrected for the 88% transparent 12 μm mylar beam splitter used for LO injection.

SIS mixer for CHAMP+ instrument for APEX telescope [25], [26]. The intended frequency range for this mixer was within the $787-950$ GHz atmospheric window, however, the mixer response is high also at frequencies down to 700 GHz for a few of the designs, see Fig.2 The DSB noise temperature for SIS mixers has been demonstrated as low as $3hf/k_B$ [26]. As illustrated in Fig. 3, the noise temperature of the dual-sideband mixer in the target frequency range is approximately 200 K. Similar to the $500-600$ GHz instrument we plan to use here at least one sideband-separating pixel. This will be based on the experience gained in the development of $800-950$ GHz two-sideband SIS mixer [27] .

C. 1080-1230 GHz channel

This is the highest frequency band in HPS instrument based on SIS mixer technology. In the past it was successfully proven for HiFi Band 4 and Band 5 [7], [28]. The DSB noise temperature for HiFi Band 4 was below 400 K for frequencies up to 1120 GHz. This indicates that it may be possible to reach the SSB noise temperature of approximately 1000 K with this receiver.

We plan to build our SIS mixer using the same technology as for the band 740-900 GHz, i.e. based on Nb-AlN-NbN junctions and with NbTiN ground electrode. To ensure the feasibility of superconducting mixer based on NbTiN within the specified frequency range, we have addressed separately the question on the quality of the produced NbTiN films at frequencies up to 1300 GHz.

We have already started the work on studying the superconducting NbTiN films which would allow us to manufacture the devices with operating frequency range up to 1.2 THz [29], [30] and to be published soon. The films fabricated were measured using terahertz time-domain spectrometer Menlo Systems Tera K15. In [29] we have found the optimal pressure of nitrogen in magnetron for obtaining the NbTiN film on silicon substrate with high energy gap Δ, critical temperature T_c and low normal-state resistivity near T_c. In [30] we studied the impact of the layers which are necessary for technological processes (buffer 100 nm-thick Al_2O_3 layer at the substrate for the surface protection in RIE-processes; 10 nm-thick anodized Al layer for protection of the bottom electrode during etching in CF_4 when SIS-junctions are formed). Only slight deterioration of the NbTiN film properties was observed for all the buffer layers. In [F3, to be published] we studied the properties of NbTiN films sputtered on quartz substrates.

Our results show that NbTiN film with gap frequency as high as 40 cm^{-1} or 1200 GHz can be obtained which corresponds to $hf_{gap} = 2\Delta = 4.82$ meV. Fig.4 depicts the spectra of transmission coefficient (a), permittivity (b) and real part of conductivity (c) for the NbTiN film on quartz substrate with Al_2O_3 buffer layer and anodized Al on the film surface. This is exactly the film that forms bottom electrode in superconductor microstrip line in receiver (see Fig.1).

The results obtained thus far instill confidence that the embedding circuit for the future $1080-1230$ GHz SIS mixer will exhibit acceptable losses, thereby enabling the desired receiver noise temperature to be achieved.

IV. HEB MIXER-BASED HETERODYNE RECEIVERS

The high sensitivity of HEB mixers ensured their application in astronomical projects of the European Space Agency: SOFIA airborne observatory with a 2.5-meter mirror capable of observing at frequencies from 1 to 5 THz. [31] and TELIS [32] for operation in the range 1.76-1.86 THz. In the HERSCHEL space observatory, the $1.4-1.9$ THz channel also used HEB mixers [7]. Among the SOFIA instruments, the 1.4, 1.9 and 2.5 THz channels of the GREAT instrument are based on single waveguide NbN HEB mixers [33]. The upGREAT tool, the successor to GREAT, is already a multi-pixel heterodyne matrix [34]. The upGREAT Low Frequency Array (LFA) consists of 2×7 pixel waveguide HEBs, two sets needed to separate polarizations at 1.9–2.5 THz. The upGREAT High Frequency Array (HFA) consists of 7 waveguide HEBs tuned to operate at 4.745 THz. The hexagonal configuration is chosen for maximum display efficiency. An LFA with a noise temperature of 600 K at the center of the frequency range was successfully commissioned to observe the [C II] line at 1.905 THz. Direct detection of atomic oxygen on

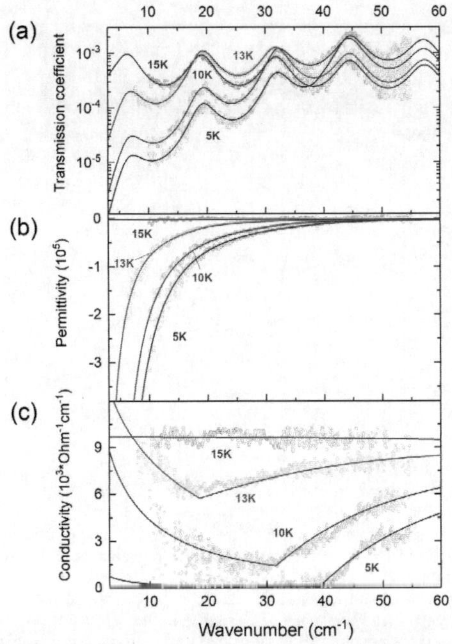

Fig. 4. Measured spectra of the transmission coefficient (a), permittivity (b) and conductivity (c) of the superconducting $NbTiN$ film on quartz substrate with 100 nm Al_2O_3 buffer layer and anodized aluminum on surface.

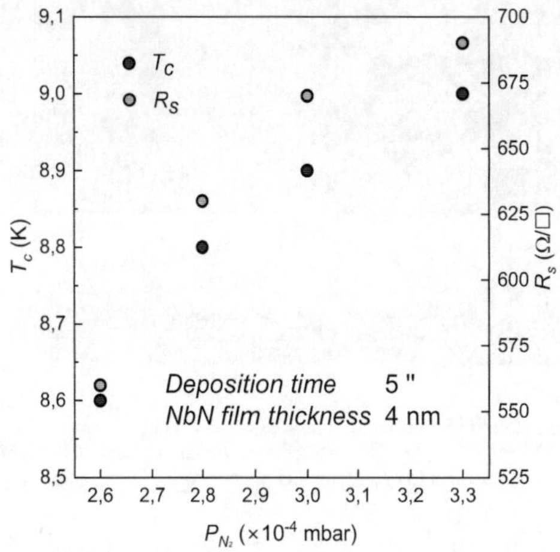

Fig. 5. Dependence of the T_c and R_s of the NbN films on the N_2 partial pressure during deposition process, each deposition took 5 sec.

the dayside and nightside of Venus [35] was demonstrated by HFA. The DOME A observatory, being developed by the Purple Mountain Observatory of the Chinese Academy of Sciences, may become the most promising ground-based observation platform for THz astronomy [36]. It is planned to build a 5-meter THz telescope, DATE5, equipped with a dual-band heterodyne receiver for atmospheric window frequencies of 0.85 THz and 1.4 THz based on an HEB mixers.

Disordered ultra thin ($3.5 - 4$ nm) superconducting NbN film is used as a sensitive element in these terahertz range HEB mixers. Due to the physical principle of the HEB mixer operation, this "ingredient" is the most crucial, the combination of extremely low film thickness and its superconducting properties determines the characteristics of the final receiver.

Modern NbN HEBs as heterodyne detectors have almost reached their sensitivity limit [33], [37], however, they have a conversion band (IF band) not exceeding $3 - 4$ GHz [38]. Further reduction of the noise temperature, an increase in the IF band as well as reduction of the required local oscillator (LO) power has a significant practical interest. Currently, each pixel of the heterodyne matrix – HEB mixer – is considered as a separate detector, requiring individual adjustment of the bias voltage and the LO power. With a matrix size of at least a ten pixels, such an individual approach is difficulty applicable. The solution of the problem could be the fabrication of an unified HEB mixers that do not require individual settings. To meet the issue we improve the technology for deposition of NbN films and the NbN HEB mixers fabrication processes to fabricate mixers with as close as possible R(T) characteristics and a minimum spread of normal resistances R_{300} for devices with same geometry within the one batch.

A. NbN HEB mixers fabrication and DC test

HEB mixers were fabricated from a bilayer $NbN - Au$ structure with a thickness of NbN film of $3.5 - 4$ nm and Au $20 - 25$ nm deposited *in-situ* using reactive magnetron sputtering onto the surface of a high-resistivity Si substrate. The substrate temperature could vary up to 900 $°C$. Before film deposition, the substrate surface was cleaned in oxygen plasma, followed by treatment in a hydrofluoric acid solution. This, along with speed optimization by reducing the discharge current to 300 μA, made it possible to achieve high uniformity of the superconducting properties (film thickness, T_c and surface resistance R_s) of the NbN film over the Si substrate surface.

The dependence of the critical temperature and surface resistance of NbN films at a deposition time of 5 seconds is presented in Fig.5. The optimal values of the critical temperature T_c and critical current density j_c were obtained with the following parameters of the NbN film deposition process: substrate temperature 800 $°C$, partial pressures of Ar and N_2 were 5×10^{-3} $mbar$ and $3 - 3.3 \times 10^{-4}$ $mbar$, current and discharge voltage were 300 μA and 300 V. NbN deposition was followed by *in-situ* deposition of an Au layer after cooling the Si substrate down to 400 $°C$. Long time cooling of the Si substrate made it possible to anneal the fabricated NbN film to improve its superconducting properties and its distribution over the substrate. The NbN film deposition rate, for given process characteristics, was determined by test the dependence of film thickness on sputtering time. Same to the NbN films, the Au layer thickness was controlled over the deposition process time based on the obtained deposition rates. The thickness of the deposited films was measured using an atomic force microscope AFM.

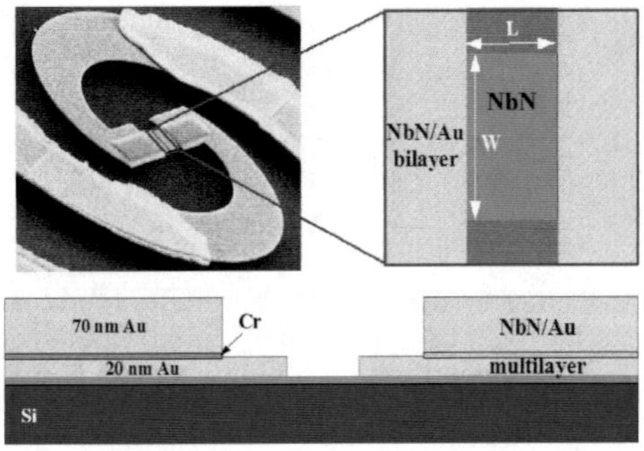

Fig. 6. False color SEM photo and crosssection sketch of central part of a typical NbN HEB device, showing the active material of NbN (blue). The yellow parts made of gold (Au) serve as ports of THz spiral antenna.

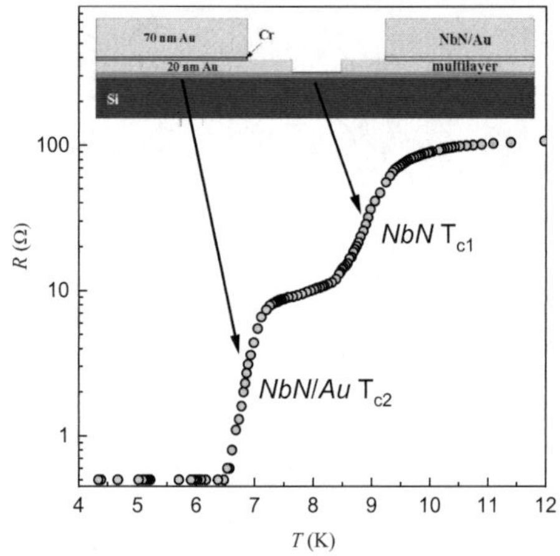

Fig. 7. Temperature dependence of the NbN HEB mixer resistance R. The appearance of the second superconducting transition at T_{c2} is caused by the proximity effect in the NbN film under $in-situ\ Au$. The value of the mixer resistance after the first transition is determined by the width NbN of the bridge and the coherence length ξ in the NbN film. Resistance appears in a superconducting NbN film due to the process of conversion of normal electrons into Cooper pairs. The conversion region is determined by the coherence length of the resulting Cooper pairs.

In the process of an HEB mixer fabrication, electron and photo lithography methods, chemical and ion etching processes, thermal metal deposition and electron beam evaporation deposition, as well as the lift off process in acetone are used. For photo and electron lithography processes, Ti/Au alignment marks are made on top of Au surface of the NbN/Au two-layer structure. This is followed by the fabrication of the spiral THz antenna and setting the width W of the future NbN mixer bridge using electron lithography and liquid and plasma-chemical etching processes. The photolithography method, thermal deposition of Ti and Au layers and the lift off process are used to form the outer part of the antenna and the mixer bond pads. At the final step of the fabrication, the length L of the NbN bridge of the mixer is set; for this purpose, electron lithography, ion and liquid etching methods are used. Central part of the mixer is covered with a layer of SiO_2 to prevent the external undesirable environment influence.

Typical sizes of superconducting bridges of HEB mixers ranged from 0.1–0.4 μm in length L, and 1–4 μm in width W. SEM photo and crosssection of the central part of the THz antenna, shown in Fig.6. With a normal resistance per square NbN film $R = 650 - 700\ \Omega/\Box$ (with a NbN thickness of $4\ nm$), the ratio of the bolometer length to its width L/W is kept constant to achieve a 65 Ω resistance in the normal state for optimal RF and IF coupling.

Optimized sequence of the HEB mixers fabrication processes, the sensitive element - the NbN bridge of the mixer - during the fabrication had been coated with $in-situ\ Au$ layer, made it possible to minimize the spread of normal resistances R_{300} of the mixer; along with the optimized process of NbN film deposition, it also allowed to minimize the spread in T_c down to 0.1 K for mixers fabricated within the same process. The dependence of the NbN HEB mixers resistance R as function of the temperature is presented in Fig.7. A conventional HEB device consists of 2 main parts. The central, bare NbN film with critical temperature T_{c1}, connected to a superconducting $NbN - Au$ layer, with critical temperature T_{c2}, which serves as THz antenna. The appearance of the second superconducting transition at T_{c2} is caused by the suppression of superconductivity in the NbN film under thin $in-situ\ Au$ due to the proximity effect. The remaining NbN film under the shunt Au antenna is in a normal state at temperatures above 4.2 K. The difference in T_c between the NbN bridge and NbN/Au bilayer for our NbN films exceeds 2K, this indirectly indicates the quality of the electrical contact, as well as the thickness of the NbN film. The proximity effect appears at sizes on the order of the coherence length ξ in a superconductor, which is on the order of 3 nm for thin NbN films.

B. NbN HEB mixers noise temperature

Double side band noise temperature ${T_n}^{DSB}$ measurements were carried out at a local oscillator frequency of 2.52 THz using the "cold" / "hot" (77K/300K) blackbody technique. A quantum cascade laser QCL was used as a local oscillator source. QCL had an output power of about 5 mW, frequency tuning of about 10 GHz, power dissipation of 1 W at a physical temperature of 17 K. To couple the mixer with the radiation, we used an elliptical Si lens. Using a 3.5 μm - thick Mylar beamsplitter, the local oscillator was combined with the blackbody. The transmittance of the beamsplitter was no less than 0.97 at a frequency of 2.52 THz. The optical window of the cryostat was made of high-density polyethylene HDPE. To prevent the direct detection effect, we used a cooled to 4 K narrow-band interference filter with a transmittance of 0.9. A cold amplifier with a gain of 35 dB had a T_n of no higher than 10 K in the $1 - 8\ GHz$ band. The estimated contribution of the optical path to the measured ${T_n}^{DSB}$ value did not exceed 100 K. Fig.8 shows the experimental results of measuring ${T_n}^{DSB}$ for mixers from three different batches. Red circles indicate the results

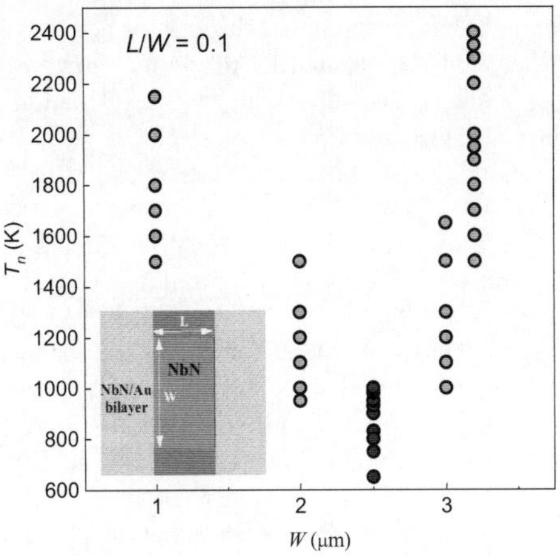

Fig. 8. NbN HEB mixer double side band noise temperature T_n as function of the NbN bridge width W measured at $2.52\ THz$ LO frequency.

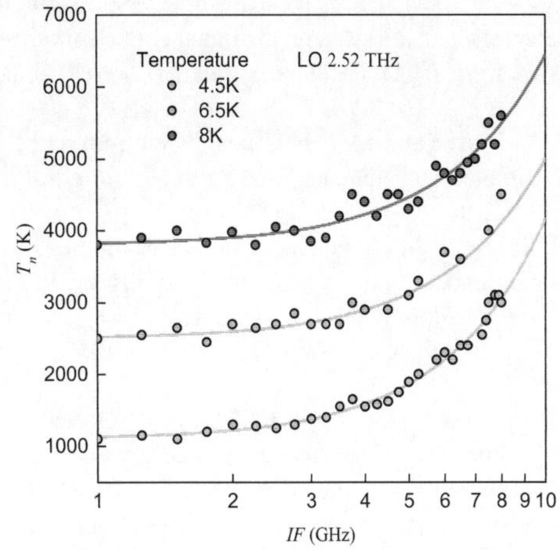

Fig. 9. NbN HEB mixer double side band $T_n@2.52\ THz$ as function of the intermediate frequency IF measured on different bath temperature T.

for mixers fabricated using optimized NbN film deposition and mixer fabrication routes. Blue circles indicate mixers fabricated using "old" non-optimized $in-situ$ and $ex-situ$ technology. The spread in $T_n{}^{DSB}$ for mixers fabricated using optimized technology does not exceed $200\ K$ with an average level of $800\ K$. This value well meets the design requirements and has a margin for losses in the future optical path of the instrument. Fig.9 shows the experimental results of $T_n{}^{DSB}$ in the intermediate frequency IF band. The measurements were carried out for three temperatures $4\ K$, $6.5\ K$ and $8.5\ K$. Since the measurements were limited to the band of the cooled amplifier, the experimental points were approximated by a curve of the form $T_n{}^{DSB}(f) = [1 + (f/\Delta f)]^2$ to determine the frequency Δf at which $T_n{}^{DSB}$ will double. During the work, it was determined that as the temperature of the NbN HEB mixer increases, $T_n{}^{DSB}$ naturally increases, but at the same time Δf also increases. This effect is caused by the design of the mixer, namely the presence in the mixer NbN bridge regions near the normal metal contacts of the antenna with suppressed superconductivity due to the proximity effect. Electron conversion regions at temperatures below T_{c2} block the diffusion of "hot" electrons into cold metal contacts.

V. Conclusion

The development of mixer, which can be used to built receivers for the High Resolution Spectrometer for Millimwtron space observatory gives us confidence to formulate the expected sensitivity levels for all the bands within the the range 0.5 -2.6 THz. For channel M1 the Nb base SIS mixers with AlN tunnel barrier will be utilised. For channels M2 and M3 SIS mixers with NbTiN wiring will be built. To implement channels $M4 - M7$ the NbN HEB mixers will be used; the achieved unification of the parameters of the fabricated mixers will make it possible to create a heterodyne matrix of 7 pixels with parameters that meet the requirements of the instrument. Since the mixers use a helical antenna that works well in the $1 - 5\ THz$ band, it will be possible to use a pair (H/V polarization) of such matrices for sequential observation in channels $M4 - M7$.

References

[1] N. S. Kardashev, I. D. Novikov, V. N. Lukash, *et al.*, "Review of scientific topics for the Millimetron space observatory," *Physics Uspekhi*, vol. 57, no. 12, 1199-1228, pp. 1199–1228, Dec. 2014. DOI: 10.3367/UFNe . 0184 . 201412c . 1319. arXiv: 1502 . 06071 [astro-ph.IM].

[2] I. D. Novikov, S. F. Likhachev, Y. A. Shchekinov, *et al.*, "Objectives of the Millimetron Space Observatory science program and technical capabilities of its realization," *Physics Uspekhi*, vol. 64, no. 4, pp. 386–419, Jul. 2021. DOI: 10.3367/UFNe.2020.12.038898.

[3] S. F. Likhahev and T. I. Larchenkova, "From the Spektr-R project to the Spektr-M project: milestones in space radio astronomy," *Physics Uspekhi*, accepted for publication. DOI: 10.3367/UFNe.2024.03.039662.

[4] M. S. Kirsanova, P. V. Baklanov, E. O. Vasiliev, *et al.*, "Origin and transfer of water in the Universe," *Physics Uspekhi*, submitted for publication.

[5] G. N. Gol'tsman, A. D. Semenov, Y. P. Gousev, *et al.*, "Sensitive picosecond nbn detector for radiation from millimetre wavelengths to visible light," *Superconductor Science and Technology*, vol. 4, no. 9, p. 453, Sep. 1991. DOI: 10.1088/0953-2048/4/9/020. [Online]. Available: https://dx.doi.org/10.1088/0953-2048/4/9/020.

[6] "Millimetron observatory website." (2024), [Online]. Available: https://millimetron.ru/en/.

[7] T. De Graauw, F. Helmich, T. Phillips, *et al.*, "The herschel-heterodyne instrument for the far-infrared (hifi)," *Astronomy & Astrophysics*, vol. 518, p. L6, 2010.

[8] J. Zmuidzinas and P. L. Richards, "Superconducting detectors and mixers for millimeter and submillimeter astrophysics," *Proceedings of the IEEE*, vol. 92, no. 10, pp. 1597–1616, 2004.
[9] J. R. Tucker and M. J. Feldman, "Quantum detection at millimeter wavelengths," *Reviews of Modern Physics*, vol. 57, no. 4, p. 1055, 1985.
[10] A. R. Kerr, M. J. Feldman, and S.-K. Pan, "Receiver noise temperature, the quantum noise limit, and the role of the zero-point fluctuations," in *Proc. of the 8th Int. Symp. on Space Terahertz Technology*, 1997, pp. 101–111.
[11] "Alma observatory website." (2024), [Online]. Available: https://www.almaobservatory.org/en/about-alma/.
[12] "The noema radio telescope website." (2024), [Online]. Available: https://www.cnrs.fr/en/press/european-radio-telescope-noema-reaches-full-power.
[13] "The submillimeter array (sma) website." (2024), [Online]. Available: https://lweb.cfa.harvard.edu/sma/index.html.
[14] "Apex - atacama pathfinder experiment telescope website." (2024), [Online]. Available: https://www.eso.org/public/teles-instr/apex/.
[15] "Iram - the 30-meter telescope website." (2024), [Online]. Available: https://iram-institute.org/observatories/30-meter-telescope/.
[16] "Herschel space observatory website." (2024), [Online]. Available: https://www.esa.int/Science_Exploration/Space_Science/Herschel_overview.
[17] Y. Delorme, M. Salez, B. Lecomte, *et al.*, "Space-qualified sis mixers for herschel space observatory's hifi band 1 instrument," *Proc. 16th ISSTT*, pp. 444–448, 2005.
[18] A. Baryshev, R. Hesper, F. Mena, *et al.*, "The alma band 9 receiver-design, construction, characterization, and first light," *Astronomy & Astrophysics*, vol. 577, A129, 2015.
[19] V. Belitsky, I. Lapkin, M. Fredrixon, *et al.*, "Sepia–a new single pixel receiver at the apex telescope," *Astronomy & astrophysics*, vol. 612, A23, 2018.
[20] R. Hesper, A. Khudchenko, A. M. Baryshev, J. Barkhof, and F. P. Mena, "A new high-performance sideband-separating mixer for 650 GHz," in *Society of Photo-Optical Instrumentation Engineers (SPIE) Conference Series*, vol. 9914, 2016, 99140G-1–99140G-11. DOI: 10.1117/12.2233065. [Online]. Available: http://dx.doi.org/10.1117/12.2233065.
[21] R. Hesper, A. Khudchenko, A. M. Baryshev, J. Barkhof., and F. P. Mena, "A high-performance 650-GHz sideband-separating mixer – design and results," *IEEE Transactions on Terahertz Science and Technology*, vol. 7, no. 6, pp. 686–693, Nov. 2017, ISSN: 2156-342X. DOI: 10.1109/TTHZ.2017.2758270.
[22] M. Y. Torgashin, V. P. Koshelets, P. N. Dmitriev, A. B. Ermakov, L. V. Filippenko, and P. A. Yagoubov, "Superconducting integrated receiver based on nb-aln-nbn-nb circuits," *IEEE transactions on applied superconductivity*, vol. 17, no. 2, pp. 379–382, 2007.
[23] M. Westig, S. Selig, K. Jacobs, T. Klapwijk, and C. Honingh, "Improved nb sis devices for heterodyne mixers between 700 ghz and 1.3 thz with nbtin transmission lines using a normal metal energy relaxation layer," *Journal of Applied Physics*, vol. 114, no. 12, 2013.
[24] A. Traini, B.-K. Tan, J. D. Garrett, *et al.*, "The influence of lo power heating of the tunnel junction on the performance of thz sis mixers," *IEEE Transactions on Terahertz Science and Technology*, vol. 10, no. 6, pp. 721–730, 2020.
[25] A. Khudchenko, A. M. Baryshev, K. I. Rudakov, *et al.*, "High-gap nb-aln-nbn sis junctions for frequency band 790–950 ghz," *IEEE Transactions on Terahertz Science and Technology*, vol. 6, no. 1, pp. 127–132, 2015.
[26] K. I. Rudakov, A. V. Khudchenko, L. V. Filippenko, *et al.*, "Thz range low-noise sis receivers for space and ground-based radio astronomy," *Applied Sciences*, vol. 11, no. 21, p. 10087, 2021.
[27] A. Khudchenko, R. Hesper, A. M. Baryshev, *et al.*, "Design and performance of a sideband separating sis mixer for 800–950 ghz," *IEEE Transactions on Terahertz Science and Technology*, vol. 9, no. 6, pp. 532–539, 2019.
[28] A. Karpov, D. Miller, F. Rice, *et al.*, "Low noise 1 thz–1.4 thz mixers using nb/al-aln/nbtin sis junctions," *IEEE Transactions on Applied Superconductivity*, vol. 17, no. 2, pp. 343–346, 2007.
[29] F. V. Khan, E. S. Zhukova, B. P. Gorshunov, *et al.*, "Characterization of microwave properties of superconducting nbtin films using tds," *IEEE Transactions on Terahertz Science and Technology*, vol. 13, no. 6, pp. 627–632, 2023. DOI: 10.1109/TTHZ.2023.3321252.
[30] E. S. Zhukova, B. P. Gorshunov, L. S. Kadyrov, *et al.*, "Impact of the buffer layers and anodization on properties of nbtin films for thz receivers," *IEEE Transactions on Applied Superconductivity*, vol. 34, no. 3, pp. 1–5, 2024. DOI: 10.1109/TASC.2024.3353139.
[31] "Stratospheric observatory for infrared astronomy (sofia)." (2024), [Online]. Available: http://www.sofia.usra.edu.
[32] R. W. M. Hoogeveen, P. A. Yagoubov, A. Maurellis, *et al.*, "New cryogenic heterodyne techniques applied in TELIS: the balloonborne THz and submillimeter limb sounder for atmospheric research," in *Infrared Spaceborne Remote Sensing XI*, M. Strojnik, Ed., International Society for Optics and Photonics, vol. 5152, SPIE, 2003, pp. 347–355. DOI: 10.1117/12.521354. [Online]. Available: https://doi.org/10.1117/12.521354.
[33] P. Pütz, C. Honingh, K. Jacobs, M. Justen, M. Schultz, and J. Stutzki, "Terahertz hot electron bolometer waveguide mixers for great," *Astronomy & Astrophysics*, vol. 542, p. L2, 2012.
[34] C. Risacher, R. Güsten, J. Stutzki, *et al.*, "First supra-thz heterodyne array receivers for astronomy with the sofia observatory," *IEEE Transactions on Terahertz Science and Technology*, vol. 6, no. 2, pp. 199–211, 2015.
[35] H.-W. Hübers, H. Richter, U. U. Graf, *et al.*, "Direct detection of atomic oxygen on the dayside and nightside

of venus," *Nature Communications*, vol. 14, no. 1, p. 6812, 2023.

[36] S.-C. Shi, S. Paine, Q.-J. Yao, *et al.*, "Terahertz and far-infrared windows opened at dome a in antarctica," *Nature Astronomy*, vol. 1, no. 1, ? 2017. DOI: 10.1038/s41550-016-0001.

[37] I. Tretyakov, S. Ryabchun, M. Finkel, *et al.*, "Low noise and wide bandwidth of nbn hot-electron bolometer mixers," *Applied Physics Letters*, vol. 98, no. 3, 2011.

[38] J. J. Baselmans, M. Hajenius, J. Gao, *et al.*, "Doubling of sensitivity and bandwidth in phonon cooled hot electron bolometer mixers," *Applied physics letters*, vol. 84, no. 11, pp. 1958–1960, 2004.

Ivan V. Tretyakov received the M.S. degree in Solid state physics from Pomor State University, Arkhangelsk, Russia, in 2006. The Ph.D. degree in Radio physics from the Moscow State Pedagogical University, Moscow, Russia, was received in 2013.

Currently he is a Senior researcher, since 2006 he has been with the Moscow State Pedagogical University, Moscow. From 2020 he joined the Astro Space Center of Lebedev Physical Institute, Russian Academy of Sciences. The main activity is related to the development of THz and infrared devices.

Andrey V. Khudchenko received the M.S. degree in applied physics and mathematics and the Ph.D. degree in radiophysics from the Moscow Institute of Physics and Technology, Moscow, Russia, in 2007 and 2009, respectively.

From 2004 to 2008, he was an Engineer and, in 2009, a Researcher with the Kotel'nikov Institute of Radio Engineering and Electronics, Russian Academy of Sciences, Moscow. From 2010 to 2015, he has been an Instrument Scientist with The Netherlands Institute for Space Research SRON. Since 2015 to 2020, Instrument Scientist at the Kapteyn Astronomical Institute, University of Groningen. Since 2020, he joined Astro Space Center of Lebedev Physical Institute. The main activity is related to the development of new heterodyne THz instruments.

Kirill I. Rudakov received B.S. and M.S. degrees in applied physics and mathematics from the Moscow Institute of Physics and Technology in 2013 and 2015, respectively. After internships as a Research Student in the ALPHA experiment at CERN in 2013 and as a Guest Scientist in the CHAMP+ experiment at a SRON in 2014, he started a Ph.D. and then a PostDoc research with the Kapteyn Astronomical Institute, University of Groningen, The Netherlands, in 2016 and 2020. His Ph.D. research is devoted to the modeling and development of elements of the array SIS receivers in terahertz range. His research interests include heterodyne waveguide SIS-based receivers and high-frequency simulations of superconducting circuits, and measurements of them.

Irina V. Ivashentseva received the B.S. degree in fundamental physics (in english) from the Moscow State Pedagogical University, Moscow, Russia, in 2023. At present moment she is a student of the Master's degree program Applied electronics and photonics, National Research University Higher School of Economics, Moscow, Russia.

Since 2020, she is research assistant in Laboratory of quantum detectors, the Moscow State Pedagogical University, Moscow, Russia. The main subject of work is thin film sputtering and fabrication and characterisation of HEB.

Ronald Hesper received the M.Sc. degree in experimental solid state physics from the University of Leiden, Leiden, The Netherlands, and the Ph.D. degree in experimental solid state physics from the University of Groningen, Groningen, The Netherlands.

Since 2000, he has been an Instrument Scientist with the Kapteyn Astronomical Institute, University of Groningen. From 2000 to 2008, he was involved in the technological development of the ALMA Band 9 receivers, including the process of industrialization, as well as related projects like the CHAMP+ mixer arrays for APEX; from 2008 to 2013, on the development of a sideband-separating mixer upgrade for the ALMA Band 9 receivers; and from 2013 to the beginning of 2015, on the industrialization of the ALMA Band 5 receivers. Currently, he is involved in the development of new heterodyne detector technologies and industrialization of the ALMA Band 2 receivers.

Andrey M. Baryshev received the Master's degree (summa cum laude) in physical quantum electronics from the Moscow Institute of Physics and Technology, Moscow, Russia, in 1993, and the Ph.D. degree from the Technical University of Delft, Delft, The Netherlands, in 2005.

He is currently a Senior Instrument Scientist. Since 1998, he has been with the SRON Low Energy Astrophysics Division and the Kapteyn Astronomical Institute, University of Groningen, Groningen, The Netherlands. Since 2000, he has been involved in a joint effort to develop the SIS receiver (600–720 GHz) for ALMA. In 2013, he became an Associate Professor of astronomical instrumentation for the far-infrared with the Kapteyn Astronomical Institute, University of Groningen. His main research interests include the areas of heterodyne and direct detectors for large focal plane arrays at THz frequencies, and quasi-optical system design and experimental verification.

Dr. Baryshev was the recipient of a Netherlands Organisation for Scientific Research-VENI grant for research on heterodyne focal plane array technology in 2008, and an EU commission Starting Researcher Grant for work on direct detector focal plane arrays in 2009.

Fedor V. Khan received the B.S. and M.S. degree in applied mathematics and physics from the Moscow Institute of Physics and Technology, Dolgoprudny, Russia, in 2021 and 2023, respectively, where he is currently working towards his Ph.D. degree in condensed matter physics. Since 2019, he has been an Engineer with the Kotelnikov Institute of Radio Engineering and Electronics of Russian Academy of Sciences, Moscow, Russia, working on characterization of the superconducting materials for THz-range electronics and modeling of superconducting devices.

Elena S. Zhukova received the M.S. degree in applied physics and mathematics and the Ph.D. degree in condensed matter physics from the Moscow Institute of Physics and Technology, Moscow, Russia, in 2007 and 2010, respectively. From 2010 to 2014, she was a junior researcher in the A.M. Prokhorov General Physics Institute, Russian Academy of Sciences, Moscow. Since 2014 she has been a senior researcher and a deputy head of the Laboratory of Terahertz Spectroscopy in the Moscow Institute of Physics and Technology (National research institute), Dolgoprudnyi, Russia. Her current scientific interests are in the field of terahertz and infrared spectroscopy of systems with correlated electronic states.

Valery P. Koshelets received the M.S. degree in physics from Lomonosov Moscow State University, Moscow, Russia, in 1973, and the Ph.D. degree in radio physics and Doctor of Sciences (Habilitation) degree in physical electronics from the Kotel'nikov Institute of Radio Engineering and Electronics of the Russian Academy of Sciences, Moscow, in 1978 and 1990, respectively. Since 1973, he has been with the Kotel'nikov Institute of Radio Engineering and Electronics of the Russian Academy of Sciences, Moscow, Russia, where he is currently the head of the Laboratory of superconducting devices for signal detection and processing.

Tatiana I. Larchenkova received the M.S. degree in physics from Moscow Engineering Physics Institute, Moscow, Russia, and the Ph.D. degree in astronomy and Doctor of Sciences (Habilitation) degree in astronomy from the Lebedev Physical Institute of the Russian Academy of Sciences, Moscow, in 2000 and 2021, respectively. Since 1999, she has been with the Lebedev Physical Institute of the Russian Academy of Sciences, Moscow, Russia, where she is currently the head of the Department of Theoretical astrophysics and cosmology.

Maria S. Kirsanova received the M.S. degree in physics from Ural State University, Ekaterinburg, Russia, and the Ph.D. degree in astronomy from the Institute of Astronomy of the Russian Academy of Sciences, Moscow, Russia in 2005 and 2009, respectively. Since 2005 she has been with the Institute of Astronomy of the Russian Academy of Sciences, where she is currently has a senior researcher position. Since 2023, she joined Astro Space Center of Lebedev Physical Institute. She is experienced in observations with mm and submm telescopes and in numerical modelling of interstellar medium.

Amplitude and phase beam pattern measurements of a waveguide-type HEBM at 1.9 THz

Y. Irimajiri[1*], I. Morohashi[1], A. Kawakami[1], M.-J. Wang[2], W.-C. Lu[2], R. Sakai[3], A. Gonzalez[3], H. Maezawa[4], N. Sekine[1]

***Abstract*— We measured beam patterns of a fabricated corrugated horn of a waveguide-type hot electron bolometer mixer (HEBM) in amplitude and phase at 1.9 THz. The HEBM block was set on a 4-K cold stage of a cryostat with an external ellipsoidal mirror. Amplifier/multiplier chain (AMC) source were used as a local oscillator and as a RF source. The beam pattern was measured from a detected beat-note by scanning the RF source horizontally and vertically. The measured data was calibrated using the data at the beam center. A far-field beam pattern was obtained by Fourier transform calculation.**

Keywords—beam pattern measurements, corrugated horn, hot electron bolometer mixer, terahertz.

I. Introduction

Measurement of an antenna beam pattern of a hot electron bolometer mixer (HEBM) is important to evaluate a receiver performance especially for astronomical observations or atmospheric measurements. We have measured beam patterns of a corrugated horn of a waveguide-type HEBM in amplitude and phase at 1.9 THz. The concept of a measurement setup was presented in the 29th ISSTT symposium [1]. The measurement setup was referred the previous work [2, 3]. Although a phase-locked THz-QCL is described in the figure of the proceedings, we are currently using an amplifier/multiplier chain (AMC) source as a local oscillator. In the future, we will try to use a phase-locked THz-QCL for the measurement at the higher frequency of 2~4.75 THz.

II. Measurement and Results

A corrugated horn at 2-THz band was developed for a waveguide-type HEBM [4]. The corrugated horn at 2-THz band was manufactured using a 15-μm blade by KMCO (Kawashima Manufacturing Co. Ltd., Japan). The width of the teeth and the notch are measured to be 18 μm and 15 μm, respectively and the depth is 37.5 μm (75 μm at the deepest point). The number of the teeth is 140. The HEBM and an external ellipsoidal mirror were installed on a 4-K cold stage of a cryostat of a pulse-tube cooler. The HEBM was pumped using an amplifier/multiplier chain (AMC) source using a mylar beam splitter. Another AMC source with a probe (WR-0.65) set on the XY stage was used as an RF source. The beam patterns in amplitude and phase were measured from a beat-note using a lock-in amplifier by scanning RF source horizontally and vertically in a range of 15.2 mm with a spatial resolution of 79 μm which corresponds to a Nyquist sampling rate of λ/2 at 1.9 THz. Figure 1 shows the block diagram of the measurement system. It is noted that the phase beam pattern was measured by subtracting the unwanted component from the oscillators of the AMC sources. The same unwanted component was generated by a microwave circuit shown in the boxed area on the left of the figure. Figure 2 shows the photograph of the measurement system. A radio wave absorber is placed around the horn to avoid the radio wave reflection.

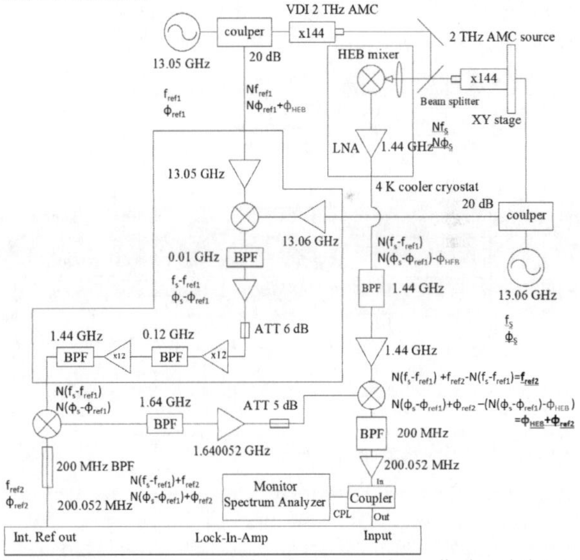

Fig. 1. The measurement setup of the beam pattern in amplitude and phase at 1.9 THz. The beat note from a 2 THz AMC source with a WR-0.65 RF probe was detected by the HEBM. The amplitude and phase of the beat note was measured by scanning the XY stage using the lock-in amplifier. It is noted that the phase beam pattern was measured by subtracting the unwanted

[1]National Institute of Information and Communications Technology, Koganei, Tokyo, 184-8795, Japan; [2]Academia Sinica, Institute of Astronomy and Astrophysics, Taipei, 10617, Taiwan; [3]National Astronomical Observatory, Mitaka, Tokyo, 184-0015, Japan; [4]Osaka Metropolitan University, Osaka, Osaka, 558-8585, Japan. *Yoshihisa Irimajiri (email: irimaji@nict.go.jp).

NOTES:

components from the oscillators, which was generated by the microwave circuit shown in the boxed area in the left of the measurement system diagram.

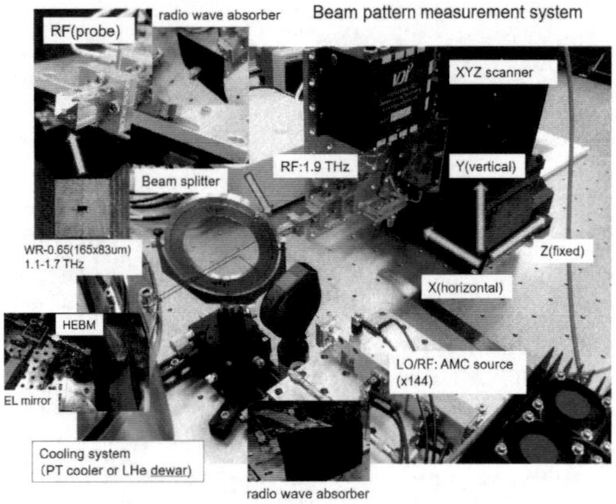

Fig. 2. A photograph of the measurement setup. The HEBM and an elliptical mirror are put on the 4 K stage of the cryostat. A LO signal was fed to the HEBM using a mylar beam splitter. The RF source which has a probe with a WR-0.65 waveguide output is set on the XY stage. A radio wave absorber is placed around the horn to avoid the radio wave reflection.

Figure 3 shows the block diagram of the data acquisition system. The XY stage is controlled by a controller for continuous motion, while data were measured by the lock-in amplifier and acquired using an AD converter by a trigger from the stage. It takes 2 hours for the full resolution mapping. This system is ~7 times faster than the method of controlling the stage and data acquisition using one by one GPIB command.

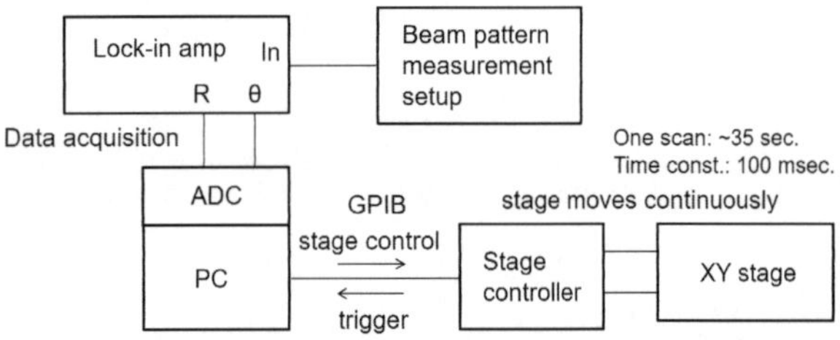

Fig. 3. The data acquisition system. The XY stage is controlled by a controller for continuous motion, while data were measured by the lock-in amplifier and acquired using an AD converter by a trigger from the stage. It takes 2 hours for the full resolution mapping.

Figure 5 (a, b) shows the measured near-field beam pattern of the corrugated horn in amplitude and phase. The phase beam pattern is slightly distorted by the instability of the receiver. In order to calibrated the data, the data at the beam center position was acquired after each horizontal scan as shown in Figure 4 (a). Figure 4 (b, c) show the acquired data at the beam center for the calibration in amplitude and the phase. It slightly changes with time due to the instability of the receiver system.

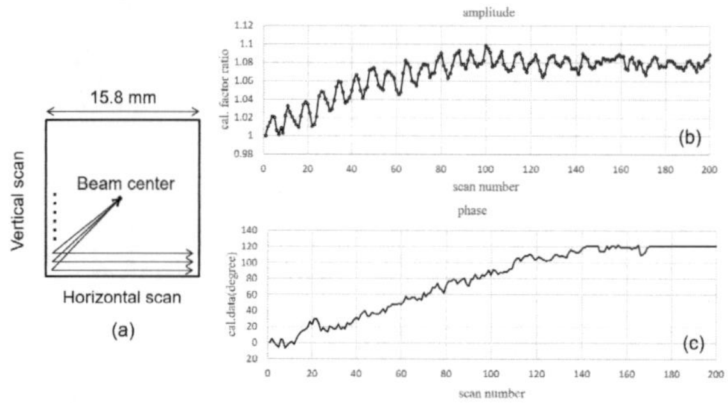

Fig. 4. (a) The calibration data was acquird by measuring the center position of the beam pattern after each horizontal scan. (b, c) The measured data at the beam center for amplitude and phase. It slightly changes with time due to the instability of the receiver system.

Figure 5 (c, d) show the beam pattern after the calibration. Especially for the phase pattern, the data has been improved to be smoothed. The data are plotted in a liner scale. The amplitude data was converted to a logarithmic scale and analyzed by fitting to a Gaussian profile which results the Gaussicity of 98 %.

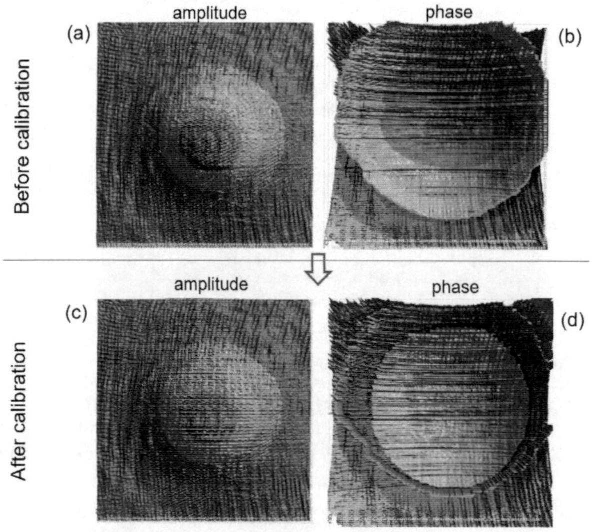

Fig. 5. The measured near-field beam pattern of a corrugated horn with an ellipsoidal mirror in amplitude (a, c) and phase (b, d) in spatial resolution of 79 μm at 1.9 THz. The data were calibrated using the data at the beam center position. The upper images (a, b) show the beam pattern before the calibration and the lower (c, d) shows that after the calibration. Especially for the phase pattern, the data has been improved to be smoothed. The data are plotted in a linear scale.

NOTES:

Figure 6 shows a far-field beam pattern obtained by Fourier transform calculation.

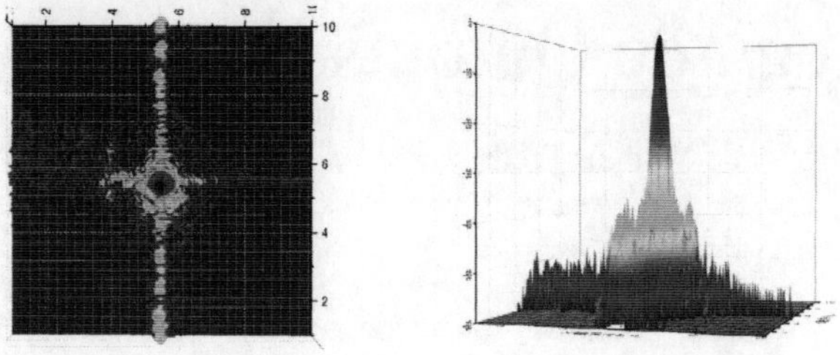

Fig. 6. The calculated far-field beam pattern of a corrugated horn in amplitude at 1.9 THz.

III. Summary

We measured beam patterns of a corrugated horn at 1.9 THz in amplitude and phase. We will also measure the beam patten in cross-polarization. In the future, we plan to measure a beam pattern at higher frequencies of 3 to 4.75 THz using a phase-locked THz-QCL as a local oscillator. At 4.75 THz, a RF source is an important issue.

References

[1] Y. Irimajiri et al., “Beam pattern measurements of a quasi-optical HEB mixer at 2 THz”, *in Proc. 29th Int. Symp. Space THz Techn.*, Mar. 2018.

[2] A. Gonzalez et al., "Reconfigurable Near-Field Beam Pattern Measurement System from 0.03 to 1.6 THz", *IEEE Trans. THz Sci. and Technol.*, vol. 6, no. 2, Mar. 2016.

[3] W. Jellema, “Optical design and performance verification of Herschel-HIFI”, doctor’s thesis, ISBN 978-90-367-7444-4(electric version), 2015.

[4] Y. Irimajiri, A. Kawakami, M.-J. Wang, W.-C. Lu, "Development of a corrugated horn at 2-THz band for a hot electron bolometer mixer", *in Proc. 47th Int. Conf. on Infrared, Millimeter, and Terahertz Waves*, Delft, Netherlands, Aug. 28–Sept. 3, 2022.

NOTES:

Wideband OMT with Modified Bøifot Layout and Co-aligned Waveguide Outputs

Igor Lapkin, Cristian López, Denis Meledin, Leif Helldner, Mathias Fredrixon, Alexey Pavolotsky, Sven-Erik Ferm, Vincent Desmaris, Victor Belitsky, *Senior Member IEEE*

***Abstract*—We present the design of an orthomode transducer, OMT, that aims for the frequency band 210 - 375 GHz. The OMT employs a modified Bøifot layout. The OMT is optimized to fit into the tight spatial constraints, e.g., of the ALMA cartridge and harmonizes the receiver cartridge components for both polarizations by allowing use of the same configuration and components in both polarization chains. The OMT features a built-in novel broadband 90-degree waveguide twist, which minimizes the insertion loss by removing the H-split waveguide while eases receiver components integration with the 2SB mixers in the ALMA cartridge. The OMT was designed and optimized using HFSS™ 3D simulation software and tested with Keysight PNA-X VNA and three VNA extensions WR5.1, WR3.4 and WR2.2 in order to cover ultra-broad RF band.**

***Index Terms*—Orthomode transducer, waveguides, ALMA receiver cartridge, dual-polarization receivers**

I. INTRODUCTION

An orthomode transducer, OMT, is a polarization diplexer that allows to physically separate the polarization components of an incoming signal. A typical place for such diplexer is between the feed, e.g., a corrugated horn, and the first stage of a dual-polarization receiver, e.g., a mixer. Consequently, the input port of an OMT should allow propagation of both orthogonal linear polarizations and the OMT RF insertion loss directly affects the receiver noise performance. A few derivatives of the Bøifot OMT layout [1] are currently used in the ALMA Band 3, 4, 5, 6, 8 receiver channels [2] providing the superior beam squint (Fig.1) at the sky with the single feed for both polarizations as compared to the receivers using polarization grids and thus employing individual feeds for each polarization, requiring careful mechanical alignment. The implementation of an OMT in the next generation of ALMA receivers, e.g., ALMA Band 7 is therefore a desirable option.

The ALMA receiver cartridges provide a very confined space for placing the different receiver components particularly considering all cold optics with bigger mirrors, especially for lower frequency bands with all-cold optics, e.g., ALMA Band 5 [3] and even Band 6.

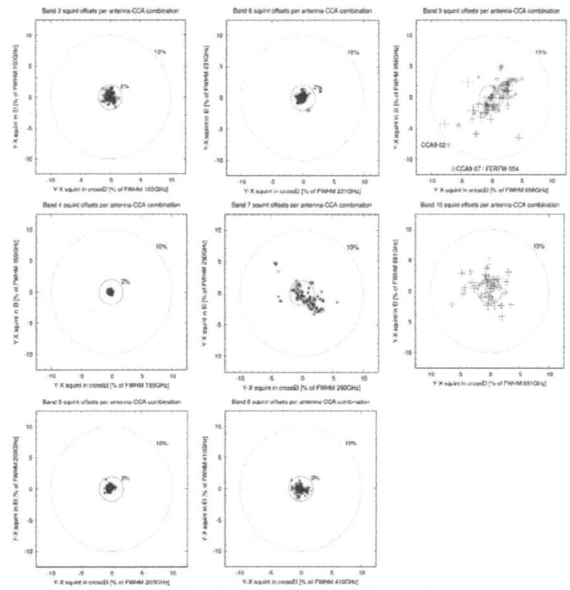

Fig. 1. On-sky Y-X polarization beam squint in cross-elevation vs. elevation coordinates in units of the beam FWHM at the observing frequency for ALMA Bands 3 to 10. Multiple measurements have been averaged per antenna-CCA combination when possible. The 10% specification limit is shown (green circle), in addition to a threshold of 2% which all existing single-horn receivers (Bands 3, 4, 5, 6, 8) could comply with and which may be a practical future goal. This figure corresponds to Figure 8 in [4].

The original Bøifot OMT's T-layout precludes using the same components and arrangement of the receivers for both polarizations, which would be a disadvantage for, e.g., ALMA receiver cartridge production. However, this difficulty could be circumvented by employing slightly more complex OMT configurations, which include additional split of the block and 90-degree H-plane turns of the waveguides at the Y-junction branch as in ALMA Band 5 OMT [5]. Though, this solution has a drawback of featuring an extended H-split waveguide for one polarization that may result in a RF leak and correspondingly an increased insertion RF loss. This is especially true for an OMT at higher frequencies, e.g., 210 - 375 GHz where the

All authors are with the Group for Advanced Receiver Development, Onsala Space Observatory, Department of Space, Earth and Environmental Sciences, Chalmers University of Technology, Gothenburg, SE-41296, Sweden. E-mail oф all authors is: name.surname@chalmers.se.

Corresponding author: Victor Belitsky, victor.belitsky@chalmers.se.

waveguide dimensions are substantially smaller and an OMT requires even higher accuracy for fabrication.

In order to avoid the unwanted H-split and risk of increasing the insertion RF loss, we propose in this paper the introduction of an integrated 90-degree wideband waveguide twist placed close to the OMT polarization splitting junction, PSJ. This allows to align OMT output waveguides and avoid using external 90 - degree twist as in [5, 6]. By harmonizing the OMT outputs, we open a possibility of using the same receiver layout and components for both polarizations.

II. OMT Layout & Simulations

The OMT design largely follows the one successfully used in the ALMA Band 5 receiver. However, in order to achieve substantially broader fractional bandwidth, 210-375 GHz or 57,6% of the center frequency and better control the internal matching, additional elements were added to the polarization splitting junction: a multi-step transformer was added to the direct output of the PSJ and in multiple circular recesses to Y-junction branch of the PSJ. Fig. 2 demonstrates the HFSS 3D model of the proposed OMT. The orange arrow points to the waveguide 90-degree 3-section twist.

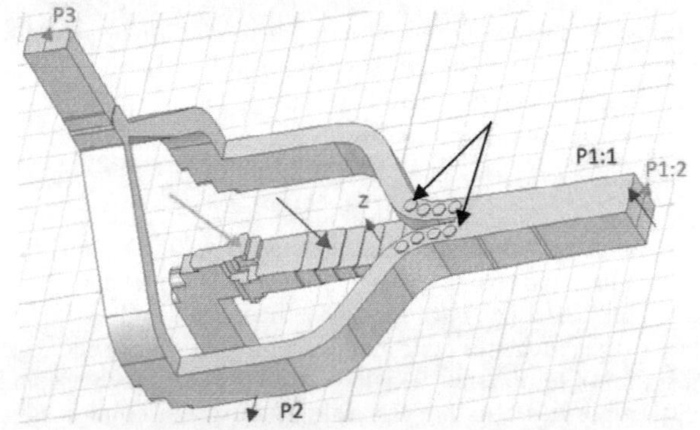

Fig. 2. HFSS optimized OMT 3D model. Yellow arrow points on the built-in twist. Polarization split illustrated by color-coded port denotations. The multi-step transformer in the direct branch of the PSJ marked by blue arrow. The multiple circular recesses in the Y-junction branch of the OMT PSJ marked by black double-arrow. The OMT to be fabricated off three parts with split going in the middle of the square input waveguide, dimensions 760x760 μm, in the direction of P1:2 vector; these two parts formed PSJ and the second split goes in the middle of the output waveguides, dimensions 380x760 μm, in the plane of vectors P3 & P2.

In order to keep the length of the waveguides inside OMT to possible minimum and thus introduce minimum insertion RF loss, we would need a compact 90-degree waveguide twist for integration inside the OMT, e.g., such as described in [7, 8, 9, 10]. The twist should also be compatible with split-block fabrication technique intended to fabricate the OMT. Another obvious requirement is that the twist should provide 56% of the fractional RF band and deliver 210 - 375 GHz operational bandwidth.

The required bandwidth could be provided by the twist suggested in [8], however its design is hardly suitable for the split-block fabrication. The twists suggested in [9, 10] could be accommodated to fabrication using split-block technique but do not provide the required RF bandwidth. Following these conclusions, we employ the twist suggested [9] in a multi-step design in order to extend the RF band. Fig. 2 illustrates the initial design of the 3-stage twist with the twist [9] in the central position while the two adjacent peripheral steps use the twist shape similar to presented in [11].

During design phase, all major waveguide components of the OMT, the polarization splitting junction, the Y-junction, the twist and H-turns were HFSS simulated and optimized separately with the goal of having input/output reflection loss well below -20 dB. Thereafter, the complete HFSS OMT model was assembled by combining these partial models and the optimization procedure was run for the complete OMT also applying realistic constraints of the fabrication using precision CNC miller. The results of the HFSS simulations are presented in Fig. 3.

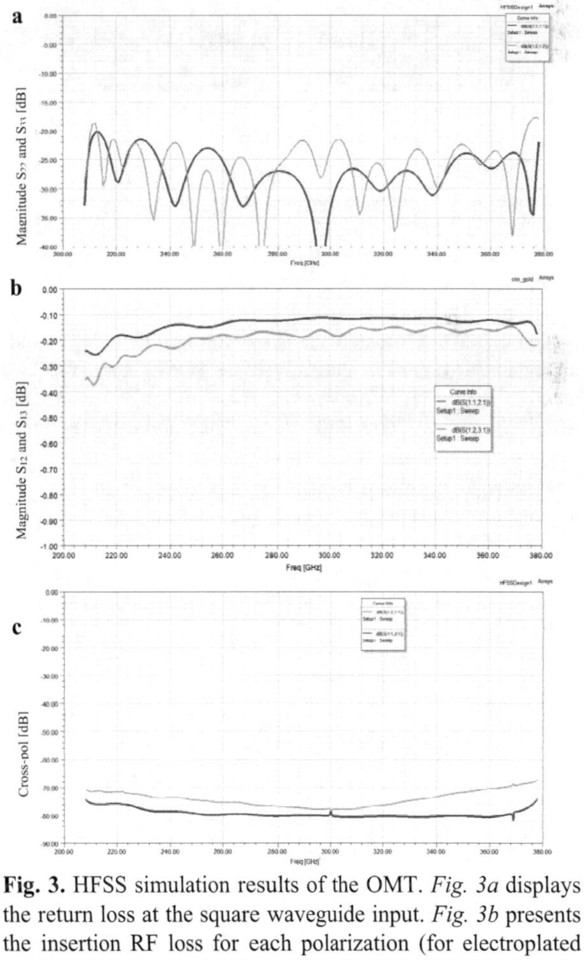

Fig. 3. HFSS simulation results of the OMT. *Fig. 3a* displays the return loss at the square waveguide input. *Fig. 3b* presents the insertion RF loss for each polarization (for electroplated gold assigned as material for the OMT in HFSS). *Fig. 3c* presents the cross-pol simulation results.

III. OMT Fabrication & Characterization

The OMT was manufactured in-house in tellurium copper using a precision CNC milling machine from Kern GmbH. Subsequently, the produced parts were electroplated with 0.5 µm of Gold. Fig. 4 shows the manufactured OMT.

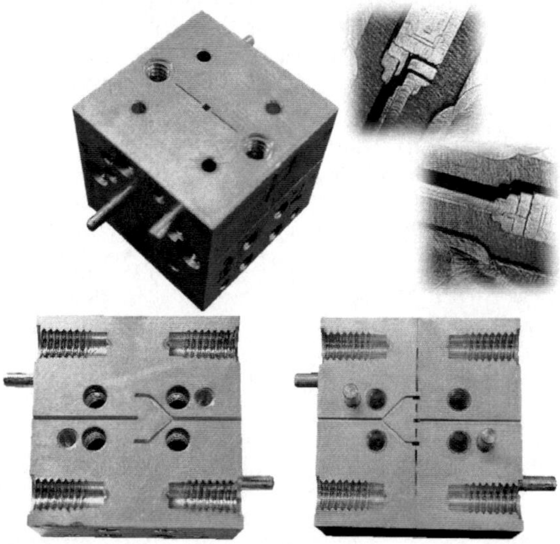

Fig. 4. Manufactured OMT. Pictures show assembled OMT with the square input waveguide, top photo. Solid part of the second split (left) and the split part that containing the PSJ, right. In this split all waveguides are divided in the middle of the broad wall (non-radiating split) and are co-aligned. The OMT is manufactured from Tellurium copper and gold-plated. The inserts show the details of the 90-degree split.

The OMT was characterized using Keysight PNA VNA and frequency extension modules WR5.1, WR3.4 and WR2.2. Because the OMT uses custom-sizes of the waveguide 380 x 760 µm and 760 x 760 µm, the measurements required adapters for rectangular waveguides and transitions rectangular-to-square from the waveguides of the VNA frequency extension modules. The contribution of the transitions used for the measurements are later de-embedded using back to back direct measurements of the transitions themselves., yielding the OMT's characterization results, depicted in Fig. 5 below.

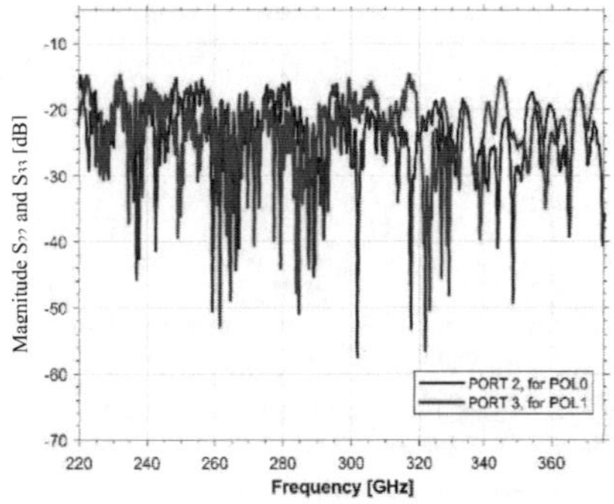

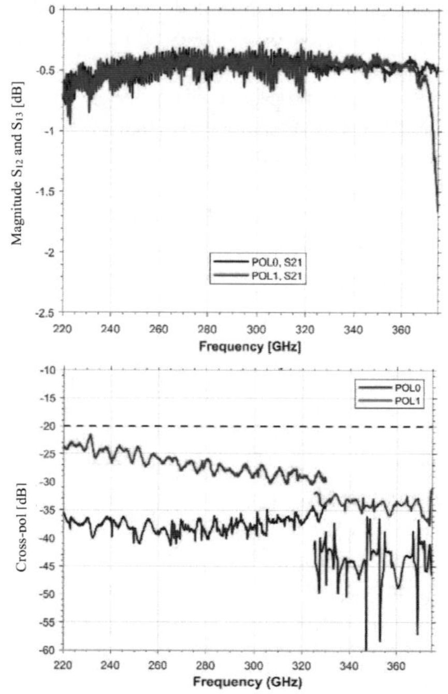

Fig. 5. Measured performance of the OMT. *Fig. 5a* displays the return loss at the square waveguide input. *Fig. 5b* presents the insertion RF loss for each polarization. *Fig. 5c* presents the cross-pol measurement results.

IV. Conclusion

We present the design and characterization of the modified Bøifot orthomode transducer that covers the frequency band 210 - 375 GHz, i.e., covering a 56% fractional bandwidth. The OMT design features built-in novel 90-degree waveguide twist that allows to minimize length of the H-split waveguides and thus minimizes a risk of RF leaks and extra insertion losses. Additionally, the twist allows placing the OMT outputs co-aligned that facilitates the following receiver components design and helps fitting into a tight space, e.g., of the ALMA receiver cartridge the receiver cartridge. The manufactured prototype performance follows well the simulations in HFSS. An early version of the OMT with an external twist was used in SEPIA345 [6] receiver that is installed at APEX telescope in Chile [13] since January 2020.

Acknowledgments

Authors would like to thank Dr. N. Phillips, ESO, for providing data and plots on ALMA receivers' beam squint on sky and used in Fig. 1. This work was supported by the Onsala Space Observatory, Swedish National Facility for Radio Astronomy, and the European Southern Observatory via Collaboration Agreement No. 94468/20/103859/HNE.

References

[1] A. M. Bøifot, "Classification of ortho-mode transducers," European Transactions on Telecommunications, vol. 2, no. 5, pp. 503–510, 1991.

[2] ALMA receiver bands, on-line https://www.eso.org/public/sweden/teles-instr/alma/receiver-bands/?lang

[3] B. Billade et al., "Performance of the First ALMA Band 5 Production Cartridge," in IEEE Transactions on Terahertz Science and Technology, vol. 2, no. 2, pp. 208-214, March 2012, doi: 10.1109/TTHZ.2011.2182220.

[4] S. Asayama, G. H. Tan, K. Saini, J. Carpenter, G. Siringo, T. Hunter, N. Phillips, H. Nagai, "Report of the ALMA Front-end & Digitizer Requirements Upgrade Working Group", ALMA Report ALMA-05.00.00.00-0048-A-REP

[5] V. Belitsky et. al., "ALMA Band 5 receiver cartridge: Design, performance, and commissioning", A&A, A98, Volume 611, March 2018, https://doi.org/10.1051/0004-6361/201731883

[6] D. Meledin, et. al., "SEPIA345: A 345 GHz dual polarization heterodyne receiver channel for SEPIA at the APEX telescope," Astronomy & Astrophysics, 668, A2, 2022, doi: https://doi.org/10.1051/0004-6361/202244211

[7] C. López, V. Desmaris, D. Meledin, A. Pavolotsky, & V. Belitsky, "Design and Implementation of a Compact 90° Waveguide Twist with Machining Tolerant Layout," in IEEE Microwave and Wireless Components Letters, vol. 30, no. 8, pp. 741-744, Aug. 2020, doi: 10.1109/LMWC.2020.3000833

[8] C. López, D. Montofré, V. Desmaris, A. Henkel, & V. Belitsky, "Ultra-Wideband 90° Waveguide Twist for THz Applications," IEEE Transactions on Terahertz Science and Technology, vol. 13, no. 1, pp. 67-73, Jan. 2023, doi: 10.1109/TTHZ.2022.3213468

[9] L. Rud, D. Kulik, A. Kirilenko, "Compact Broadband 90° Twist Based on Square Waveguide Section with Two Stepped Corner Ridges", Microwave and Optical Technology Letters, Vol. 51, No. 3, March 2009, DOI 10.1002/mop

[10] L. Zeng, C. E. Tong, S. N. Paine and P. K. Grimes, "A Compact Machinable 90° Waveguide Twist for Broadband Applications," in IEEE Transactions on Microwave Theory and Techniques, vol. 68, no. 7, pp. 2515-2520, July 2020, doi: 10.1109/TMTT.2020.2987800.

[11] A. Kirilenko, D. Kulik, L. Rud, "Compact 90° Twist Formed by a Double-Corner-Cut Square Waveguide Section", IEEE Transactions on Microwave Theory and Techniques, Vol. 56, No. 7, pp. 1633-1637, July 2008.

[12] A. Gonzalez and K. Kaneko, "High-Performance Wideband Double-Ridged Waveguide OMT for the 275–500 GHz Band," in IEEE Transactions on Terahertz Science and Technology, vol. 11, no. 3, pp. 345-350, May 2021, doi: 10.1109/TTHZ.2021.3062451.

[13] APEX telescope, https://www.apex-telescope.org/ns/ .

Design of Octave-Band Magic-T Using Stepped Ridges and Posts

N. Tasouji[1*], D. Henke[2], T. Sieverding[3], S. Salem Hesari[2], J. Bornemann[1], and A. Densmore[2]

Abstract— **In this paper, the design of a wideband ridged waveguide magic-T is presented. The magic-T has three branches of single-ridged waveguide which is connected to a standard double ridged waveguide at the output. Utilizing an L-shaped septum and posts aid in achieving a return loss (RL) better than 20 dB over a frequency range of 20 GHz to 40 GHz. Additionally, the proposed Magic-T exhibits high E-H port isolation (>60 dB). This innovative design is well-suited for utilization in applications related to radio astronomy.**

Keywords— **Magic-T, matching structure, rectangular waveguide, octave- band, ridged waveguide, radio astronomy.**

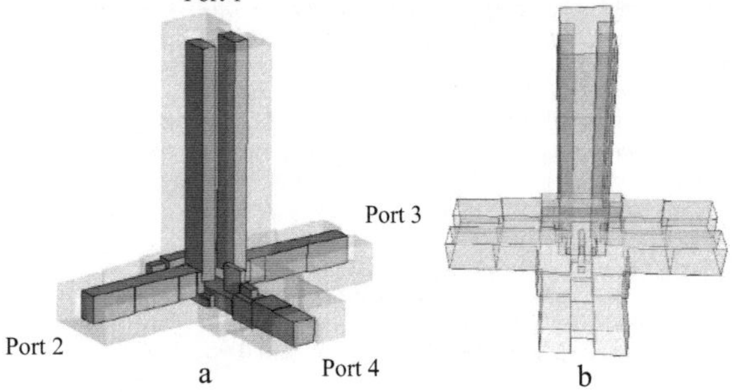

Fig. 1. Prototype of the proposed magic-T (a) CST, (b) μWaveWizard

I. Introduction

THE magic-T is a passive microwave device that features four ports. Due to the limitation of regular three-port networks, which cannot exhibit loss-lessness, reciprocity, and perfect matching across all ports simultaneously, the four-port magic-T addresses this challenge by integrating a combination of E-plane and H-plane T-junctions [1]. Upon introducing a signal to any of the ports, it divides equally between two output ports in terms of amplitude. Additionally, if the signal is applied to the H-plane (sum arm) port, the outputs will be in phase, while they will be 180 degrees out of phase if the signal originates from the E-plane (difference arm) port. The remaining port stays isolated and the level of isolation in a magic-T is significantly influenced by the matching structures employed in the junction. Usually, this configuration includes capacitively placed posts within the H-plane and inductively positioned irises near the E-plane [2]. Also, the symmetry in structure, particularly at the junctions, is a crucial factor that needs careful consideration during the design [3].

The primary benefit of the magic-T, distinguishing it from other T junctions, is that it ensures isolation among all ports regardless of the port which power is introduced. This characteristic makes the magic-T particularly well-suited for various applications like duplexers, mixers, isolators, and power combiners [4]. It can also improve the performance of an orthomode transducer and decrease the possibility of higher order mode excitation [5], [6], [7]. This design aims to create an octave band magic-T with excellent isolation, minimal loss, compatible with standard double-ridge waveguide.

II. Magic-T Design

The advancement of millimeter-wave instrumentation has created a high demand for magic-Ts that have wide bandwidth and can be integrated with other RF components specially in the field of radio astronomy [6], [8]. Nevertheless, it has been observed when designing a magic T, to achieve full 2:1 frequency bandwidth operation while also simplifying the fabrication and assembly process remain a challenging task. Adding ridges to waveguides is a way to lower the cutoff frequency and increase the total bandwidth [9]. The matching structures are crucial to address the inherent mismatches in Magic-Ts. Various types of matching structures have been proposed in earlier studies to fulfill this requirement. For example, in [3], metallic cones at the junction's center, a septum in the first port, and conducting posts in the second, third, and fourth ports enable a 36% bandwidth with RL < 20 dB. However, this design is tailored for lower frequencies, and challenges arise in machining the sharp cone edge at higher frequencies. The presented design in [10] incorporates stepped waveguides and a ridge at the sum port to achieve the desired performance within the frequency range of 28 to 36 GHz. In [11], a magic T is created with two rectangular posts, an L-shaped septum, and a cylindrical platform, operating in the 24 to 28 GHz frequency range for satellite communication. In [12], two designs utilizing Double-Ridged Waveguide (DRWG) and Pyramidal Ridged Waveguide (PRWG) are presented. The DRWG magic-T operates within a bandwidth of 27% (12.2 to 16 GHz) but requires assembly after separate processing. In contrast, the PRWG magic T, with a narrower bandwidth of 8.6%, meets printing requirements without overhang concerns, allowing for direct printing as a single piece. In [2], a magic-T operates from 7.5 to 18 GHz, utilizing a double ridged waveguide and two cylindrical posts. While achieving desired results for ports 1,2 and 3, there is no reported information on the return loss of port 4 and the isolation between the sum and differential ports.

Fig. 1 illustrates the design of the suggested wideband magic T, featuring four ports arranged in two planes: E-plane (first, second, and third ports) and H-plane (second, third, and forth port). The second, third, and forth ports each adopt a single ridged waveguide configuration, while the first port utilizes a double ridged WRD180 waveguide for the convenience of a standard interface. As previously indicated, the effectiveness of the magic-T relies heavily on achieving substantial isolation

[1] Department of Electrical and Computer Engineering, University of Victoria, Victoria, BC, V8P 5C2, Canada; [2] NRC Herzberg Astronomy and Astrophysics Research Centre, Victoria, BC, V9E 2E7, Canada.
[3] Mician GmbH, Bremen, Schlachte 21, 28195, Germany; *Corresponding author (email: nasrintasouji@uvic.ca).

NOTES:

between port 1 and port 4, also between port 2 and port 3. On the other side, minimizing insertion loss between port 2 (port3), and ports 1 and 4 is crucial so different impedance matching components, such as L-shaped metal elements, steps, and metallic segments, must be employed collectively to function as inductive and capacitive elements in arms and junctions. While modal analysis has been conducted for a basic magic-T structure [13], there are currently no explicit solutions available for designing a wideband symmetric magic-T [14]. In this design, it has been noted that the rectangular step located at the midpoint of the junction near the E-arm output significantly influences the achievement of the desired outcomes for port 1. Additionally, the inclusion of an L-shaped septum in the sum arm is crucial, serving to separate the RF waves and achieve the desired isolation between the ports.

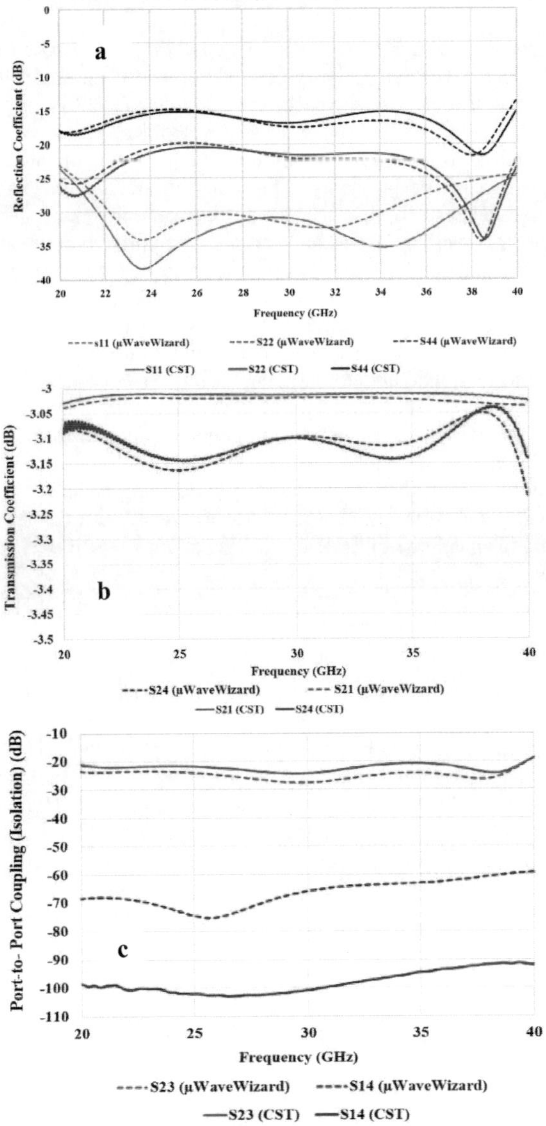

Fig. 2. Simulation results of the magic-T (a) Reflection coefficients at Ports 1, 2 and 4. (b) Transmission coefficients (Insertion loss) from Port 1 to 2 and from Port 4 to 2. (c) Isolation between port 2 and 3and between port 1 and 4.

NOTES:

III. Simulation Result

The proposed magic-T was designed and simulated using both µWave Wizard and CST Microwave Studio to confirm its functionality. Dashed lines represent simulation results from µWaveWizard, while solid lines depict simulation results from CST. The results of the simulated S parameters, shown in Fig. 2, demonstrate consistent and satisfactory agreement between the two software tools throughout the entire frequency range of the octave band. Because of structural symmetry, the results for both port 2 and port 3 are identical. Hence, only the results for one of them are showcased in Fig. 2.

IV. Conclusion

This paper explores a waveguide magic-T designed to attain one octave bandwidth, effective isolation, machinable, and a compact size. Two high-frequency software tools were employed to analyze the design. The use of stepped ridged waveguide and posts has enhanced the bandwidth in comparison to prior designs. Simulation results indicate that the suggested magic-T exhibits promising potential for integrated front-end receiver designs.

References

[1] D. M. Pozar, *"Microwave Engineering."* Hoboken, NJ: Wiley, 2012.

[2] C. Yuan, Y. Luo, F. Meng, and G. Chen, "A full-frequency band matching structure of double-ridge magic T," in *Proc. 2nd Int. Conf. Advances Energy, Environment and Chemical Engr*, (AEECE), Singapore, 2016.

[3] E. Ameri, M. Khalaj-Amirhosseini, H. Sedighy, Y. Azizi, "Design of new wideband waveguide magic-T for X-band applications," Microw. Opt. Technol. Lett., vol. 65, pp. 2530–2534 2023.

[4] L. Guo, Y. Shi Y, G. Wang, and Y. Tian, "An X-band coplanar magic-T with high power handing," Microw. Opt. Technol. Lett., vol. 64, pp. 1387-1393, 2022.

[5] D, Henke, and C. Groppi, "Focal Plane Array Concept for ALMA 2030,", ALMAFED2021, Oct. 1, 2021. doi.org/10.5281/zenodo.5548445.

[6] D. Ch. Son, M. A. Elmansouri, L. B. Boskovic, and D. S. Filipovic, "Spline-Based Aspheric Dielectric Lens-Corrected Quad-Ridge Horn Antenna," *IEEE Microw. Wirel. Compon. Lett.*, pp. 1-5, Nov 2023, doi.org/10.1109/LAWP.2023.3332251.

[7] D. Henke, D. Atherton, I. Wevers, and A. Densmore, "A turnstile OMT using magic-tees and integrated noise-injection couplers,", Proc. 33rd Int. Symp. Space Terahertz Technol., Charlottesville, Virginia, USA, Apr. 7–11, 2024.

[8] W. Shan, S. Ezaki, H. Kang, A. Gonzalez, T. Kojima, and Y. Uzawa, "A compact superconducting heterodyne focal plane array implemented with HPI (hybrid planar integration) scheme," *IEEE Trans. Terahertz Sci. Technol.*, vol. 10, no. 6, pp. 677-689, Nov 2020.

[9] W.J.R. Hoefer, M.N. Burton, "Closed-Form Expressions for the Parameters of Finned and Ridged Waveguides," *IEEE Trans. Microw. Theory Techn.*, vol. 30., pp. 2190 - 2194, Dec 1982.

[10] Y. J. He, D. Y. Mo, Q. S. Wu, and Q. X. Chu, "A Ka-band waveguide magic-T With coplanar arms using ridge-waveguide transition," *IEEE Microw. Wirel. Compon. Lett.*, vol. 27, no. 11, pp. 965 - 967, Nov 2017.

[11] V. Senthil Kumar and D. G. Kurup, "A new broadband magic Tee design for Ka-band satellite communications," *IEEE Microw. Wirel. Compon. Lett.*, vol. 29, pp. 92 - 94, Feb 2019.

[12] J. Wu, C. Wang, and Y. Guo, "Ridged waveguide magic Tees based on 3-D printing technology," IEEE Trans. Microw. Theory Techn., vol. 68, no. 10, pp. 4267 - 4275, Oct. 2020.

[13] T. Sieverding and F. Arndt, "Modal analysis of the magic Tee," *IEEE Microw. Guided Wave Lett.*, vol. 3, no. 5, pp. 150-152, May 1993.

[14] W. W. Siekanowicz and R.W. Paglione, "Broadband Double-Ridge Waveguide Magic Tee," *United State Patent.*, Dec 1971.

Investigating Pin-Holes Issues in Josephson Junction Travelling Wave Parametric Amplifiers Requiring Large Area of Dielectric Layer

Javier Navarro Montilla[1], Nikita Klimovich[1], Arnaud Barbier[2], Eduard F. C. Driessen[2] and Boon-Kok Tan[1]

***Abstract*—Microwave superconducting Josephson Travelling Wave Parametric Amplifiers (JTWPAs) exploit the non-linear inductance of a long superconducting metamaterial line formed by thousands of Josephson junctions to achieve broadband parametric gain with quantum limited added noise. Nevertheless, pin-holes in the dielectric (spacer) layer required for fabricating these superconducting transmission lines (STLs) represent a challenge for JTWPAs fabrication. In this paper, we explore two pin-holes mitigation techniques, which shown promising results with DC characterisation of a suite of test structures at cryogenic temperatures. When implemented for actual JTWPA designs with much longer length, they have shown to improve the fabrication yield albeit some pin-holes still seems to exist over the large wafer area. This indicates that further mitigation effort is required to completely eradicate the pin-holes issue for applications requiring large area of dielectric layer such as microwave JTWPAs.**

***Keywords*—travelling wave, parametric amplifier, quantum-limited, Josephson junctions, metamaterial, microwave.**

INTRODUCTION

Travelling Wave Parametric Amplifier (TWPA) can generate high amplification gain over a large bandwidth in the microwave regime with quantum-limited noise performance [1-3]. TWPAs use superconducting transmission lines (STL) with high non-linear inductance to promote the parametric processes required to transfer energy from a pump wave (ω_p) to a weak signal (ω_s). In this process, an idler tone (ω_i) is generated. The non-linear inductance can be sourced from high kinetic inductance superconducting films (termed K-TWPA) [1, 4, 5], or metamaterial films comprising many Josephson junctions (J-TWPA) [2, 3]. TWPAs find applications in many fields ranging from quantum computing [6], dark matter axion search [7, 8, 9] to astrophysics; replacing the semiconductor-based readout amplifiers for millimetre (mm) and sub-mm receiver to minimise heat dissipation and enhance sensitivity [10, 11].

JTWPAs are challenging to fabricate because they require a large number of junctions with low critical currents. Most recent JTWPAs have utilised shadow-evaporated aluminium junctions with high fabrication yield [2, 12]. However, the use of aluminium-based junctions restricts their operation to mK environments and frequencies below 90 GHz, which limits some potential applications for JTWPAs. Niobium/aluminium-aluminium oxide/niobium ($Nb/Al\text{-}AlO_x/Nb$) junctions offer an alternative solution to elevate the operational temperature and frequency into the mm/sub-mm range, as well as expediating JTWPA on-chip integration with astronomical receivers such as superconductor-insulator-superconductor (SIS) mixers.

Trilayer deposition techniques for fabricating $Nb/Al\text{-}AlO_x/Nb$ junctions have been extensively explored and matured for the development of SIS mixers. Utilising them for JTWPAs, on the other hand, is a relatively new development. One of the challenges in fabricating niobium-based JTWPAs is related to the existence of pin-holes defects in the inherent dielectric layer, rather than just the need for many junctions. This dielectric layer, referred as the spacer, is required to define the trilayer-junctions and to form parallel plate capacitors (PPCs) needed in some JTWPA designs. These pin-holes can result in electrical connections between the bottom niobium layer of the trilayer and the top wiring layer, creating an unwanted short-circuits to ground effect, rendering the device unusable.

In this paper, we explore two mitigation techniques to reduce the number of pin-holes in the spacer layer. We measure the performance of these techniques by fabricating dedicated test devices that we characterised using DC signal analysis. Then, we implement one of these techniques in our JTWPA designs, that we fabricated and tested at cryogenic temperatures. The results from the preliminary tests are presented and discussed.

PIN-HOLES MITIGATION TECHNIQUES

The intended metamaterial transmission line designed for our JTWPAs is presented in Fig.1. This elementary unit cell representing a small section of length 'a' in the STL, is cascaded numerous times to create the entire device. Often, the unit cell includes one or several junctions and a parallel plate capacitor (PPC) to increase the shunt capacitance to the ground C_s, ensuring a characteristic impedance $Z_0 = 50\ \Omega$ of the line.

[1]University of Oxford, Oxford, OX1 3RH, United Kingdom; [2]Institut de Radioastronomie Millimétrique (IRAM), Saint Martin d'Hères, 38406, France.

NOTES:

Furthermore, resonators can be added to the transmission line [13] to correct for the phase mismatch caused by the cross and self-phase modulation of the tones propagating in the JTWPA, increasing the gain-bandwidth product.

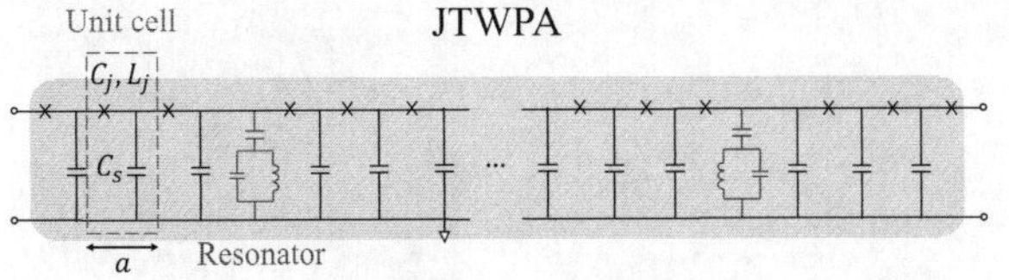

Fig. 1. Schematic circuit diagram of a JTWPA. The unit cell is composed of a single Josephson junction denoted with a cross, with capacitance C_j and inductance L_j, and the shunt capacitance C_s to the ground. The unit cell is cascaded to create the metamaterial nonlinear transmission line. Resonators are periodically added to the line for resonant phase matching (RPM), to enhance the gain and bandwidth.

For fabricating the tunnel junction required for the JTWPAs, we modify the trilayer fabrication process recipe originally developed for SIS mixers. As shown in Fig.2(a), the cross-section topology of our transmission line comprising the junctions is composed of a Nb/Al-AlO_x/Nb trilayer, a dielectric spacer layer surrounding the junction and a niobium wiring layer contacting the top electrode of the junction. For a detailed summary of the fabrication steps, we refer the reader to [14].

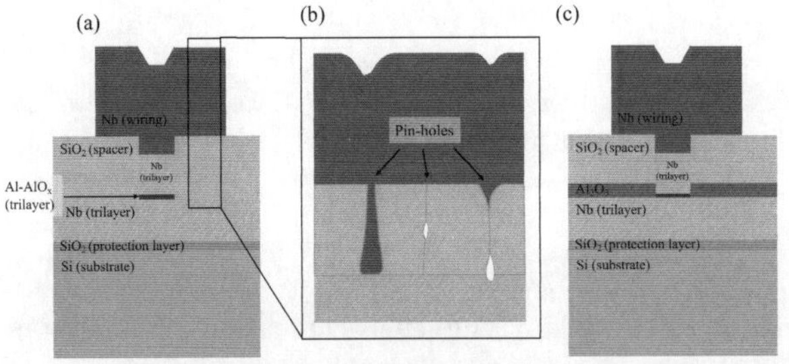

Fig. 2. Cross-section diagram of a trilayer niobium Josephson junction. (a) standard layer distribution. (b) Zoom-in the spacer layer, showing the formation of pin-holes. (c) Pin-hole mitigation strategy consisting of adding an extra layer of Al_2O_3 in the spacer layer. The diagrams are not to scale.

In principle, any dielectric material could be used for the spacer layer, although silicon dioxide (SiO_2) is our preferred choice due to its electromagnetic properties and the availability of existing equipment in our cleanroom facility to grow them on niobium. However, RF-sputtered SiO_2 layers suffer from the formation of pin-holes potentially leading to short-circuits between the trilayer bottom niobium and the wiring layer (Fig.2(b)). The issue is less critical for relatively small devices operating at mm/sub-mm frequency, but for a microwave JTWPA measuring several mm^2 in size, this could be a challenge. Numerous factors could contribute to the formation of these unwanted pin-holes, such as the substrate surface conditions, substrate's temperature during sputtering, a low RF power used and pressure fluctuations in the chamber. The density of these pin-holes decreases exponentially with the dielectric thickness [15], therefore, a thick dielectric layer is a viable mitigation strategy. Furthermore, the pin-holes density is also inherently related to the deposition technique and the material choice e.g., evaporated Al_2O_3 (aluminium oxide) have shown to have a lower pin-hole density, hence our alternative mitigation approach. We acknowledge that dielectric layer formed using plasma-enhanced chemical vapor deposition (PECVD) or atomic layer deposition (ALD) technique could also help to resolve the pin-hole issue, but unfortunately, we do not have access to the required facilities to perform such deposition. Hence, we focus on exploring possible solutions to RF-sputtered SiO_2 for our JTWPA development here.

We have explored both pin-holes mitigation strategies described above: (a) increasing the RF-sputtered SiO_2 spacer layer from 200 to 400 nm, and (b) adding a protective layer of 20 nm evaporated Al_2O_3 underneath the 200 nm RF-sputtered SiO_2, as shown in Fig.2(c). We compare the performance of these two approaches by fabricating a suite of test devices composed of a niobium coplanar waveguide (CPW) embedded with 100, 500 and 1,000 PPCs coupled to the transmission line as shown in Fig.3. To ease the fabrication and reduce the complexity, the trilayer of these test devices was substituted with simply a bottom niobium layer, and no junctions are added to the line. Mitigation technique (a) and (b) are implemented in two different wafers, denoted Wafer 1 and Wafer 2 respectively. We use a 280-um thick high resistivity silicon wafers for the substrate.

These test devices were DC-screened at cryogenic temperatures using the 2-wires measurement shown in Fig.4. The DC screening setup allows us to measure the current-voltage (IV) curve through the trace of the devices and from the trace to the ground, by rewiring the experiment at room temperature. At temperatures below 9 K, the device's resistance is negligeable and we measured an average resistance ($R_{\text{throu}} = 31.1\ \Omega$) originated from the cooper wires used to connect the current source and voltmeter. When measuring the resistance from the trace of the device to the ground, we found an average value of $R_{\text{gnd}} = 19.4\ \Omega$ for short-circuited test devices and 2.22 kΩ for non-faulty devices. The two orders of magnitude difference in the resistance allows us to easily identify devices with pin-holes issues that result in electrical shorts to ground.

NOTES:

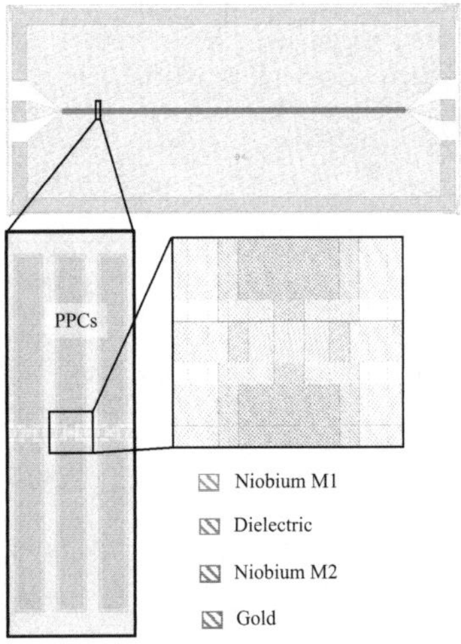

Fig. 3. Layout of the test devices used to study the performance of the two pin-holes mitigation techniques. The parallel plate capacitors are connected to the trace of the CPW.

To further confirm the validity of our measurements, we measure the RF transmission of two non-faulty devices with 1,000 and 100 PPCs, plotted in Fig.5(a) and (b) respectively. Note that the devices' characteristic impedance $Z_0 \neq 50\ \Omega$ (due to the absent of the tunnel junctions) results in a Fabry-Perot-like effect in the transmission profile, caused by the cavity modes. The periodicity of this transmission variation is inversely proportional to the length of the lines, hence the ten times more rapid periodicity for the 1,000 PPCs sample compared to the 100 PPCs sample.

A total of 16 test-structure devices were tested, 12 at cryogenic temperatures and 4 at room temperature, where the effect of the pin-holes can equally be observed through the DC screening process. The results from these measurements are summarised in Fig.5(c). Despite the limited number of fabricated test devices, the larger number of 'good' test devices in Wafer 2 indicates that increasing the dielectric thickness is the preferred pin-holes mitigation strategy. Although this may be applied for JTWPAs with purely bare junctions without PPCs, the solution may not be applied to cases that need PPCs. The capacitance of the PPCs decreases linearly with the thickness of the dielectric layer, therefore, to achieve the same shunt capacitance values, the PPCs will now need to have a much larger area than originally intended. The larger PPC area required not only increases the footprint of the JTWPA device, but will equally increase the chances of having a pin-hole, hence potentially limiting the yield. On the other hand, the addition of a 20 nm evaporated Al_2O_3 will barely modify the size of the required PPCs in our design, while showing promising improvements in the reduction of pin-holes. Therefore, we chose the pin-holes mitigation strategy (b) for our devices.

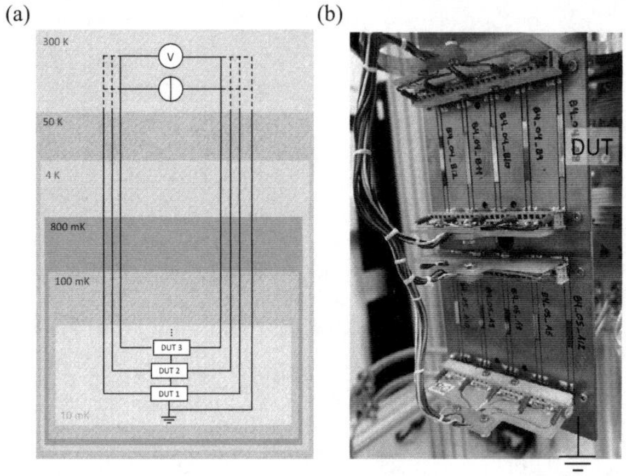

Fig. 4. DC-screening experimental setup. (a) Schematics of the experimental setup. The dotted lines indicate possible reconfiguration options to measure the different devices, as well as the connection from the device to the ground. (b) Picture of the DUT mounted on the DC-screening board inside the cryostat.

JTWPA DESIGN AND EXPERIMENTAL RESULTS

Using the chosen pin-hole mitigation technique presented in the previous section, we fabricated JTWPAs composed of 2,024 Nb/Al-AlO_x/Nb junctions and a targeted critical current density J_c = 1.4 and 0.9 kA/cm^2 (for two different designs), where the spacer layer is composed of the stack of 200 nm SiO_2 and 20 nm Al_2O_3. The spacer layer is equally used as the dielectric layer for a large number of PPCs deployed to match the impedance of our JTWPA to 50 Ω. Furthermore, 23 resonators (comprising a meandered inductive line and a PPC) are added to the transmission line for phase-matching purpose. More details on the design can be found in [16].

Fig.6 shows the IV curve measured for the two JTWPAs: Device A with J_c = 1.4 kA/cm^2, and Device B with J_c = 0.9 kA/cm^2. The measurements were taken using the DC-screening setup presented in Fig.4. The measured IV curves for Device A and B are plotted in the top and bottom panels respectively, with the left column measured through the trace while the right column from the trace to the ground.

NOTES:

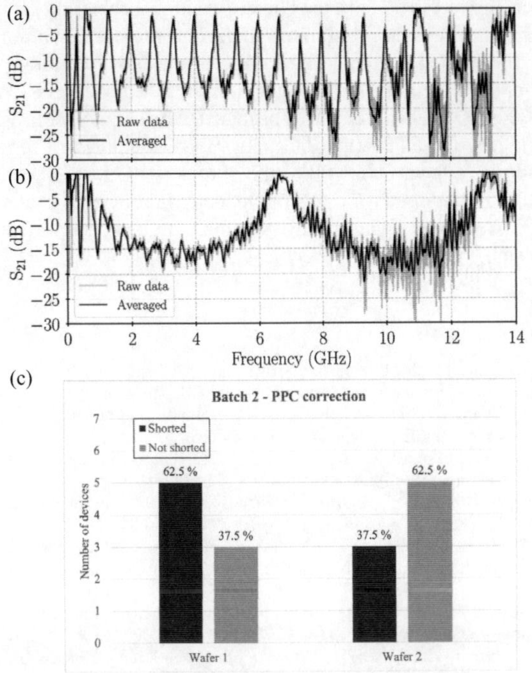

Fig. 5. RF transmission profile at T = 4 K calibrated with a passive feedthrough for (a) a test device with 1,000 PPCs and (b) a test device with 100 PPCs. The data is averaged with a 43.2 MHz window. (c) Number of short-circuited (red) and non-faulty (green) test devices after DC-screening. The spacer layer is formed of 20 nm evaporated Al_2O_3 and 200 nm RF-sputtered SiO_2 (Wafer 1), and 400 nm RF-sputtered SiO_2 (Wafer 2).

Assuming no short-circuits to the ground in our device, we would expect an IV characteristic similar to a bare single Josephson junction, but with a gap voltage 2,024 times higher than the V_{gap} of a single junction, since the junctions are connected in series. However, from Fig.6 (a) and (c), we observe that the transition to normal state happened at a value smaller than the expected $V_{\mathrm{gap}} \approx 6.4$ V. We believe this is due to the presence of several short-circuits in the line as illustrated in Fig.7(a), where the current partly flows through the ground plane, bypassing a number of junctions. When measuring the IV curve to the ground (Fig.6(b) and (d)), we notice the same effect for a smaller value of V_{gap}. This is consistent with our theory, since in this case, the current only sees the junctions before the first short-circuit in the device, without returning to the trace, as shown in Fig.7(b).

Despite the existence of short-circuits in the line, the data in Fig.6 can be used to estimate the I_c of the junctions. From the V_{gap}, we estimate the number of junctions seen by the current to $N_{\mathrm{jj}} = 74 \pm 7$, 218 ± 10 and 23 ± 2 for Fig.6(a), (c) and (d) respectively (we exclude Fig.6(b) due to the low quality of the V_{gap} transition). Measuring the normal resistance value and divideding by the number of junctions, we can make use of the Ambegaokar-Baratoff formula to estimate the average I_c of the junctions, resulting in $I_c = 7.25 \pm 0.68$ μA and $I_c = 2.85 \pm 0.23$ μA for device A and B respectively. These results are close, in terms of order of magnitude, to the targeted value of $I_c = 7$ and 4 μA for device A and B respectively.

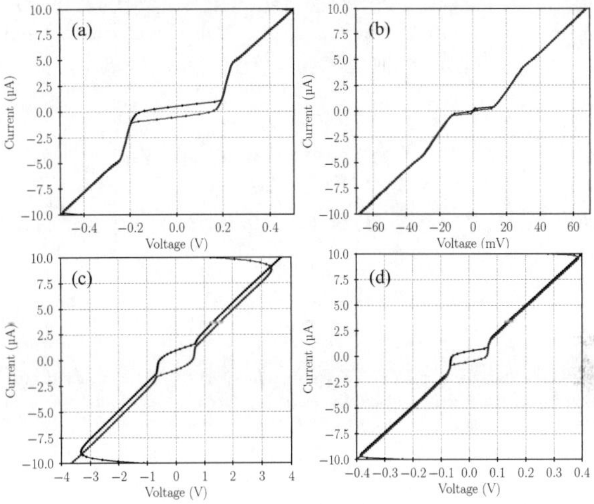

Fig. 6. Current-voltage (IV) curves of our JTWPAs measured at T = 10 mK for Device A (a) through the device and (b) from the device trace to the ground, and for Device B (c) through the device and (d) from the device trace to the ground. The non-linear behaviour from (b) and (d) indicates a possible short-circuit in the line. The current is swept from -10 to 10 μA (black) and from 10 to -10 μA (red).

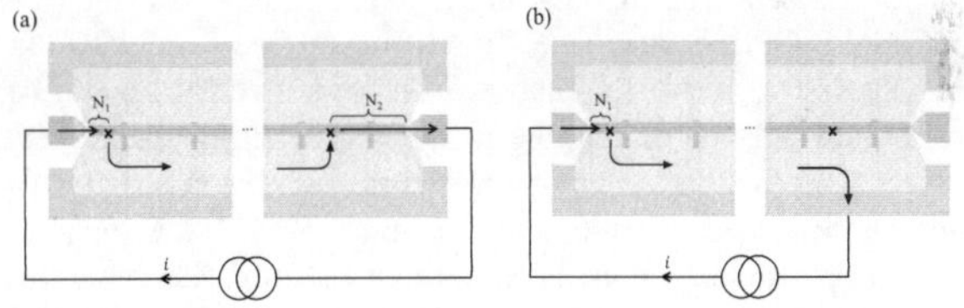

Fig. 7. Schematic representation of the current flowing in our device when several short-circuits to the ground are presented in the line, (a) when the current is applied through the device, and (b) from the device to the ground.

We found similar results for the most of the JTWPA devices measured using this DC-screening technique, suggesting pin-holes issues still persisted despite the yield improvement, most likely due to the much larger area of dielectric layer used in these devices. However, we did manage to measure a satisfactory transmission profile for some of the devices.

NOTES:

The transmission measured for one of these devices is shown in Fig.8. We notice several high-Q resonant dips around 6 GHz, that we associated to the inclusion of the resonators originally designed for a resonance frequency at 7.4 GHz. The shift and spread in resonant frequency could indicate another fabrication issue, related to the fabrication tolerance of the photolithography technique used to define the resonators. Furthermore, the wider dip appears around 8.5 GHz is probably due to the loading effect of the resonators in the transmission line. Nevertheless, the device shows a satisfactory overall transmission, with a moderate expected frequency dependent loss originating from the dielectric layer in the PPCs.

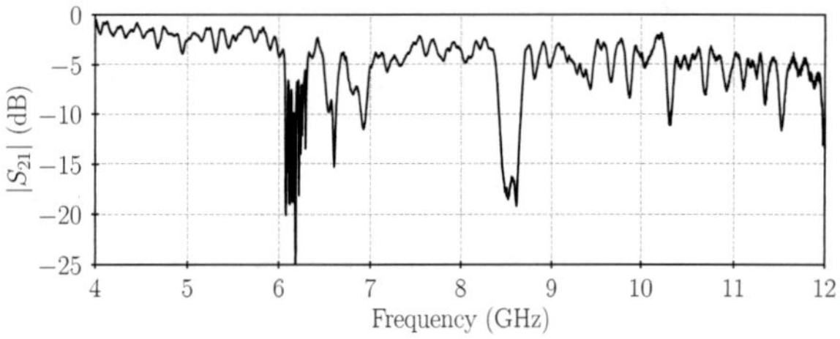

Fig. 8. RF transmission at T = 10 mK measured for the only satisfactory JTWPA, calibrated with a blank. We associate the resonance around 6 GHz to the resonators, and the dip around 8.5 GHz to the periodic loading effect of the resonators in the transmission line. We see a moderate loss with frequency as expected from the dielectric used for the PCCs.

Conclusion

Pin-holes formed in the dielectric used as the spacer layer in defining the trilayer-junction of a microwave JTWPA are a technical concern. They can result in a short-circuit to the electrical ground, rendering the device unusable. We explored two techniques to reduce the number of pin-holes in our RF-sputtered SiO_2 dielectric layer fabrication recipe: (a) increasing the SiO_2 layer, and (b) adding a protective layer of 20 nm of evaporated Al_2O_3 below the SiO_2 layer. Both techniques showed promising results on fabricated test structures. Nevertheless, when implemented in our JTWPA design with much larger spacer areas, it seems that the pinholes have not been completely eradicated, although we observe an improvement in fabrication yield. Therefore, we believe that further efforts are needed to eliminate them completely for devices requiring large area of RF-sputtered dielectric layer, such as our JTWPA design.

Acknowledgement

This research is supported by the European Research Council (ERC) under the European Union's Horizon 2020 research and innovation programme with grant agreement No. [803862] (Project [SPA4AstroQIT]), the UKRI Quantum Technology for Fundamental Physics programme under the project Quantum Sensor for Hidden Sector, and the UK Science and Technology Facilities Council (STFC) Quantum for Science Grant (ST/Y004973/1). J. Navarro Montilla's D.Phil. studentship is supported by the STFC and the Foley-Bejar Scholarship from the Balliol College, Oxford.

References

[1] N. Klimovich et al., "Demonstration of a quantum noise limited traveling-wave parametric amplifier," *ArXiv:2306.11028*, June. 2023.

[2] Luca Planat et al., "Photonic-Crystal Josephson Traveling-Wave Parametric Amplifier", *Physical Review X*, April. 2020.

[3] C. Macklin et al., "A near-quantum-limited Josephson traveling-wave parametric amplifier", *Science,* October. 2015.

[4] Byeong Ho Eom et al., "A wideband, low-noise superconducting amplifier with high dynamic range", *Nature Physics,* July. 2012.

[5] S. Shu et al., "Nonlinearity and wide-band parametric amplification in a (Nb,Ti)N microstrip transmission line", *Physical Review Research*, June 2021.

[6] L. Ranzani et al., "Kinetic inductance traveling-wave amplifiers for multiplexed qubit readout", *Applied Physics Letters,* December. 2018.

[7] LC. Bartram et al., "Dark Matter Axion Search Using a Josephson Traveling Wave Parametric Amplifier", *ArXiv:2110.10262,* October. 2021.

[8] C. Braggio et al., "An Haloscope Amplification Chain based on a Travelling Wave Parametric Amplifier", *ArXiv:2205.02053*, May. 2022.

[9] J. M. Navarro et al., "Optimising the design of a broadband Josephson junction TWPA for axion dark matter search experiments", *Proc. SPIE 11881, Quantum Technology: Driving Commercialisation of an Enabling Science II*, October. 2021.

[10] M. Malnou et al., "Performance of a Kinetic Inductance Traveling-Wave Parametric Amplifier at 4 Kelvin: Towards an Alternative to Semiconductor amplifiers", *Physical Review Applied*, April. 2022.

[11] J.N. Montilla et al., "Design of high compression point Josephson junction travelling wave parametric amplifiers for readout of millimetre and sub-millimetre astronomical receivers", *Proc. SPIE 12190, Millimeter, Submillimeter, and Far-Infrared Detectors and Instrumentation for Astronomy XI*, August. 2022.

[12] A. Ranadive et al., "Kerr reversal in Josephson meta-material and traveling wave parametric amplification," *Nature Communications*, April. 2022.

[13] K. O'Brien et al., "Resonant phase matching of Josephson junction traveling wave parametric amplifiers", *Physical review letters*, October. 2014.

[14] M. Gurvitch et al., "High quality refractory Josephson tunnel junctions utilizing thin aluminum layers", *Applied Physics Letters*, March. 1983.

[15] T. Hattori et al., "Detection of pinholes in R.F.-diode-sputtered SiO_2 films". *Thin Solid Films,* November. 1982

[16] J. N. Montilla et al., "Preliminary Characterisation of Microwave Travelling Wave Parametric Amplifiers Comprising Josephson Junction Meta-material Line". (Under review)

NOTES:

Development of a Phase-modulating Beam Multiplexer for a THz Local Oscillator

B. Pedroni[1*], Y. Shao[1], D. Ren[1], W. Ji[1], J. R. G. Silva[2], A. J. L. Adam[1], J. R. Gao[1,2*]

Abstract— **We propose a novel approach to multiplex a single Gaussian beam from a THz source into seven beams arranged in a hexagonal configuration, matching an array of Hot Electron Bolometer (HEB) mixers. This configuration is necessary for the High-Resolution Receiver (HiRX) instrument on the proposed NASA Single Aperture Large Telescope for Universe Studies (SALTUS) space mission. The beam splitter relies on a reflector that introduces a phase shift to the incident Gaussian beam; after propagation, the desired amplitude distribution is achieved at the mixer array plane. Unlike existing THz multiplexers, our method does not use a phase grating based on the repetition of a unit cell. Instead, we employ an iterative phase reconstruction (Gerchberg-Saxton) algorithm to retrieve the required phase shift. This paper discusses the scientific motivation, current state of the art, design methodology, simulation outcomes, and experimental validation of the reflector.**

Keywords— **Heterodyne, Local Oscillator, mixer array, multiplexer, phase retrieval, SALTUS.**

I. Introduction

Electromagnetic radiation is exploited for the most different purposes, from medical imaging to information transmission, from renewable energy to optical sensing. To fully characterize electromagnetic waves, both amplitude and phase must be considered. Amplitude determines brightness, while phase describes the wave's position in its oscillation cycle. Measurement setups rely on translating wave information into an electric, mechanical, thermal or chemical signal. In most cases, the response time of physical devices is too slow to keep up with the rapid changes of a system it is intended to monitor. Detectors average out the rapidly oscillating wavefront over time; the output is its average intensity over the measurement period. In this process, the phase information, which encodes the precise timing and spatial variation of the wave oscillations, is unavoidably lost. The result is that many of those cutting-edge technologies that leverage electromagnetic radiation for the most ambitious goals are still using just half of the information that waves carry. The problem complicates if such waves are in the "Terahertz gap", laying in the range between microwave and infrared, where the realms of electronics and optics overlap. Implementing technologies for generating and detecting Terahertz radiation is particularly challenging [1]. However, the pace of progress in THz technology has accelerated significantly in recent years, witnessing innovative laser sources and highly sensitive detectors. The initial aim of this project was the development of a reflective Fourier grating to generate a THz local oscillator with seven uniform beams. However, after delving more deeply into the topic, we decided to undertake a different route. The idea is to explore a novel way to manipulate radiation, exploiting the relatively simple principles of phase reconstruction and reflection. The challenge is that two still relatively unexplored worlds, that of phase retrieval techniques and that of Terahertz radiation, are meeting. The drive is that, if a proof of concept is obtained, we propose a simple, general yet powerful method which can be exploited for a wide range of applications. Thus, the implementation of our beam multiplexer is not just a fulfillment by itself, but it also tests the feasibility of a broad technique to manipulate electromagnetic fields.

II. Scientific Motivation

The terahertz spectral region is essential for understanding star formation and the composition of dense interstellar clouds; the latter contain ionized atoms and complex molecules, detectable through their emission and absorption lines. Mapping these spectral lines provides insights into the structure, temperature, density, and chemical composition of astrophysical objects [2]. SALTUS (Single Aperture Large Telescope for Universe Studies) is a mission proposed to NASA, designed to address these scientific challenges by deploying a far-IR space observatory [3]. If accepted, the launch is scheduled currently in 2031. The mission aims to explore the evolution of galaxies and their supermassive black holes, trace the formation of cosmic structures, and evaluate habitability by mapping astrochemical signatures in protoplanetary disks. SALTUS will employ a 14-meter inflatable, lightweight telescope achieving arcsecond-scale angular resolution, a record for far-IR space observatories. This high-resolution capability is crucial for overcoming the spatial confusion limit, where densely packed celestial sources blur together.

The High-Resolution Receiver (HiRX) instrument on board of SALTUS [4] will include cryogenic heterodyne receivers across four frequency bands: Band 1 (455-575 GHz), Band 2 (1.1-2.1 THz), Band 3 (2.475-2.875 THz), and Band 4 (4.744 and 5.35 THz). The HiRX instrument is designed to detect and resolve significant astrophysical lines, such as those from water

[1]Department of Imaging Physics, Delft University of Technology, Delft, the Netherlands; [2]SRON Netherlands Institute for Space Research, Leiden/Gronigen, the Netherlands; [*]Corresponding author (email: barbarapedroni00@gmail.com , j.r.gao@sron.nl).

NOTES:

and deuterium hydride (HD), with a high spectral resolution (ranging from 10^5 to 10^7). To detect faint signals from distant sources, high-sensitivity and high-resolution technologies like heterodyne receivers are essential. Such receivers can bring unprecedented spectral resolution ($\nu/\Delta\nu > 10^7$), providing information about the type of emitter, as well as the temperature, density, and motion of gases [2]. Their working principle relies on down-converting the THz signal to the GHz range, where low-noise amplifiers are available, by mixing it with a Local Oscillator (LO) signal. Bands 2 to 4 will use superconducting Hot Electron Bolometers (HEBs), which offer coherent detection with low noise levels and no upper frequency limit. To increase mapping speed, multiple pixels can be simultaneously employed. Band 4 in the HiRX instrument includes a seven-pixel array arranged in a hexagonal pattern, with one pixel in the middle. Due to difficulties in achieving the same frequency from all Quantum Cascade Laser (QCL) sources and due to power dissipation, it is necessary to use a single QCL and distribute its output into multiple beams.

The goal of this project is to develop a beam multiplexer to split one incoming THz beam from a single source onto multiple detectors, ensuring a high efficiency, equal power distribution, and Gaussian-shaped beams. The design will focus on two frequencies: 4.7 THz, required for the mission, and 1.627 THz, for a proof of the concept purpose.

III. A NOVEL APPROACH

Commonly, phase gratings (Fig.1) are employed in heterodyne receiver systems as LO beam multiplexers [5].

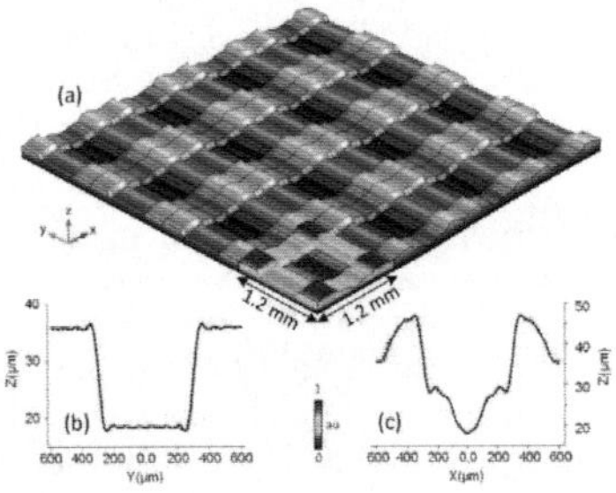

Fig. 1. (a) 3D drawing of 16 unit cells of a phase grating realized for a 4.7 THz source (height is out of scale) [6]. The unit cell is in color. (b)(c) Calculated and manufactured 2D cross sections of a unit cell along x and y directions. The dashed black and solid red curves are the manufactured and calculated profiles respectively.

Such reflective devices are based on the periodic repetition of a single unit cell. Based on diffraction, a beam incident on the grating is split among multiple orders; the size of the unit cells determines the angular separation between modes, while its shape affects power distribution. So-called "pseudo-2D" phase gratings, obtained by orthogonally superimposing two 1D gratings, can only generate rectangular patterns [7]; moreover, a fraction of the input power is unavoidably lost in unwanted diffraction modes.

In this project, we explore a novel approach to design a beam multiplexer by dropping the periodicity assumption on which phase gratings rely. The goal is to design a reflector which induces a position-dependent phase shift to the incident wavefront; after reflection and subsequent propagation, the desired amplitude is obtained at a specified distance from the reflector. If the surface is designed properly, the target distribution at the mixer array plane is achieved without additional collimating optics. This method aims to minimize power in unwanted modes and demonstrate that phase reconstruction can be a powerful and general tool for manipulating radiation for various applications.

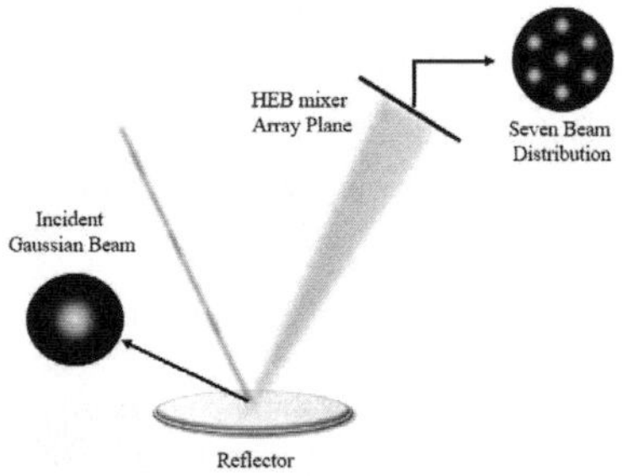

Fig. 2. Working principle of the reflector. The incident beam acquires a position-dependent phase shift through reflection by the surface; after propagation, the seven-beam amplitude is obtained at the HEB mixer array.

IV. DESIGN AND SIMULATION

The design of the surface lies in determining the necessary phase shift to apply to the initial wavefront; this translates into solving a phase retrieval problem [8]. When an electromagnetic wave reflects off the surface, each point on the wavefront undergoes a different optical path, resulting in a position-dependent phase shift. By carefully controlling this phase shift, the reflected wave can be manipulated to produce a desired far-field pattern, in this case, seven Gaussian beams at the HEB mixer array. To recover the required phase information, various computational techniques can be employed. One common approach is the use of iterative algorithms that update an initial estimate of the phase based on measured amplitude data and some prior knowledge or constraints. These algorithms aim to minimize a cost function that quantifies the discrepancy between the measured and reconstructed wave. The Gerchberg-Saxton (GS) algorithm (Fig.3) is a widely used method for phase retrieval [9]; an initial guess of the phase at the initial

NOTES:

plane is multiplied by the constrained initial beam amplitude. The resulting field is propagated to the image plane by performing a Fourier transform, resulting in a complex field. The amplitude of this complex field is then replaced by the desired target amplitude while keeping the phase component unchanged. An inverse Fourier transform is applied to return to the initial plane, and the process is repeated iteratively.

The algorithm converges when the difference between the target intensity distribution and the current one is minimized.

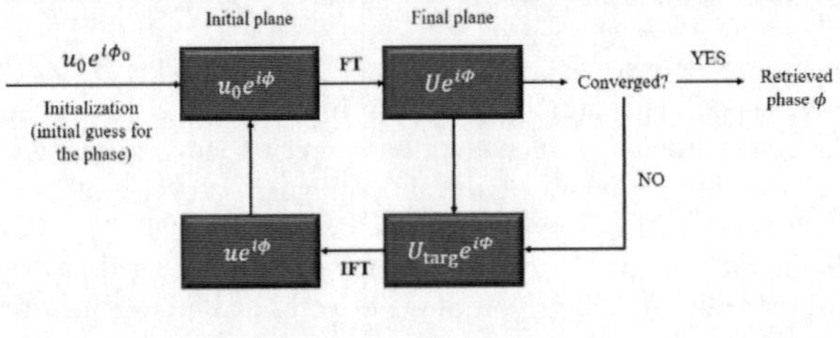

Fig. 3. Flow diagram of the principle of the Gerchberg-Saxton algorithm.

Such method was implemented on Python. The initial single Gaussian beam amplitude and the target seven-beam amplitude were defined as constraints in the algorithm. After convergence, phase patterns were obtained; adding such phase shifts to an incident Gaussian amplitude and propagating it for 10 cm yields amplitude distributions similar to Fig.4.

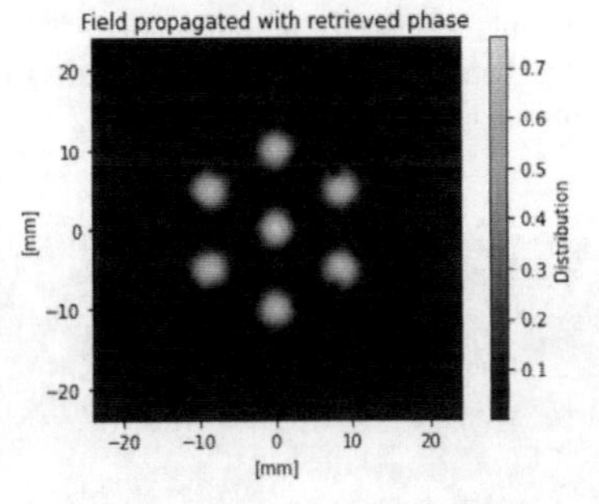

Fig. 4. Seven-beam distribution obtained by adding the position-dependent phase shift to the single Gaussian beam distribution and propagating it for 10 cm (1.627 THz).

The simulated efficiency, i.e. the ratio between the power inside the seven beams and the overall power, for the 1.627 THz and the 4.7 THz reflectors, is 62.6 % and 71% respectively.

To translate phase information into a surface profile, we can apply the ray optics relation $d = \lambda\Delta\phi/(4\pi\cos\theta)$, where d is the reflector groove depth, $\Delta\phi$ is the phase modulation, λ is the working wavelength and θ is the angle of beam incidence. The latter must be considered so that elements in the experimental setup do not block the beam path and the laser emitting source and the detector don't overlap. 3D plots of the reflectors are shown in Fig.5, while Table I summarizes the design parameters.

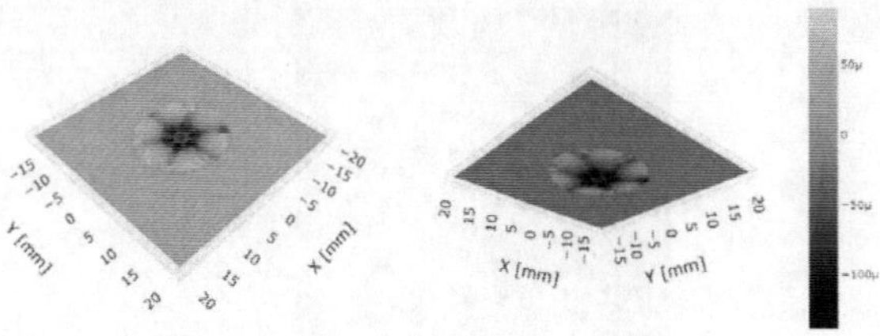

Fig. 5. Seven-beam distribution obtained by adding the position-dependent phase shift to the single Gaussian beam distribution and propagating it for 10 cm (1.627 THz). The color bar shows the scale in the z-direction (in μm)

TABLE I – PARAMETERS FOR THE REFLECTORS

Parameter	1.627 THz	4.7 THz
Reflector dimensions	Δx = 19.16 mm Δy = 18 mm	Δx = 19.16 mm Δy = 18 mm
Reflector-scanning plane distance	10 cm	10 cm
Angle of beam incidence (with respect to the normal to the reflector)	20°	20°
Design beam waist (radius)	4 mm	5 mm
Minimum Radius of Curvature (MRC)	1.23 mm	0.55 mm
Coordinates of MRC	x = -6.96 mm y = 6.04 mm	x = 4.67 mm y = 7.75 mm

V. FABRICATION

Prototypes for both 1.627 THz (Fig.6) and 4.7 THz frequencies were fabricated at the Fraunhofer Institute for Applied Optics and Precision Engineering IOF in Jena, Germany.

The reflectors, made from Aluminum RSA6061 with dimensions of 48 mm x 15 mm, were produced using diamond machining in fast tool servo mode. Surface smoothness was enhanced by removing outlier points; according to simulations, the seven-beam pattern would not be visibly affected. The 1.627 THz prototype had significant surface smoothing, which may affect performance. Three additional prototypes for the 1.627 THz surface were created at the SRON Netherlands Institute for Space Research in Groningen, using CNC machining with increasingly finer tools. The 1.627 THz prototype manufactured

NOTES:

in Germany has a smoother surface, while the latter exhibits a horizontal line pattern and sharper features due to the CNC process. The reflector with a lower surface roughness would suggest better experimental performance.

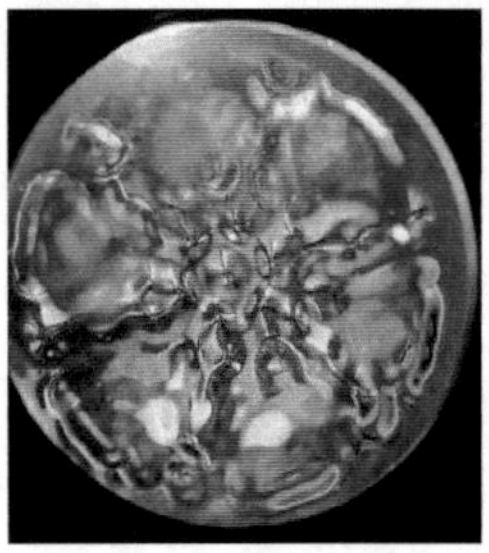

Fig. 6. Optical micrography of the prototype at 1.627 THz manufactured via diamond machining at Fraunhofer IOF (Jena, Germany).

VI. EXPERIMENTAL RESULTS

Characterization of the reflector at 1.627 THz was conducted at SRON Netherlands Institute for Space Research in Groningen, where a Far-Infrared (FIR) gas laser at such frequency was available. A scheme of the experimental setup is shown in Fig.7.

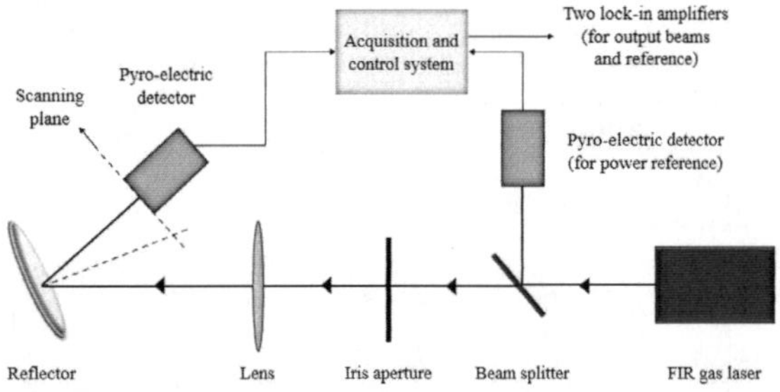

Fig. 7. Scheme of the experimental setup for characterization of the reflector at 1.627 THz. The laser beam is emitted by a Far-Infrared gas laser source at 1.627 THz; through a 6 μm beam splitter, a fraction of the power is sent to a pyro-electric detector for reference. To obtain a Gaussian profile, the transmitted beam is adjusted by an iris aperture; through a lens, we get the proper beam waist size (4 mm) at the location of the reflector. The reflected field intensity is measured through another pyro-electric detector mounted on a xy stage. The data is sent to an acquisition and control system. Two lock-in amplifiers are needed for simultaneous measurement of the output beams and the reference beam.

The laser beam is emitted by a Far-Infrared gas laser source at 1.627 THz; through a 6 μm beam splitter, a fraction of the power is sent to a pyro-electric detector for reference. To obtain a Gaussian profile, the transmitted beam is adjusted by an iris aperture; through a lens, we get the proper beam waist size (4 mm) at the location of the reflector. The reflected field intensity is measured through another pyro-electric detector mounted on a xy stage. The latter is placed inside a metal box, which includes an opening for the light. Such opening (diameter: 2 mm), together with the step size of the scanning stage, will determine the resolution in the 2D scan. The data is sent to an acquisition and control system. Two lock-in amplifiers are needed for simultaneous measurement of the output beams and the reference beam.

First, the reflector fabricated at Fraunhofer IOF (Prototype I) was tested. The initial scans at a 10 cm distance from the reflector reveal a clear pattern with seven beams, showing a central beam with significantly higher power (39.5%) compared to the others (ranging from 8.7% to 11.7%). The horizontal line pattern visible from the scans is caused by the offset between the position of the detector and the moment of acquisition from the measurement system.

Following adjustments aimed at correcting power imbalances among the upper and lower parts of the hexagonal ring; a subsequent scan shows improved power distribution in favor of the lower-right beams. This highlights the impact of alignment on the power distribution of the beams. Increasing the incident beam waist, we aimed to mitigate the power discrepancy between the central and outer beams. To achieve this, we replaced the lens to achieve a larger beam waist of approximately 5.5 mm (compared to a design beam waist of 4 mm). The dominance of the central beam persisted, but an overall balanced distribution among the six beams in the hexagon was obtained (Fig.8).

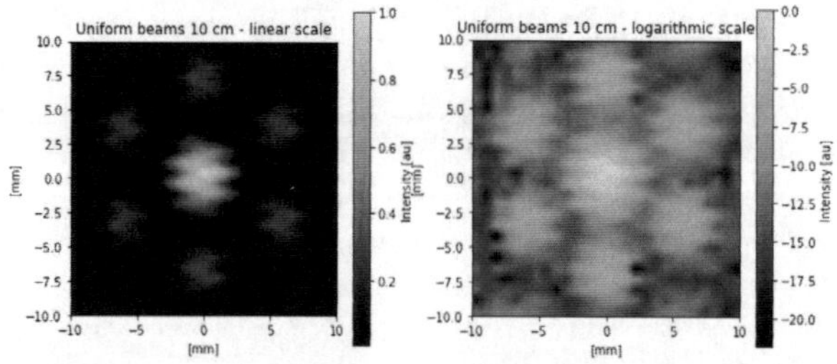

Fig. 8. Scan of the output pattern from Prototype I after adjusting the optical setup and increasing the beam waist of the incident beam (reflector-scanning plane distance: 10 cm, step size: 0.25 mm) in linear (left) and logarithmic (right) scale.

We further performed scans of the output beams at a greater distance (15 cm); although a hexagonal pattern is still discernible, the individual beam shapes do not resemble a Gaussian profile; this emphasizes the sensitivity of the performance to distance parameters.

NOTES:

The next phase involved testing Prototype II, the reflector manufactured at SRON. The experimental setup remained unchanged from the previous measurements, with an input beam waist of approximately 5.5 mm. A distinct 7-beam pattern is achieved through this reflector, with improved power uniformity among the six beams in the ring compared to Prototype I (Fig.9). Prototype II also exhibits a higher concentration of power within the hexagonal area.

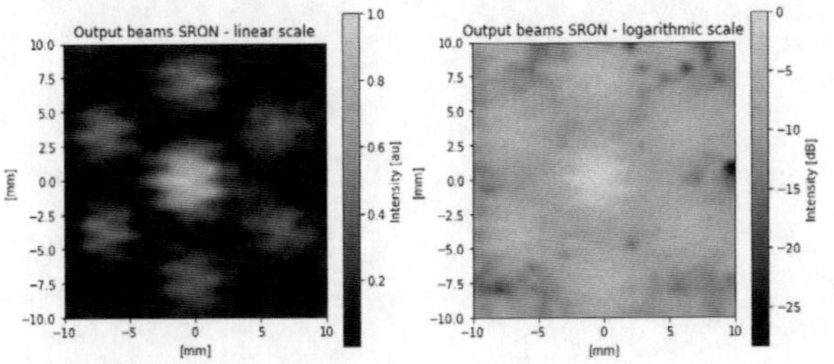

Fig. 9. Scan of the output pattern obtained through Prototype II (reflector-scanning plane distance: 10 cm, step size: 0.25 mm) in linear (left) and logarithmic (right) scale.

To determine the efficiency, the total power of the beam reflected by a flat mirror was initially scanned at a 20° angle of incidence before placing the reflector. Due to the instability of the laser source, normalization of scans was crucial. This involved adjusting each data point by the corresponding value from a reference pyro-electric detector, normalized to its maximum intensity to mitigate laser fluctuations. Support functions were defined to isolate areas for efficiency calculations. For the single Gaussian beam, efficiency was evaluated within a circle four times larger than its beam waist; for the seven-beam pattern, comparisons were made using either a seven-circle support or a single circle enclosing the entire hexagon. Table II reports efficiency estimations, where values are corrected for the reference power.

TABLE II – EFFICIENCY CALCULATIONS

Reflector Prototype (1.627 THz)	Seven-circle support	Single circle support
Prototype I	41.2%	42.4%
Prototype II	60.2%	62.9%

The analysis focuses on two main results: the disparity in power distribution between central and outer beams, and the different performances from Prototype I and Prototype II.

The simulations were run keeping into account only the phase-modulation effect. The idea was that, after reflection, each point in the field has some constructive or destructive interference with the others; this is made in such a way that, at 10 cm, the combination of constructive and destructive interference among points has maximum amplitude in the seven beams, minimum elsewhere. However, if only a fraction of the input beam undergoes this process, while the rest is reflected as by a simple mirror, we would expect the central part of the output pattern to be more intense, as it corresponds to the peak of the Gaussian beam. In other words, the resulting pattern could be a superposition of a single Gaussian beam and the desired pattern obtained through phase manipulation. That could justify the enhanced power of the central beam. The partial phase modulation might be caused by the accuracy of approximations made in the design.

Second, discrepancies in reflector performance between the prototypes may arise from the fabrication process. While simulations only considered phase modulation effects, practical reflector implementations can deviate due to fabrication imperfections. Variations in groove depth across the reflector's surface might influence phase modulation efficiency and affect the output beam distribution. It turns out that surface corrections and interpolation between points during the fabrication process can have an impact on reflector performance.

CONCLUSIONS

We have investigated a novel method for multiplexing a single Gaussian beam into seven beams using a reflector that introduces a position-dependent phase shift. This phase modulation, determined through the Gerchberg-Saxton algorithm, ensures the desired output pattern is achieved at a specific distance. Two reflector prototypes were fabricated for 1.627 THz and 4.7 THz. Prototype I, manufactured via diamond machining, exhibited higher surface smoothness, while Prototype II, produced using CNC machining, maintained design fidelity despite a visibly less polished finish. Experimental testing at 1.627 THz revealed an imbalance in power distribution between the central and outer beams, potentially due to incomplete phase modulation of the field distribution. Adjustments to the optical setup alignment and input beam waist improved power uniformity among the hexagonally distributed beams. The efficiency of the two reflectors was approximately 42% for Prototype I and 61% for Prototype II, compared to the simulated expectation of 67.6%. It appears that point smoothing can negatively impact efficiency.

To further improve the multiplexer for LO applications, we propose the following research activities: a) Increasing phase

NOTES:

modulation efficiency, potentially by enhancing the angle of incidence, to minimize the imbalance between central and outer beams; b) Implementing rigorous diffraction treatments for higher angles of incidence to refine phase retrieval algorithms, as discussed in [10]; c) Adding phase constraints to ensure Gaussian phase profiles of the seven output beams, which is crucial for optimal coupling to the HEB mixer array; d) Characterizing and further investigating the feasibility of our method with the already manufactured 4.7 THz reflector.

Acknowledgements: The authors would like to thank Ralf Steinkopf and Andreas Gebhardt at Fraunhofer Institute for Applied Optics and Precision Engineering IOF, Jena, Germany for fabricating the reflectors (Prototype I), Erik van der Meer at SRON-Groningen for fabricating the reflectors (Prototype II), and Yuner Gan, Behnam Mirzaei, Wouter Laauwen and Willem Jan Vreeling for sharing their knowledge and very useful discussions. TU Delft Space Institute is acknowledged for the support.

REFERENCES

[1] Pawar, Ashish Yashwantrao et al. "Terahertz technology and its applications." *Drug Invention Today* 5 (2013): 157-163.
[2] C. Walker, "Terahertz Astronomy", *CRC Press* (2015)
[3] Chin, Gordon et al. "Single Aperture Large Telescope for Universe Studies (SALTUS): Science Overview." (2024).
[4] J. R. Silva et al., "High-resolution receiver (HiRX) for the Single Aperture Large Telescope for Universe Studies (SALTUS)"
[5] Graf, U.U., & Heyminck, S. (2001). "Fourier gratings as submillimeter beam splitters". *IEEE Transactions on Antennas and Propagation, 49*, 542-546.
[6] Mirzaei, B., Silva, J.R., Hayton, D.J., Groppi, C., Kao, T., Hu, Q., Reno, J.L., & Gao, J.R. (2017). "8-beam local oscillator array at 4.7 THz generated by a phase grating and a quantum cascade laser". *Optics express, 25 24*, 29587-29596.
[7] Graf, U.U. (2018). "Enhanced diffraction efficiency of two-dimensional phase gratings". Optics express, 26 25, 32739-32742.
[8] Taylor, L.S. (1981). The phase retrieval problem. *IEEE Transactions on Antennas and Propagation, 29*, 386-391.
[9] Zalevsky, Z., Mendlovic, D., & Dorsch, R.G. (1996). "Gerchberg-Saxton algorithm applied in the fractional Fourier or the Fresnel domain". *Optics letters, 21 12*, 842-4.
[10] Matsushima, K., Schimmel, H., & Wyrowski, F. (2003). Fast calculation method for optical diffraction on tilted planes by use of the angular spectrum of plane waves. *Journal of the Optical Society of America. A, Optics, image science, and vision, 20 9*, 1755-62.

NOTES:

S-parameter Measurements of ALMA Band 2 Orthomode Transducer using Cryogenic System at Room Temperature

Sho Masui[1*], Ryo Sakai[1], Takafumi Kojima[1,2], and Keiko Kaneko[1]

Abstract— **This paper describes a verification of our cryogenic measurement system of insertion and return losses for the ALMA Band 2 orthomode transducer (OMT). This system consists of a vector network analyzer, two frequency extenders in 67-116 GHz band, vacuum feedthroughs (FTH), and thermal insulation waveguides (TIWG). In this cryogenic system, the waveguides and their interfaces including the FTHs and TIWGs significantly affect the accuracy of the *S*-parameter measurement. Thus, we have attempted to perform *S*-parameter calibration of the measurement system at the output port of TIWGs. When measuring the *S*-parameters under cooling with this system, long time interval is required between each calibration and measurements, which causes the drift of the extender characteristics to degrade the measurement accuracy. Therefore, in this study, we adopted a method to perform the *S*-parameter calibration at the input/output ports of the frequency extenders in every calibration and measurement step. To verify our calibration method, we compared two different *S*-parameter measurements of the OMT at room temperature between the direct method and our method performed at reference planes inside the cryostat. The result confirmed that our measurements were consistent except on the low-frequency side.**

Keywords— **ALMA, cryogenic, measurement system, orthomode transducer, S-parameter, waveguide, W-band.**

I. Introduction

The development of Atacama Large Millimeter/submillimeter Array (ALMA) Band 2 receiver [1] has been led by European Southern Observatory (ESO) in collaboration with several countries under contract/agreement. The radio frequency (RF) range covers 67–116 GHz band, which is challenging development for RF components in terms of wide bandwidth. In this development, National Astronomical Observatory of Japan (NAOJ) has contributed to the development, production, and testing of wideband optical components including the waveguide orthomode transducer (OMT) [2]. The OMT should be very low insertion loss because of the design and its operating temperature. However, it is essential to understand their characteristics at cryogenic temperatures because the losses directly affect the receiver performance. Nevertheless, theoretical conductivity and surface roughness model have been normally applied to estimate component losses without verifications [3]. Recently, measurements of waveguide attenuation constant at cryogenic temperature were reported in [4], which improve the estimation accuracy for insertion loss of waveguide components under cooling. However, this system does not allow to directly measure frequency dependence of *S*-parameters of waveguide components. In this research, aiming to measure the insertion and return losses of ALMA Band 2 OMT at cryogenic temperature, we established a cryogenic measurement system in 67–116 GHz band.

II. Measurement System

We established a cryogenic measurement system as shown in Fig. 1. This system consists of a measurement instrument at room temperature with a vector network analyzer (VNA) and two frequency extenders in 67-116 GHz band, and a cryogenic system with vacuum feedthroughs (FTH) and thermal insulation waveguides (TIWG). Since the insertion loss of the OMT is approximately 0.2–0.3 dB for both polarizations at room temperature, a high-accuracy cryogenic measurement

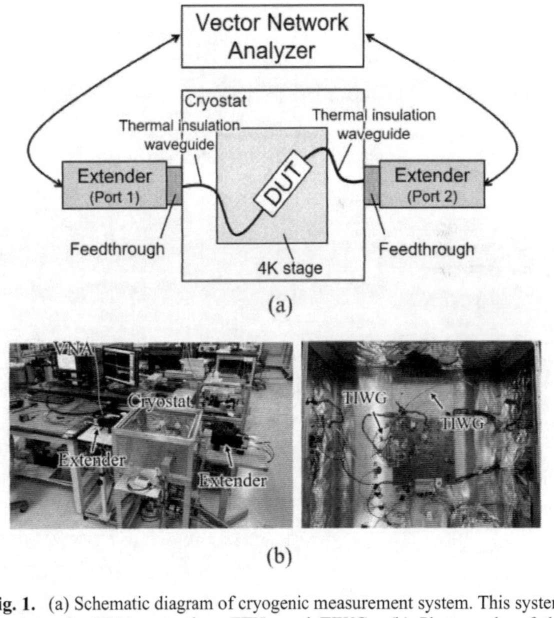

Fig. 1. (a) Schematic diagram of cryogenic measurement system. This system consists of a VNA, extenders, FTHs, and TIWGs. (b) Photographs of the measurement system for (left) overview and (right) inside.

[1]Advanced Technology Center, National Astronomical Observatory of Japan (NAOJ), Mitaka, Tokyo 181-8588, Japan, [2]Graduate University of Advanced Studies (Sokendai), Mitaka, Tokyo 181-8588, Japan; [*]Corresponding author (sho.masui@nao.ac.jp).

NOTES:

system should be necessary. To realize high-accuracy measurements, we performed Thru-Reflect-Line (TRL) calibration [5] at the output port of TIWGs, thereby de-embedding the error terms of FTHs and TIWGs from measurement results. However, TRL calibration requires at least three vacuum cooling cycles, thus it is necessary to have repeatability of transmission characteristics in cryogenic systems. In addition, the characteristics of the extenders are not stable in long time, thus TRL calibration cannot be properly performed in this measurement system, which takes 12 hours for one cooling cycle. To avoid above extender's drift, we calibrated at output ports of the extenders before all calibrations and measurements. This calibration method makes it possible to eliminate extender's drift.

III. *S*-PARAMETER MEASUREMENT AT ROOM TEMPERATURE

To verify that our calibration method is worked properly, we compared the results with our method and the results measured by directly connecting the OMT to the extender. The measured insertion losses were shown in Fig. 2. In these results, we have

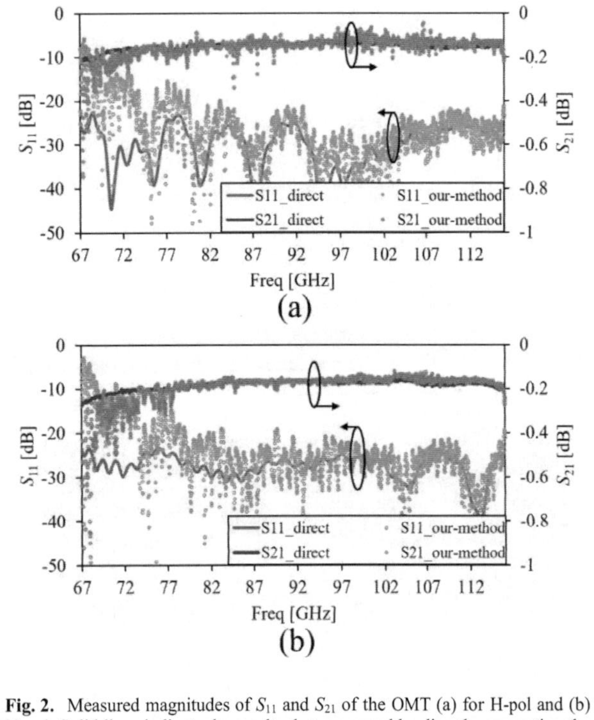

Fig. 2. Measured magnitudes of S_{11} and S_{21} of the OMT (a) for H-pol and (b) V-pol. Solid lines indicate the results that measured by directly connecting the OMT, and open circles indicate those that measured the OMT with our method.

de-embedded the error terms of FTHs and TIWGs and compensated the insertion loss of the waveguide transition that is necessary for the OMT measurements from the measured results. For the magnitude of S_{11}, these results are compatible at the higher frequency. However, at lower frequency, there are large differences between our method and direct measurements. We suspect that the assumption, the $|S_{11}|$ is less than $|S_{11} - S_{21} \cdot S_{12}/S_{22}|$, does not work well due to the high insertion and return losses of the error terms at lower frequency and gives rise to the limitation of our calibration method and this measurement system. On the other hands, the magnitude of S_{21} with our method are consistent with that of direct measurements except for the lower frequency. Based on this result, the insertion loss above 80 GHz could be measured with high accuracy.

IV. *S*-PARAMETER MEASUREMENT AT CRYOGENIC TEMPERATURE

The insertion and return losses of the ALMA Band 2 OMT were also measured at cryogenic temperature. These results have been included in an extended paper submitted to the IEEE Transactions on Terahertz Science and Technology. This paper notes a practical issue of the OMT that fabricated with split block techniques.

REFERENCES

[1] P. Yagoubov, et al., "Wideband 67–116 GHz receiver development for ALMA Band 2," A&A, 634, A46 (2020). https://doi.org/10.1051/0004-6361/201936777.

[2] A. Gonzalez and S. Asayama, "Double-Ridged Waveguide Orthomode Transducer (OMT) for the 67–116-GHz Band." J Infrared Milli Terahz Waves 39, 723–737 (2018). https://doi.org/10.1007/s10762-018-0503-5.

[3] S. Masui, et al., "Development of a new wideband heterodyne receiver system for the Osaka 1.85 m mm–submm telescope: Receiver development and the first light of simultaneous observations in 230 GHz and 345 GHz bands with an SIS-mixer with 4–21 GHz IF output," Publ. Astron. Soc. Japan, vol. 73, no. 4, Pages 1100–1115, Aug. 2021, doi: 10.1093/pasj/psab046

[4] J. D. Garrett and C. -Y. E. Tong, "Measuring Cryogenic Waveguide Loss in the Terahertz Regime," in IEEE Transactions on Terahertz Science and Technology, vol. 12, no. 3, pp. 293-299, May 2022, doi: 10.1109/TTHZ.2022.3155725.

[5] G. F. Engen and C. A. Hoer, "Thru-Reflect-Line: An Improved Technique for Calibrating the Dual Six-Port Automatic Network Analyzer," IEEE MTT, vol. 27, no. 12, pp. 987-993, 1979, doi: 10.1109/TMTT.1979.1129778.

NOTES:

Improvement of the Polarization Performance of ALMA Band 9

Realini S.[1*], Hesper R.[1], Baryshev A.[1], Barkhof J.[1]

***Abstract*—In the framework of the project "Study Towards a Producible ALMA2030-Ready Band 9 CCA" funded by ESO, we present the results of the feasibility of improving the beam squint of ALMA Band 9 receivers by adjusting the mechanical alignment of the optical components on an existing Cold Cartridge Assembly.**

***Keywords*— ALMA Band 9, Beam squint, Polarization, Submillimeter astronomy, Grid, Optical analysis.**

I. Introduction

The ALMA Band 9 (600–720 GHz) receivers are dual channel heterodyne systems capable of detecting orthogonally polarized signals using a grid.

The cross–polar performance of the existing receivers is ~-18 dB and it does not meet the requirements specified for the ALMA channels, i.e. a cross-polar level lower than -23 dB. Moreover, they show a relatively large beam squint, which makes this channel not suitable for extended-source polarimetry. However, polarization in Band 9 is essential to better understand the physics in planet-forming disks and to better constrain the properties of the magnetic fields in very dense molecular environments around evolved stars and high mass star-forming regions [1-2].

In a previous study, we found that the limitation of the current polarimetric performance can be explained by the deviation of the grid mounting angle for the beam squint and by the presence of the grid in combination with mirrors for the cross-polar level [3-4]. For these reasons, one of the goals of our project "Study Towards a Producible ALMA2030-Ready Band 9 CCA" [5] is to reduce the worst-case beam squint by an order of magnitude by shimming the grid, based on measured beam squint data. If this is found sufficient for polarimetric observation, it would be the most cost efficient way to address the issue and it would also avoid the higher loss associated with the use of an orthomode transducer to split the two orthogonal polarizations.

II. Measurement Setup

We constructed a rotating test source to obtain high-quality dual-polarization near-field beam patterns (amplitude and phase). From the near-field scans, we can derive far-field beam patterns, co-alignment of the two orthogonal beams on the sky (beam squint) and the common aperture, polarization and focus efficiencies.

The test setup incorporate several measures and corrections to obtain reliable data:

- phase correction due to electronic, thermal and mechanical drifts during scans based on a center-field phase reference sample before each scan line;
- standing wave compensation by scanning two planes spaced at $\lambda/4$;
- spurious reflections minimization by careful baffling;
- correction for probe misalignment by measuring one polarization at 0° and 180°; the probe offset for the 90° measurement can then be inferred;
- active cable phase correction by measuring the round-trip phase delay of an out-of-band pilot signal (plus phase-stable cables suspended with low-strain).

Fig. 1. Rotating test source at 0° (left) and 90° (right).

III. Data Reduction Pipeline

We developed a data analysis pipeline to calculate the beam squint from the measured complex field. We define the beam squint as the distance of the beam pointing relative to the Full-Width Half-Maximum (FWHM) on-sky. This is equivalent to the distance of the foci in the focal plane relative to the FWHM in focal plane. The steps of the analysis are:

1. compute the far field beam pattern from the complex fields measured;
2. consider field truncation at the secondary mirror and amplitude/phase distribution (asymmetry);
3. subtract a spherical phase front with center (x, y, z) from the far field phase over the extent of the secondary mirror;
4. calculate the overlapping integral between the resulting far field and a top hat function with the size of secondary mirror to evaluate aperture efficiency;
5. repeat the process for different (x, y, z) values until a local maximum of the aperture efficiency is found to determine the nominal telescope focus position;
6. find the focus for the two polarizations individually with a gaussian fit.

NOTES:

IV. Beam Squint Analysis

We measured the beam pattern at 12 frequencies from 620 to 708 GHz at 0°, 90° and 180°. The signal to noise ratio of the measurements is ~−72 dB. The value of the beam squint calculated from the measurements is shown in Fig. 2, where is compared with the measurements performed on the same CCA in the same way as during Band 9 production phase. Although there are some significant differences, the overall shape and magnitude of the deviations are comparable.

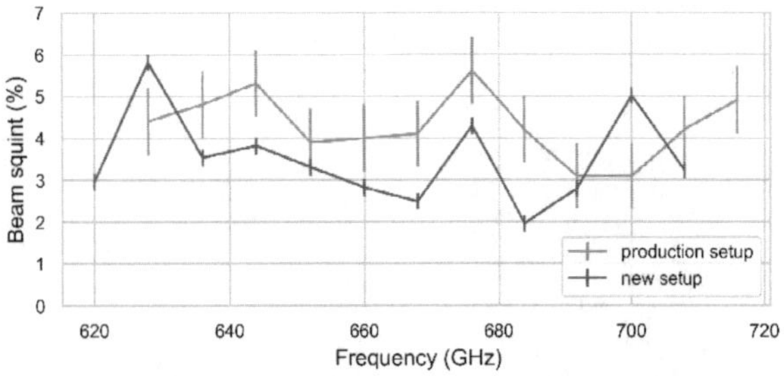

Fig. 2. Beam squint measured and analyzed in the same way as during Band 9 production phase (green) and with the new setup and data reduction (blue).

We notice that the data show rather strong frequency dependence, which suggests that a single correction of the grid is not an option for beam squint mitigation. This effect can be explained by the accuracy of the machined parts, especially the shape of the mirror surfaces. This was verified by means of optical simulation performed with GRASP.

V. Results of Mechanical Alignment

Despite the apparent infeasibility of grid adjustment, we still performed a test of the grid adjustment procedure to perform a proof of concept. We modified the original grid holder brackets with small adjustment screws and removed the original reference surfaces. With this arrangement, the grid can be adjusted over three degrees of freedom: azimuth (angle around vertical axis), inclination (angle around horizontal axis) and offset normal to grid surface.

For the initial reference measurement, the grid was set close to its nominal position as before the modification of the grid holder. After the first run, the beam squint was determined as described in section III. The results are shown in Fig. 3, blue points, where we observed again a large spread over frequency. We decided to correct the squint halfway in the inclination direction. The azimuth angle, and therefore the normal offset, were virtually unchanged. After this adjustment, the beam squint measurements were repeated (red points). For both point clouds the average position and standard deviation were determined (crosses and circles in the figure, respectively). The magnitudes of the beam squints, before and after adjustment,

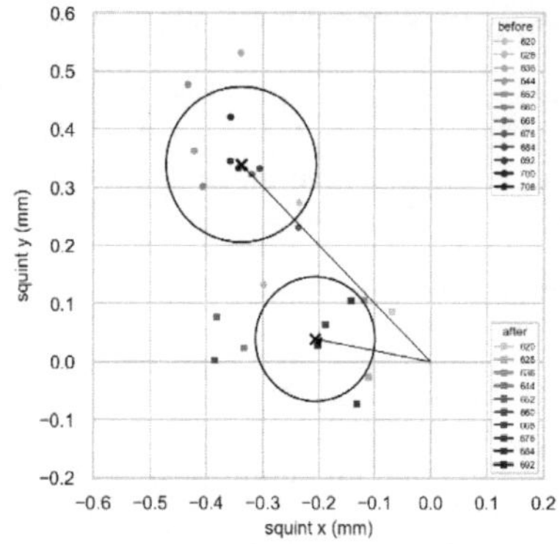

Fig. 3. Beam squint results in the x-y plane before (blue) and after (red) adjusting the grid angle by about half the desired value.

are plotted in Fig. 4 as function of frequency. The results show that the magnitude of the correction is consistent with the amount of correction applied.

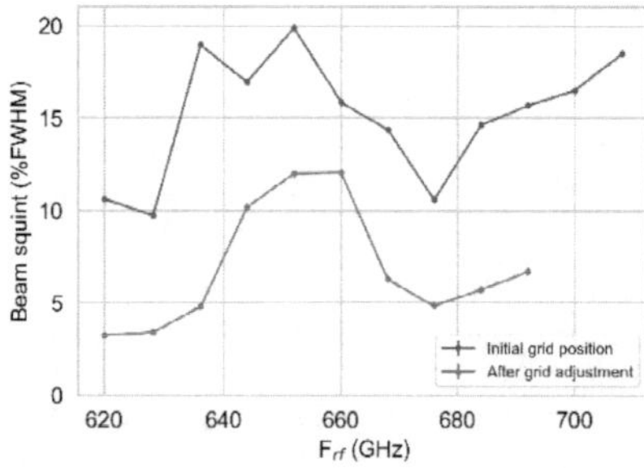

Fig. 4. Magnitudes of the beam squints, before and after adjustment of the grid, as function of frequency.

VI. Conclusion

We conclude that the beam squint can be minimized by grid angles and offset shimming with the limitation that the effect on the pointing angle to the ALMA secondary mirror should be minimized at the same time. Nevertheless, the large frequency dependence of the beam squint probably makes it impossible to do over the full band with the current optics

References

[1] Guillet, et al., "Polarized emission by aligned grains in the Mie regime: Application to protoplanetary disks observed by ALMA," A&A, 2020.

[2] Hirota, et al., "ALMA observation of the 658 GHz vibrationally excited H2O maser in Orion KL source I," The Astrophysical Journal, 2016.

[3] Hesper, et al., "ALMA Band 9 sideband separating upgrade - study report," ALMA document FEND- 40.02.09.00-1974-C-REP, 2023.

[4] Realini, et al., "Further optical analysis of the ALMA Band 9 front end for a possible upgrade", SPIE Proc., 2022.

[5] De Breuck, "Statement of work for *Study Towards a Producible ALMA2030-Ready Band 9 CCA*," 2023-10-12, document nr. ESO-526058.

NOTES:

A 200 GHz Fully Integrated Quasi-Optical Detector Using Orthogonal Heterostructure Backward Diodes with Improved Performance

Yu Shi[1*], Yijing Deng[1], Peizhao Li[1], Patrick Fay[1], and Lei Liu[1]

***Abstract*— In this paper, we report the updated results of a 200 GHz fully integrated quasi-optical detector based on orthogonal heterostructure backward diodes. The newly fabricated devices and detector circuit have been characterized through on-wafer dc and RF tests. An improved voltage responsivity as high as 500 V/W has been projected for the integrated detector.**

***Keywords*— Backward diode, on-wafer measurement, terahertz detector.**

I. Introduction

TERAHERTZ (THz) detector has drawn increasing attentions for potential applications in radio astronomy, biomedical imaging, and remote detection. Most detectors can only sense the amplitude or the phase of the THz wave. However, the polarization of the THz wave can also provide important information about the object under test, which accelerates the development of highly sensitive polarization resolved THz detector.

In our previous paper [1], a fully integrated dual polarization quasi-optical detector based on zero-bias heterostructure backward diodes (HBDs) was proposed and a reduced voltage responsivity was reported. The polarization imaging capability of such device has also been demonstrated [2]. In order to analyze the results and improve the detector performance, additional on-wafer test structures have been fabricated and characterized. An improved voltage responsivity as high as 500 V/W can be projected for the newly developed detector based on the on-wafer measurement results.

II. Device Characterization

The design and fabrication of the proposed detector has been described in [1]. In order to evaluate the performance of the detector, additional HBDs with a ground-signal-ground (GSG) probe pad have been fabricated along with the integrated detector and then characterized.

The devices were tested first through on-wafer dc measurement (see Fig. 1(a)), from which the IV curve can be obtained. The IV curve of the 1.5 μm × 1.5 μm HBD, as well as the corresponding curvature coefficient, are shown in Fig. 1(b). It can be seen that a curvature coefficient of -27 V^{-1} has been achieved for the HBD at zero bias. The curvature coefficient is higher than the results reported in [1], leading to an improved voltage responsivity.

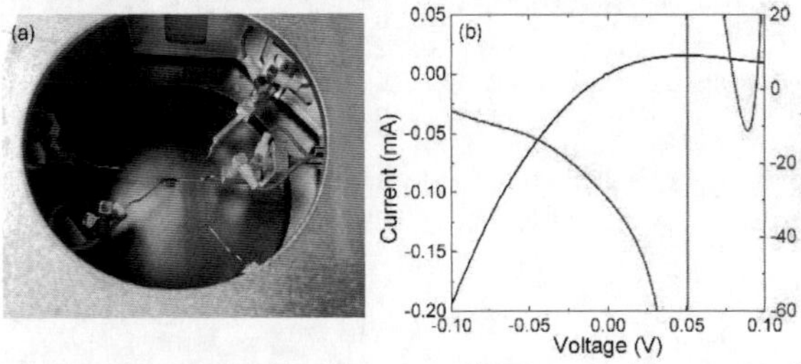

Fig. 1. (a) on-wafer dc measurement setup; (b) IV curve and the corresponding curvature coefficient of the 1.5 μm × 1.5 μm HBD.

For the small signal on-wafer RF measurements at 200 GHz, the same HBD with a GSG probe pad was characterized. The experiment setup is shown in Fig. 2(a). An Agilent E8361C vector network analyzer (VNA) with WR-5.1 band extenders (from OML Inc.) was used. An additional piece of AlN absorber was placed in between the chip and the chuck of the probe station to minimize the reflection.

In order to transfer the reference plane to the tip of the GSG probe, on-wafer calibration was first performed with the off-chip Line-Reflect-Reflect-Match (LRRM) standards. Additional on-chip Through-Reflect-Line (TRL) standards, which were fabricated along with the devices, were used to de-embed the GSG probe pad so that the intrinsic S-parameters of the HBD can be obtained. The intrinsic S_{11} from 140 GHz to 220 GHz is shown in Fig. 2(b).

The lumped element model of the device after de-embedding is shown in Fig. 3(a). In addition to the junction resistor (R_j), junction capacitor (C_j), and series resistor (R_s) from the intrinsic HBD, the model also includes a parasitic inductor L_p from the metal airbridge and the warped ground, and a parasitic capacitor C_p from the metal contacts. Their values can be obtained from

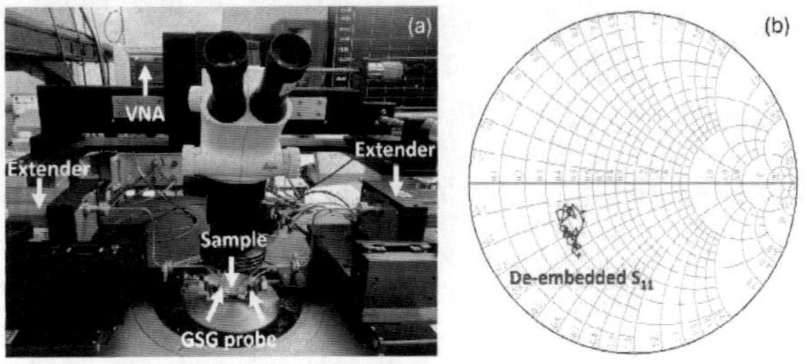

Fig. 2. (a) The experiment setup of the on-wafer small signal S parameter measurement; (b) The intrinsic S_{11} of the 1.5 μm × 1.5 μm HBD.

[1]Department of Electrical Engineering, University of Notre Dame, Notre Dame IN, 46556. *Corresponding author (email: yshi5@nd.edu).

NOTES:

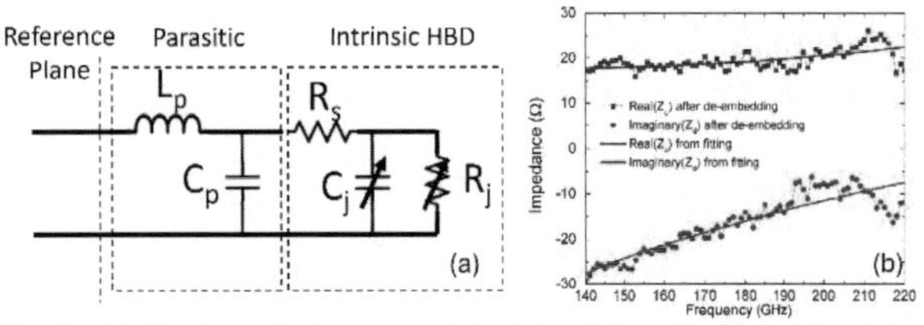

Fig. 3. (a) The lumped element model of the device; (b) The real and the imaginary part of the device impedance (Z_d) from measurement (dotted line) and from fitting (solid line).

TABLE 1. HBD Key Parameter Comparison

	This work	Ref [3]	Ref [4]	Ref [5]
Curvature (V^{-1})	-27	-47	-25	-39
Area (μm^2)	1.5 × 1.5	0.4 × 0.4	1 × 1	2 × 2
Barrier thickness (nm)	1.47	1.1	1	3.2
R_j ($\Omega\cdot\mu m^2$)	14.6×10^3	814	1045	55.6×10^3
C_j ($fF/\mu m^2$)	11.3	15	8.5	7.2
R_s ($\Omega\cdot cm^2$)	4.3×10^{-7}	2.0×10^{-7}	2.2×10^{-7}	2.2×10^{-7}
Cut-off frequency (GHz)	328	644	850	740

fitting the device impedance (Z_d) after de-embedding. The impedance from the lumped element model, as well as that from the measurement after de-embedding, are shown in Fig. 3(b). It can be seen that the results matched well with each other, showing the effectiveness of the lumped element model and the fitting process.

The extracted key device parameters of the HBD, as well as those from the previous work [3][4][5] are summarized in Table 1. It can be seen that the R_j increased with the thickness of the barrier, as expected. The R_s also became larger than the previous results, leading to a degraded cut-off frequency. The additional parasitic elements also lead to power reflection and decrease the detector responsivity since the matching condition is not satisfied.

III. CIRCUIT CHARACTERIZATION

In addition to the HBD, the annular slot antenna, as well as the orthogonal detector circuit, were also characterized through the same on-wafer S-parameter measurement. In order to obtain the intrinsic S-parameters, the same de-embedding process as mentioned earlier has been performed.

The S_{11} of the antenna after de-embedding are shown in Fig. 4(a). The S_{11} from the full wave simulation are also plotted for comparison. It can be seen that the measurement results have an offset of around 5 GHz (2.5%) compared with the simulation, which is due to the feeding signal line that was left after de-embedding the GSG probe pad.

The S-parameters of the entire detector circuit including the shorted stubs, the interdigitated dc block, and the low pass filter were also obtained. Both the measurement and the simulation results are shown in Fig. 4(b). It can be seen that the on-wafer

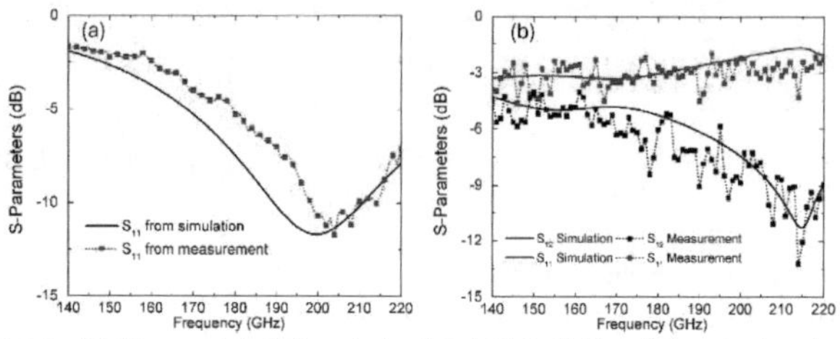

Fig. 4. (a) Measured (solid) and simulated (dotted) S_{11} of the annular slot antenna; (b) Measured (solid) and simulated (dotted) S_{11} and S_{12} of the detector circuit in one polarization.

measurement results are close to the simulation, with discrepancy likely from the noise during the measurement.

On the basis of the on-wafer dc and RF measurement results, the circuit model of the entire detector was extracted. The detector voltage responsivity can therefore be projected to be as high as 500 V/W, much higher than the results reported in [1]. This improvement in responsivity can also benefit the polarization imaging capability of the proposed integrated detector since the angular resolution of the imaging system can increase [2]. The performance could be further improved by reducing the insertion loss of the detector circuit, as well as eliminating the impedance mismatch that came from the parasitic elements.

IV. CONCLUSION

In this paper, the measurement results of the HBDs and detector circuits through on-wafer dc and RF tests were presented. An improved responsivity of the integrated detector can therefore be projected. The results show the promises of the integrated detector for THz polarization detection in remote sensing, biomedical imaging, and radio astronomy.

REFERENCES

[1] Y. Shi, Y. Deng, P. Li, P. Fay and L. Liu, "*A 200 GHz Fully Integrated, Polarization-Resolved Quasi-Optical Detector Using Zero-Bias Heterostructure Backward Diodes*," IEEE Microwave and Wireless Components Letters, vol. 32, no. 7, pp. 891-894, July 2022.
[2] Y. Shi, Y. Deng, P. Li, P. Fay and L. Liu, "*Polarization-Resolved THz Imaging with Orthogonal Heterostructure Backward Diode Detectors,*" in IEEE Transactions on Terahertz Science and Technology, vol. 13, no. 3, pp. 286-296, May 2023.
[3] Z. Zhang, R. Rajavel, P. Deelman, and P. Fay, "*Sub-Micron Area Heterojunction Backward Diode Millimeter-Wave Detectors with 0.18 pW/Hz1/2 Noise Equivalent Power*," IEEE Microw. Wireless Components Lett., vol. 21, no. 5, pp. 267-269, 2011.
[4] Z. Zhang, "*Sb-heterojunction backward diodes for direct detection and passive millimeter-wave imaging*," Ph.D. dissertation, Dept. Elect. Eng., Univ. Notre Dame, Notre Dame, IN, USA, 2011
[5] N. Su, R. Rajavel, P. Deelman, J. N. Schulman and P. Fay, "*Sb-Heterostructure Millimeter-Wave Detectors With Reduced Capacitance and Noise Equivalent Power*," in IEEE Electron Device Letters, vol. 29, no. 6, pp. 536-539, June 2008

NOTES:

An Initial Concept of A Resonance Phase Matched Junction-Loaded Travelling Wave Parametric Tripler

B-. K. Tan

***Abstract*—In this paper, we investigate the possibility of utilising a tunnel-junction loaded transmission line as high efficiency parametric frequency multiplier. Through the interaction between the injected primary tone and the nonlinear medium, higher harmonic tones can be generated through wave-mixing process. Here, we aim to maximise the third harmonic wave generation. We first establish a theoretical framework outlining the mechanism for generating the third harmonic component from a single pump wave propagating in a nonlinear transmission line. We begin by demonstrating that strong third harmonic generation is possible with the resonance phase matching technique, albeit with an extremely narrow operational bandwidth. To broaden the bandwidth, we modify the dispersion engineering element of our circuit and show that broadband operation is achievable, while preventing unwanted harmonic tone growth. We extend this calculation from the microwave to the millimetre and sub-millimetre regimes and demonstrate that by adjusting the parameters of the junctions and the dispersion engineering circuits, we can achieve high conversion efficiency close to 1 THz.**

***Index Terms*—Superconductor-Insulator-Superconductor (SIS), Heterodyne Mixers, Focal Plane Array, Circular Waveguide, Probe Antenna**

I. Introduction

GENERATION of high spectral purity tones at high frequencies with high output power and low noise properties is crucial for many applications, particularly for generating the local oscillator (LO) signal for astronomical millimetre (mm) and sub-millimetre (sub-mm) heterodyne receivers. Traditionally, this is achieved through a chain of frequency multipliers utilising semiconductor Schottky diodes [1], [2] in a waveguide structure in conjunction with a microwave source, for operation below 2 THz. However, these multipliers are often bulky, power-hungry, dissipate significant heat, and require complex optical arrangements for injection into the main detector mounted in the cryogenic stage, limiting their use in ultra-sensitive experiments. For higher frequency operation, a quantum cascade laser [3], [4] is generally preferred, albeit with limited tunability, lower power output, higher heat dissipation, and slightly inferior spectral purity. Other methods, such as beating two infrared lasers with photodiode [5], utilising plasmonic resonance of superconducting tunnel junctions [6], or using distributed superconductor-insulator-superconductor (SIS) mixers [7], all have their pro and cons, including limitations in terms of output power, spectral purity, added noise, or bandwidth.

In this context, we propose a new mechanism using the travelling-wave parametric amplifiers (TWPAs) technology

Department of Physics (Astrophysics), University of Oxford, Denys Wilkinson Building, Keble Road, OX1 3RH, Oxford, UK. E-mail: boonkok.tan@physics.ox.ac.uk

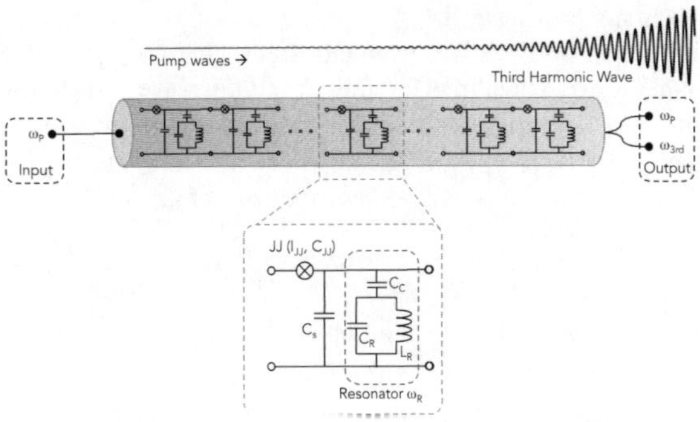

Fig. 1. Resonantly phase-matched travelling wave parametric tripler. Third harmonic photons are generated and amplified through a nonlinear interaction with the strong input pump wave as they propagate along the transmission line with a lattice period of $a = 10\,\mu$m. In each unit cell, a capacitively-shunt Josephson junction (the nonlinear inductors) is capacitively coupled to an LC resonators. The circuit parameters are $C_j = 280$ fF, $L = 100$ pH, $C_s = 50$ fF, $C_c = 10$ fF, $C_R = 7.036$ pF, $L_R = 100$ pH, $x = 2000a$, and $I_0 = 3.29\,\mu$A.

[8], [9] to generate mm and sub-mm wave signals. TWPAs offer the potential for low noise, stable wideband operation with minimal heat dissipation, and their compact design allows for integration with other circuit components. Mounting these high-frequency sources in-situ at the cryogenic stage near the main detector minimises noise leakage from room temperature sources. Additionally, integrating the compact high-frequency source with the main detector circuit enables a new operation regime, allowing the construction of a compact integrated heterodyne miniature receiver with an integrated LO source next to the mixer. This opens up the possibility of constructing an array with frequency tunability for each pixel, creating a multi-band heterodyne array, which was not feasible with conventional heterodyne receiver construction.

In this paper, we will first present the formulation of the framework necessary for predicting the generation of pump harmonics. Subsequently, we address the fundamental issues associated with employing existing TWPA technology for achieving broadband operation before introducing our proposed solution. We illustrate the solution using a practical microwave example, before extending the design to the mm/sub-mm regime, and demonstrate that by adjusting the parameters of the junctions and dispersion engineering circuits, we can theoretically achieve close to 1.5 µW of output power near 1 THz with a cascade of several parametric multipliers.

II. Theoretical Model

A meta-material transmission line loaded with a series of Josephson junctions has been successfully used as a travelling wave parametric amplifier [8], [10] to generate exponentially high gain before. This is achieved by adding additional resonance circuits [11], [12], [13] along the transmission line to ensure that the total dispersion is near zero within the operational band. In this operational regime, it has been shown that third harmonic generation from the strong pump wave is negligible due to the junction resonance [11], [14]. However, when using this junction-loaded transmission line as a frequency tripler, the third harmonic phase mismatch needs to be minimised instead.

Here, we follow the treatment from [11] to derive the coupled wave equations for the travelling wave parametric multiplier (TWPaM) in the tripler mode. The nonlinear wave equation for a Josephson junction loaded long transmission line depicted in Fig 1 can be described as [15]:

$$C_0\frac{\partial^2\phi}{\partial t^2}-\frac{a^2}{L}\frac{\partial^2\phi}{\partial x^2}-C_j a^2\frac{\partial^4\phi}{\partial x^2\partial t^2}=\frac{a^4}{2I_0^2L^3}\frac{\partial^2\phi}{\partial x^2}\left(\frac{\partial\phi}{\partial x}\right)^2 \tag{1}$$

Taking the ansatz that the solutions to be forward propagating waves of the form:

$$\phi=\frac{1}{2}[A_p(x)e^{i(k_px+\omega_pt)}+A_3(x)e^{i(k_3x+\omega_3t)}+\text{c.c.}] \tag{2}$$

where A_m is the slowly varying amplitude, k_m is the wave vector, ω_m is the angular frequency and $m=p,3$ representing the pump and third harmonic tones respectively. Here, we assume that all harmonics higher than the third will be suppressed by the plasma resonance frequency cutoff of the Josephson junctions. Substitute the above expression (Eq. 2) into the nonlinear wave equation (Eq. 1) with the following approximations:

1) neglect the second derivatives of the slowly varying amplitudes using the slowly varying envelope approximation: $\left|\frac{d^2A_m}{dx^2}\right|\ll\left|k_m\frac{dA_m}{dx}\right|$, and
2) neglect the first derivatives of the slowly varying amplitudes on the right side of the nonlinear wave equation: $\left|\frac{dA_m}{dx}\right|\ll|k_mA_m|$,

and defining the wave vector as $k_m=\frac{\omega_m\sqrt{C_0L}}{a\sqrt{1-C_jL\omega_m^2}}$, Eq. 1 can be simplified to:

$$\frac{-iC_0\omega_m^2}{k_m}\frac{\partial A_m(x)}{\partial x}e^{i(k_mx+\omega_mt)}=\frac{a^4}{2I_0^2L^3}\frac{\partial^2\phi}{\partial x^2}\left(\frac{\partial\phi}{\partial x}\right)^2 \tag{3}$$

Replacing Eq. 2 into Eq. 3, and separating out the terms that oscillate at the pumps and the third harmonic frequencies, we get the following coupled equations:

$$\frac{\partial A_p(x)}{\partial x}-\frac{ia^4k_p^3}{16C_0I_0^2L^3\omega_p^2}\left[k_p^2A_p(x)^2A_p^*(x)+k_3(2k_p-k_3)A_p^{*2}(x)A_3(x)e^{-i\Delta k_Lx}\right]=0 \tag{4a}$$

$$\frac{\partial A_3(x)}{\partial x}-i\frac{a^4k_p^2k_3}{16C_0I_0^2L^3\omega_3^2}\left[2k_3^2A_pA_p^*A_3-k_p^2A_p^3e^{i\Delta k_Lx}\right]=0 \tag{4b}$$

where $\Delta_{kL}=3k_p-k_3$. This can be further simplified to:

$$\frac{\partial A_p(x)}{\partial x}-i\alpha_{p1}A_p(x)^2A_p^*(x)-i\alpha_{p2}A_p^{*2}(x)A_3(x)e^{-i\Delta k_Lx}=0 \tag{5a}$$

$$\frac{\partial A_3(x)}{\partial x}-i\alpha_{31}A_pA_p^*A_3+i\alpha_{32}A_p^3e^{i\Delta k_Lx}=0 \tag{5b}$$

where the coupling coefficients are defined as:

$$\alpha_{p1}=\frac{a^4k_p^5}{16C_{0,p}\omega_p^2I_0^2L^3},\quad\alpha_{p2}=\frac{a^4k_p^3k_3(2k_p-k_3)}{16C_{0,p}\omega_p^2I_0^2L^3} \tag{6a}$$

$$\alpha_{31}=\frac{a^4k_p^2k_3^3}{8C_{0,3}\omega_3^2I_0^2L^3},\quad\alpha_{32}=\frac{a^4k_p^4k_3}{16C_{0,3}\omega_3^2I_0^2L^3} \tag{6b}$$

Solving the coupled amplitude equations by making the substitutions $A_p(x)=a_pe^{i\phi_px}$ and $A_3(x)=a_3e^{i\phi_3x}$, where $\phi_p=\alpha_{p1}a_pa_p^*$ and $\phi_3=\alpha_{31}a_pa_p^*$, we have:

$$\frac{\partial a_p}{\partial x}-i\alpha_{p2}a_p^{*2}a_3e^{-i\Delta_kx}=0 \tag{7a}$$

$$\frac{\partial a_3}{\partial x}-i\alpha_{32}a_p^3e^{i\Delta_kx}=0 \tag{7b}$$

where $\Delta_k=\Delta k_L+3\phi_p-\phi_3$. If we proceed with the assumption that the third harmonic wave is much weaker than the pump wave (the undepleted pump approximation) i.e., $A_p(x)=a_p(0)e^{i\alpha_px}$, $A_3=a_3e^{i\alpha_3x}$ and $\alpha_3=\frac{a^4k_p^4k_3^2}{8C_{0,3}\omega_3^2I_0^2L^3}a_p(0)a_p^*(0)$, the above coupled amplitude equations can be solved analytically, resulting in

$$a_3=\frac{(1-e^{i\Delta_k'x})\kappa}{\Delta_k'} \tag{8}$$

where

$$\Delta_k'=3k_p-k_3+3\alpha_p-\alpha_3 \tag{9a}$$

$$\alpha_p=\frac{a^4k_p^5}{16C_{0,p}\omega_p^2I_0^2L^3}a_p(0)a_p^*(0) \tag{9b}$$

$$\text{and}\quad\kappa=\frac{a^4k_p^4k_3^2}{16C_{0,3}\omega_3^2I_0^2L^3}a_p^3(0). \tag{9c}$$

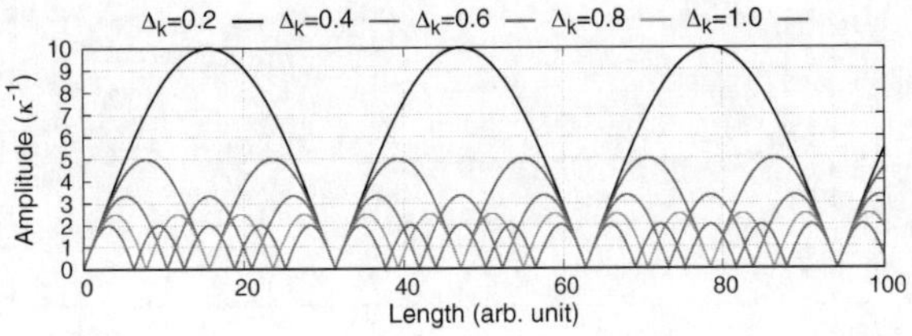

Fig. 2. Plot showing the behaviour of Eq. 8.

Eq. 8 shows that the amplitude of the third harmonic wave is periodically dependent on the length x, and the periodicity and the maximum amplitude is Δ_k dependent, as shown in Fig. 2. This means that if $\Delta_k \to 0$, then at short length, the third harmonic growth is almost linear-like. In this and the following examples, we use the circuit parameters depicted in Fig. 1. For a general case without the undepleted pump approximation, we solve Eq. 7 numerically using the Runge-Kutta method.

III. Travelling Wave Parametric Tripler

As shown in Eq. 8, the growth of the third harmonic wave depends heavily on Δ_k. Similar to the case of the TWPA, we can therefore utilise the resonance phase matching technique to engineer the TWPaM so that $\Delta_k \to 0$. In Fig. 3 (a), we show the dispersion curve of Δ_k plotted against the input pump frequency. One immediately noted that there is a strong resonance near 30 GHz. This is due to the natural resonance of the junction $f_j = 1/2\pi\sqrt{L_j C_j}$. Another weaker resonance appears at a third of this frequency, near 10 GHz, as the third harmonic's wave vector is three times higher than the pump's wave vector. Below the 10 GHz resonance, the two sharp resonance spikes were induced by the additional resonator circuits (one seen by the pump wave and the lower frequency resonance by the third harmonic), introduced to match the total phase Δ_k. The dispersion curve near this region is zoomed in Fig. 3 (b) for clarity purposes. It is clear now that near these resonance frequencies, the conversion efficiency from the pump to the third harmonic wave increase drastically, as shown in the lower panel of Fig. 3 (a).

This can be illustrated more clearly from Fig. 4. On the bottom panels, the pump frequency was set to be away from the resonance frequency at 5.994 GHz. As can be seen, only a small amount of energy is converted to the third harmonic. However, if the pump frequency is set closer to the resonance frequency at 5.99471 GHz i.e., the total phase difference is closed to zero, one can improve the power transfer from the pump to the third harmonics significantly. This is shown in the top panels, where one can see that near $I_p \approx 0.76 I_0$, the amplitude of the third harmonic wave is now higher than the pump's amplitude, meaning that more than half of the energy is now transferred to the third harmonic wave. Another important observation here is that as shown in the top right panel, the maximum power transfer is periodically oscillating along the length of the TWPaM, just as predicted via Eq. 8. Noted that these plots were made without the undepleted pump approximation depicted in Eq. 7.

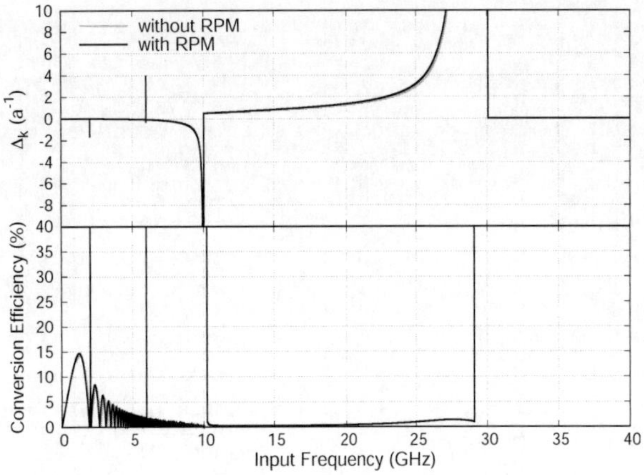

(a)

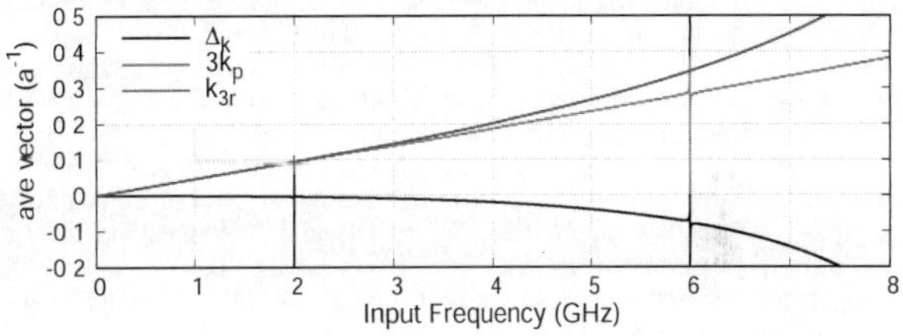

(b)

Fig. 3. (a) Top panel: Phase mismatched between the pump and the third harmonic. In this case, the junction capacitance was set to be 280 fF, L_J = 100 pH, giving a natural resonance at 30 GHz. This also creates a resonance at 10 GHz, a third of the natural resonance frequency, as the third harmonic wavelength is three times smaller than the pump. The two resonances created by the additional resonators are visible at 2 GHz and 6 GHz. This is zoomed in sub-plot (b) for clarity. Here, it is clearly seen that the 2 GHz resonance in the Δ_k curve was due to the wave vector k_{3rd}, while the intended 6 GHz resonance by the pump's wave vector k_p. Bottom panel: The conversion efficiency clearly shown that there are some conversion below 10 GHz, where at resonances, the corresponding Δ_k curve close to zero and the conversion efficiency improved up significantly. Above 30 GHz, Δ_k curve becomes purely imaginative, hence no conversion is possible. Note that these curve was plotted with the pump depletion assumption, but the underlying principle remains the same for the pump depletion case.

However, from Fig 3 (a), it is clear that the exponential growth of third harmonic wave only happens at a particular set of frequencies. In other words, with this configuration, the operational bandwidth is extremely narrow (note how the minute difference between the two frequencies used to plot Fig. 4 have on the conversion efficiency). In order to broaden the operational bandwidth, one can utilise a tunable planar resonance circuit (via changing the kinetic inductance of a superconducting line using bias current below the critical current) to compensate for the bandwidth performance, but this could add to the complexity of fabricating such devices.

IV. Broadband Microwave Third Harmonic Generator

Nevertheless, it is easy to show that this intrinsically narrow operational bandwidth is simply due to the fact that the pump needs to be very close to the resonance frequencies, which was set with very high quality factor in the case

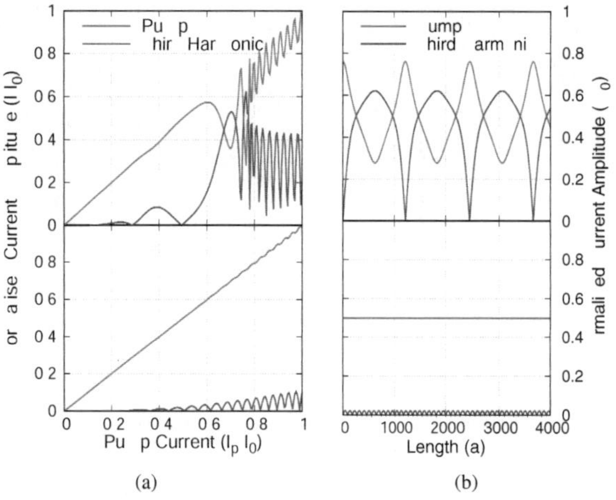

Fig. 4. (a) The normalised amplitude of the resulting pump and third harmonic wave relative to the critical current of the Josephson junction (I_0), in relation with the strength of input pump current (I_p). The bottom panel shows the case where the pump frequency is away from the resonance frequency at 5.994 GHz, while the top panel shows the case where the pump frequency is very close to the resonance at 5.99471 GHz (A similar case can be found near the 2 GHz resonance as well, but apart from that, conversion efficiency is negligible, as is clearly shown in Fig 3.). (b) The normalised amplitude of the resulting pump and third harmonic wave along the transmission line. Bottom panel shows the 5.994 GHz case, where $I_p/I_0 = 0.5$, while top panel shows the resonance case with $I_p/I_0 = 0.76$.

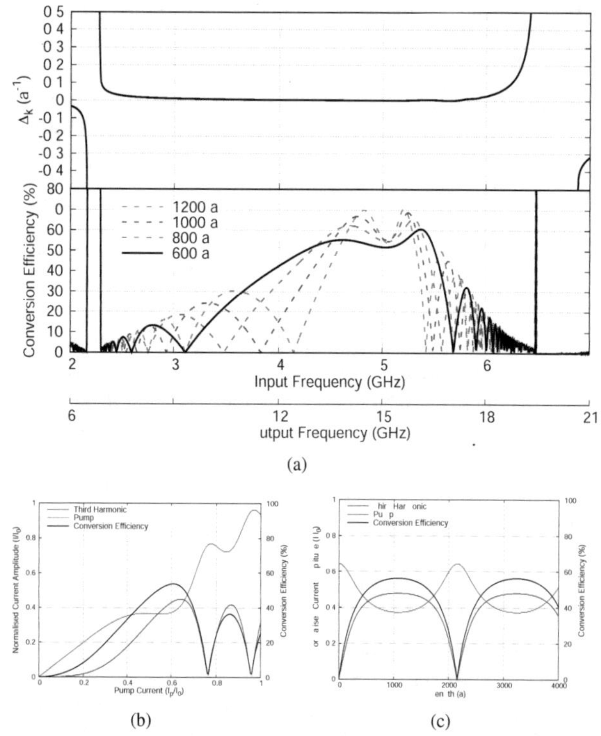

Fig. 5. (a) Top panel: Phase mismatched between the pump and the third harmonic. In this case, the resonator parameters were set to be $C_R = 4.6$ fF and $L_R = 41.3$ nH, giving resonances around 2.2 GHz and 6.6 GHz. Here, it is clearly seen that by lowing the Q-factor of the resonator circuit, we managed flatten the Δ_k curve approaching zero in between the two resonance frequencies. Bottom panel: This broaden the operational bandwidth of the third harmonic generator, with higher than 40% conversion efficiency from about 4–5.5 GHz. The asymmetry in the conversion curve was due to the the Q-factor of the 2.2 GHz resonance is naturally higher than the 6.6 GHz resonance. These plots were made by using $x = 600a$ and $I_p/I_0 = 0.645$. (b) The normalised amplitude of the resulting pump and third harmonic wave relative to the critical current of the Josephson junction (I_0), in relation with the strength of input pump current (I_p) with $f_p = 5$ GHz. (c) The normalised amplitude of the resulting pump and third harmonic wave along the transmission line.

shown above. Therefore, in order to broaden the operational bandwidth, we can reduce the Q-factor of the resonator. As described earlier, due to the third harmonic wave generated along the transmission line, the resonator will induced a lower resonance at a third of the natural resonator frequency but flip in polarity against frequency. This in effect altered the total Δ_k value between the two resonances. If the Q-factor value is low enough, the two resonance curves overlapped at larger intermittence frequencies, and there would be a region where Δ_k is very close to zero. This is shown in the Fig. 5 (a). It is worthwhile noting that the phase relation Δ_k in this case is also rely heavily on the natural resonance of the Josephson junctions. It is therefore preferable that the junction resonance is at a higher frequency, so that the Δ_k curve is flatter near the operational frequencies region, and hence it's easier to make $\Delta_k \to 0$ at a broader bandwidth.

In the bottom panel of Fig. 5 (a), we show the conversion efficiency curve of a TWPaM using the low-Q resonators to compensate for the phase difference. As can be seen, the operational bandwidth is now much wider than the previous case with a conversion efficiency better than 40% from 4–5.5 GHz, resulting in the generation of output signal from about 12–16.5 GHz.

V. Broadband Millimetre Wave & THz Third Harmonic Parametric Tripler

In order to operate the third harmonic generator at higher frequency, it is evident that we need to shift the junction resonance much further away from the operational band, since no wave can propagate beyond the junction resonance frequency. In the following case, we assume the junction inductance and capacitance is much smaller than the previous case, and by shifting the junction resonance upward in frequency, we show in Fig. 6 a similar broadband millimetre tripler can be realised using the same method presented above. In this case, the Josephson junction is resemblance more of an SIS junction with low junction capacitance.

For operation at even higher frequencies, into the sub-THz regime, the junction resonance need to be shifted to even higher frequency regime. This implies that the Josephson junction would needed to be replaced by a high current density SIS junction, such as NbN/AlN/NbN tunnel junction. The need for high current density junction here is three folds: to shift the junction resonance higher, to increase the Q-factor (therefore flatter dispersion curve at lower frequencies) and most importantly to improve the power handling. The latter requirement is due to the fact that the input pump wave cannot be stronger than the critical current density of the junction used to amplify the harmonic wave. Hence, there is a natural limit on the power handling of these devices.

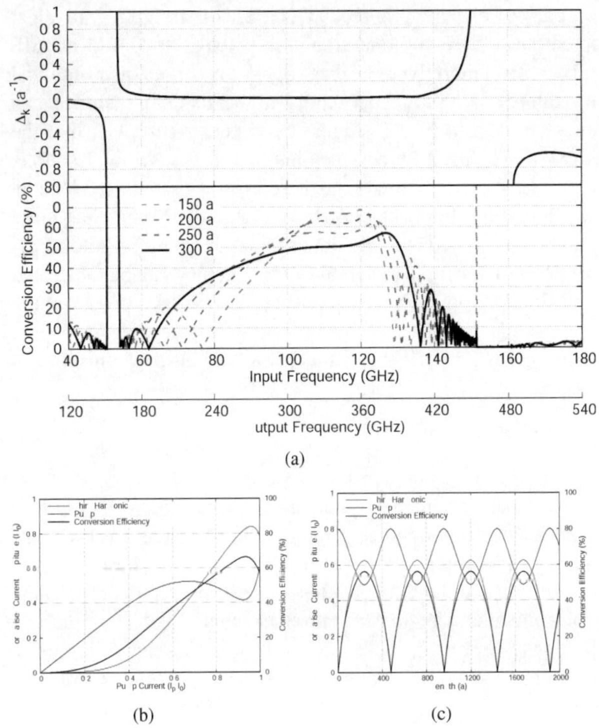

Fig. 6. (a) Top panel: Phase mismatched between the pump and the third harmonic. In this case, the junctions were set to be $C_j = 50\,\text{fF}$, $L_j = 1\,\text{pH}$, with $I_0 = 329\,\mu\text{A}$. The resonator parameters were set to be $C_R = 6.75\,\text{fF}$ and $L_R = 66\,\text{pH}$, giving resonances around 50 GHz and 150 GHz. Here, it is clearly seen that by lowing the Q-factor of the resonator circuit, we managed flatten the Δ_k curve approaching zero in between the two resonance frequencies. Bottom panel: This broaden the operational bandwidth of the third harmonic generator, with higher than 40% conversion efficiency from about 85–130 GHz. These plots were made by using $x = 250a$ and $I_p/I_0 = 0.8$. (b) The normalised amplitude of the resulting pump and third harmonic wave relative to the critical current of the Josephson junction (I_0), in relation with the strength of input pump current (I_p) with $f_p = 100\,\text{GHz}$. (c) The normalised amplitude of the resulting pump and third harmonic wave along the transmission line.

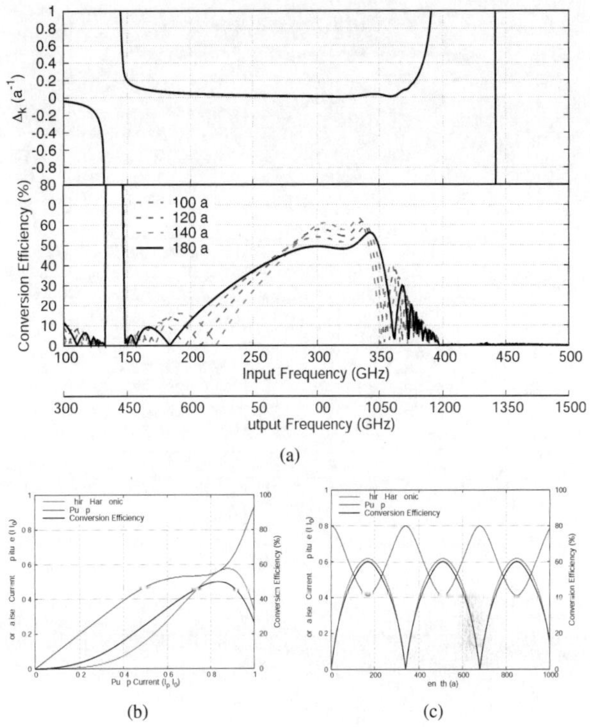

Fig. 7. (a) Top panel: Phase mismatched between the pump and the third harmonic. In this case, the junctions were set to be $C_j = 50\,\text{fF}$, $L_j = 0.22\,\text{pH}$, with $I_0 = 1.5\,\text{mA}$. The resonator parameters were set to be $C_R = 0.45\,\text{fF}$ and $L_R = 15.42\,\text{pH}$, giving resonances around 150 GHz and 450 GHz. Here, it is clearly seen that by lowing the Q-factor of the resonator circuit, we managed flatten the Δ_k curve approaching zero in between the two resonance frequencies. Bottom panel: This broaden the operational bandwidth of the third harmonic generator, with higher than 40% conversion efficiency from about 250–350 GHz. These plots were made by using $x = 100a$ and $I_p/I_0 = 0.8$. (b) The normalised amplitude of the resulting pump and third harmonic wave relative to the critical current of the Josephson junction (I_0), in relation with the strength of input pump current (I_p) with $f_p = 300\,\text{GHz}$. (c) The normalised amplitude of the resulting pump and third harmonic wave along the transmission line.

This is illustrated in Fig. 7. Here we assume a junction current density of $75\,\text{kA/cm}^2$, $2\,\mu\text{m}^2$ junctions, resulting in $I_0 = 1.5\,\text{mA}$ and $L_j = 0.22\,\text{pH}$. This current density value is possible to achieve using a full NbN tunnel junction with AlN barrier [16]. In this example, we show that it is possible to convert more than 40% of the pump energy to generate and amplify the third harmonic wave from 750–1050 GHz, an extremely broadband operation for a tripler operating at this frequency range.

VI. THz LO Chain

Although the above two examples show that it is possible to achieve a broadband tripler in mm and sub-mm wavelength, the power handling capability of these devices is still rather limited. This is due to the fact that the junction would lost its superconductivity above the critical current value of the junction I_0. Furthermore, although the conversion efficiency is very high, between 40–70%, it is very sensitive to the input power of the pump wave. This can be seen clearly from Fig. 6 (b) and Fig. 7 (b). To achieve such high conversion efficiency, the pump current needs to be at least $0.6I_0$. Therefore, if the first tripler is capable of convert only 40% of the pump power to the third harmonic wave, which in turn used to feed into the second tripler, this input power would be much less than $0.4I_0$. Hence the conversion efficiency drops significantly. Therefore, to construct a fully operational THz LO using the two TWPaMs with a W-band pump input, the two triplers need to be designed in sync.

Fig. 8 (a) shows an example of a full THz LO chain using two TWPaMs along with a W-band signal generator. The junctions and resonators parameters were altered from the example given earlier, in order to achieve decent conversion efficiency in the final stage of the frequency conversion. In this case, the conversion efficiency of the 300 GHz tripler is maximised to about 80%, with a penalty on a slightly narrower operational bandwidth. The output power from this tripler is then used to feed the 900 GHz tripler. As can be seen, with this combination, we are able to generate close to $0.5I_0$ of

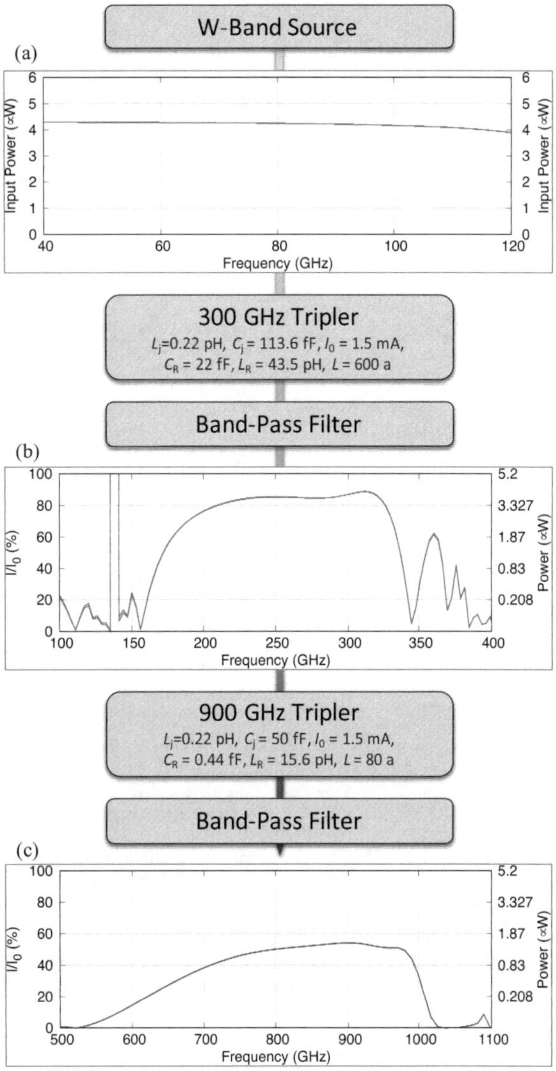

(a)

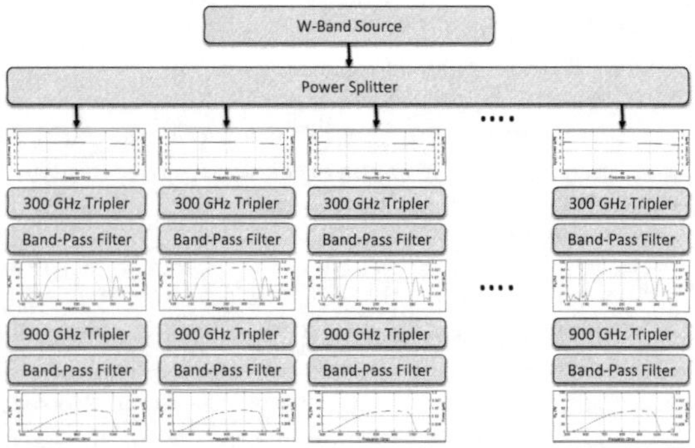

(b)

Fig. 8. (a) An example showing how the 300 GHz and 900 GHz tripler can be designed such that a maximum power output at the THz regime can be achieve with broadband operation. In this case, the output power is about $0.5I_0$, or in other word, it can be used to pump an SIS mixer up to half of its critical current value. (b) Diagram illustrating how a single W-band source can be used to feed multiple THz LO chain build using these TWPaMs.

output power from 700–1000 GHz. This is roughly equivalent to about 1.5 μW of output power, enough to pump an SIS mixer for heterodyne mixing operation in an astronomical instrument. Although one could argue that this scheme can only be used to feed a single SIS mixer, but the fact that the input power used here is negligible for any W-band source, a single W-band source can therefore be used to feed many of these LO chain, as shown in Fig. 8 (b). Noted that these planar circuit devices are much smaller than the traditional semiconductor frequency multiplier, and they can be fabricated with mass production. Hence, they can be incorporate along with the detector circuit, housed within the cryogenic system, with only a single W-band source to feed all the mm/sub-mm mixers.

Conclusion

We have presented an innovative method for developing a superconducting frequency multiplier using the TWPA technology. The travelling wave parametric tripler have a broad operational bandwidth with close to 50% conversion efficiency. They are compact in size, produce negligible heat and can be easily reproduce with standard planar circuit lithography technique. We have shown that the same technology can be use for designing low noise triplers in microwave, millimetre and sub-millimetre frequency regime. These triplers can also be cascaded together to construct a compact planar circuit THz LO chain which could be housed within the same block as the SIS mixer, or potentially integrated directly with the mixer's superconducting circuits.

Acknowledgment

This research was funded in part by the European Research Council (ERC) under the European Union's Horizon 2020 research and innovation programme with grant agreement No. [803862] (Project SPA4AstroQIT) and the MERAC Foundation. For the purpose of Open Access, the author has applied a CC BY public copyright license to any Author Accepted Manuscript version arising from this submission.

References

[1] J. Hesler, H. Song, and T. Nagatsuma, "Terahertz schottky diode technology," in *Handbook of terahertz technologies: Devices and applications*. Pan Stanford Publishing, 2015, no. 5, pp. 104–131.

[2] I. Mehdi, J. V. Siles, C. Lee, and E. Schlecht, "Thz diode technology: Status, prospects, and applications," *Proceedings of the IEEE*, vol. 105, no. 6, pp. 990–1007, 2017.

[3] M. S. Vitiello and A. Tredicucci, "Physics and technology of terahertz quantum cascade lasers," *Advances in Physics: X*, vol. 6, no. 1, p. 1893809, 2021.

[4] A. Delga, "Quantum cascade detectors: A review," *Mid-infrared Optoelectronics*, pp. 337–377, 2020.

[5] A. Leitenstorfer, A. S. Moskalenko, T. Kampfrath, J. Kono, E. Castro-Camus, K. Peng, N. Qureshi, D. Turchinovich, K. Tanaka, A. G. Markelz *et al.*, "The 2023 terahertz science and technology roadmap," *Journal of Physics D: Applied Physics*, vol. 56, no. 22, p. 223001, 2023.

[6] J. D. Garrett, H. Rashid, G. Yassin, V. Desmaris, A. B. Pavolotsky, and V. Belitsky, "A nonlinear transmission line model for simulating distributed sis frequency multipliers," *IEEE Transactions on Terahertz Science and Technology*, vol. 10, no. 3, pp. 246–255, 2020.

[7] H. Rashid, S. Krause, D. Meledin, V. Desmaris, A. Pavolotsky, and V. Belitsky, "Frequency multiplier based on distributed superconducting tunnel junctions: Theory, design, and characterization," *IEEE Transactions on Terahertz Science and Technology*, vol. 6, no. 5, pp. 724–736, 2016.

[8] C. Macklin, K. O'brien, D. Hover, M. Schwartz, V. Bolkhovsky, X. Zhang, W. Oliver, and I. Siddiqi, "A near–quantum-limited josephson traveling-wave parametric amplifier," *Science*, vol. 350, no. 6258, pp. 307–310, 2015.

[9] B. H. Eom, P. K. Day, H. G. LeDuc, and J. Zmuidzinas, "A wideband, low-noise superconducting amplifier with high dynamic range," *Nature Physics*, vol. 8, no. 8, pp. 623–627, 2012.

[10] L. Planat, A. Ranadive, R. Dassonneville, J. Puertas Martínez, S. Léger, C. Naud, O. Buisson, W. Hasch-Guichard, D. M. Basko, and N. Roch, "Photonic-crystal josephson traveling-wave parametric amplifier," *Physical Review X*, vol. 10, no. 2, p. 021021, 2020.

[11] K. O'Brien, C. Macklin, I. Siddiqi, and X. Zhang, "Resonant phase matching of josephson junction traveling wave parametric amplifiers," *Physical review letters*, vol. 113, no. 15, p. 157001, 2014.

[12] T. White, J. Mutus, I.-C. Hoi, R. Barends, B. Campbell, Y. Chen, Z. Chen, B. Chiaro, A. Dunsworth, E. Jeffrey *et al.*, "Traveling wave parametric amplifier with josephson junctions using minimal resonator phase matching," *Applied Physics Letters*, vol. 106, no. 24, 2015.

[13] J. M. Navarro and B.-K. Tan, "Optimising the design of a broadband josephson junction twpa for axion dark matter search experiments," in *Quantum Technology: Driving Commercialisation of an Enabling Science II*, vol. 11881. SPIE, 2021, pp. 139–148.

[14] J. N. Montilla and B.-K. Tan, "Design of high compression point josephson junction travelling wave parametric amplifiers for readout of millimetre and sub-millimetre astronomical receivers," in *Millimeter, Submillimeter, and Far-Infrared Detectors and Instrumentation for Astronomy XI*, vol. 12190. SPIE, 2022, pp. 1244–1253.

[15] O. Yaakobi, L. Friedland, C. Macklin, and I. Siddiqi, "Parametric amplification in josephson junction embedded transmission lines," *Physical Review B*, vol. 87, no. 14, p. 144301, 2013.

[16] Z. Wang, A. Kawakami, A. Saito, and K. Hamasaki, "1/f noise in high current density nbn/aln/nbn tunnel junctions," *IEEE transactions on applied superconductivity*, vol. 11, no. 1, pp. 84–87, 2001.

FYST CCAT Heterodyne Array Instrument Precursor

K. Vynokurova[1], U.U. Graf[1], I. Barrueto[1], N. Honingh[1], M. Justen[1], B. Klein[2], H. Krüger[1], A. Mahmood[1], B. Schmidt[1], J. Stutzki[1], S. Varghese[1,2], L. Weikert[1]

***Abstract*—This abstract offers an overview of the FYST CCAT Heterodyne Array Instrument Precursor, underlining its practical dual-frequency capabilities, intricate optics path design, and potential for incremental advancements in our understanding of celestial phenomena.**

***Keywords*—Heterodyne receiver, dual-frequency, optics path.**

I. Introduction

SPECTROSCOPY holds significant importance in improving our understanding of the Universe, offering a window into celestial bodies' composition, temperature, and dynamics. In the microwave range, it allows astronomers to probe, among other things, molecular clouds, providing insights into the star formation processes, and the physics of the interstellar medium (ISM) through detailed mapping of molecular transitions, its distribution, and properties.

The aforementioned phenomena will be observed by the Fred Young Submillimeter Telescope (FYST), a future 6m diameter telescope working in the millimeter and submillimeter wavelength regions. It is being built in the Parque Astronomico de Atacama in northern Chile, at 5600m elevation [1]. The telescope is being created as a versatile platform, allowing multiple instruments to be used for observations. One of them is the CCAT Heterodyne Array Instrument (CHAI), which will operate within 455 – 495 GHz (LFA) and 800-820 GHz (HFA) frequency ranges. In its final configuration, CHAI will feature 64 pixels in each frequency band, totalling 128 pixels (TABLE 1). The receiver allows simultaneous observations of neutral atomic carbon fine structure lines ([CI] $^3P_1 \rightarrow {}^3P_0$ in LFA and $^3P_2 \rightarrow {}^3P_1$ in HFA) and rotational lines of carbon monoxide (CO $J=4 \rightarrow 3$ and $J=7 \rightarrow 6$ respectively). [2].

TABLE I. MAIN SPECIFICATIONS OF CHAI

	LFA	HFA
RF range [GHz]	455-495	800-820
Noise temperature [K]	<100	<200
IF band [GHz]	4 – 8	4 - 8
Resolution [kHz] /[km/s]	100 / 0.06	100 / 0.04
Velocity coverage [km/s]	2500	1500
Beam size ["]	26	15
Field of View [' x ']	7.5 x 7.5	4.5 x 4.5

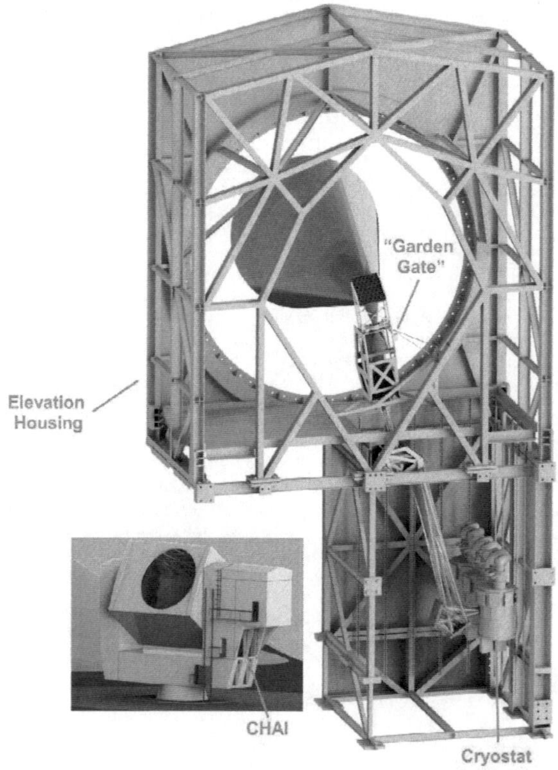

Fig. 1. The Optics path from FYST elevation bearing (large round opening), where the "Garden gate" component is located, down to the instrument space, where the instrument's cryostat is placed.

However, during preliminary tests and commissioning of the telescope, namely, holography measurements, as well as for the first light of the FYST, a less complex precursor of CHAI will be used, a dual-frequency single-pixel receiver (mini-CHAI).

II. Configuration of the Instrument

The instrument's location within the FYST introduces a unique optics path complexity, featuring a retractable

[1] I. Physikalisches Institut der Universität zu Köln, Zülpicher Straße 77, 50937 Köln, Germany, email: vynokurova@ph1.uni-koeln.de & graf@ph1.uni-koeln.de & barrueto@ph1.uni-koeln.de & honingh@ph1.uni-koeln.de & justen@ph1.uni-koeln.de & krueger@ph1.uni-koeln.de & mahmood@ph1.uni-koeln.de & bschmidt@ph1.uni-koeln.de & stutzki@ph1.uni-koeln.de & varghese@ph1.uni-koeln.de & weikert@ph1.uni-koeln.de; [2] Max-Planck-Institut für Radioastronomie, Auf dem Hügel 69, 53121 Bonn, Germany, email: bklein@mpifr-bonn.mpg.de .

NOTES:

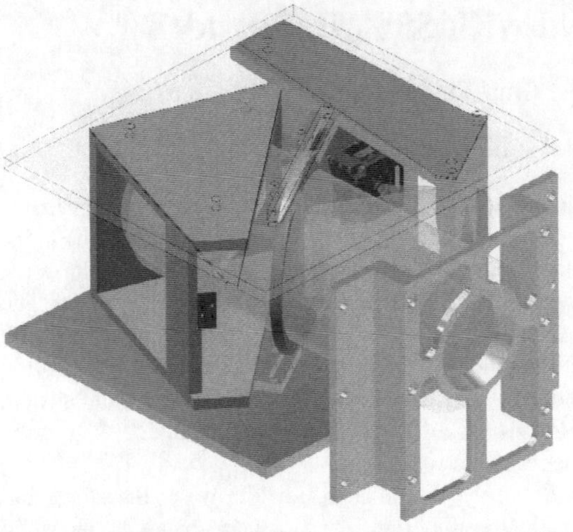

Fig. 2. CHAI precursor optics inside the cryostat. 475 GHz (orange) and 807 GHz (blue) beam paths lead from the cryostat window to the balanced SIS CHAI mixer and band 2 SIS Nb-AL2O3-Nb HIFI mixer respectively. The beams are split using polarising filter foil.

component used to direct the sky signal to our device's cryostat or enable observations with other instruments (Fig.1) [3].

Within the mini-CHAI cryostat, the double-frequency sky signal is split by a polarising filter foil. The transmitted component of the signal is then fed through the 475 GHz spline-profile diagonal horn [4] to the SIS-balanced mixer (Fig.2).

The reflected component (with polarization perpendicular to the orientation of the foil) is then fed to a second, single-ended SIS mixer, identical to the band 2 Heterodyne Instrument for the Far-Infrared (HIFI) mixer, launched onboard ESA's Herschel Space Observatory in May 2009 [5].

This single-pixel instrument not only simplifies the commissioning and initial operation of the FYST telescope but also offers a platform for field verification of CHAI components.

References

[1] G. J. Stacey, *et al*, "CCAT-Prime: science with an ultra-widefield submillimeter observatory on Cerro Chajnantor", presented at SPIE Millimeter, Submillimeter, and Far-Infrared Detectors and Instrumentation for Astronomy IX, June 14th, 2018

[2] U. U. Graf, *et al*. "CHAI, the CCAT-Prime Heterodyne Array Instrument," In Proc. Thirtieth International Symposium On Space Terahertz Technology, April 15-17, 2019, pp.77

[3] U. U. Graf, *et al.* "The CCAT-Prime Heterodyne Array Instrument (CHAI)," In Proc. 7th Chile-Cologne_Bonn-Symposium: Physics and Chemistry of Star Formation, September 26-30, 2022, pp.339-340

[4] H. J. Gibson *et al.*, "A spline-profile Diagonal horn with low cross-polarization and sidelobes, suitable for THz split block machining" *IEEE Transactions on Terahertz Science and Technology*, vol. 7, no. 6, pp. 657-663, Nov. 2017

[5] Th. Graauw, *et al.* "The Herschel-Heterodyne Instrument for the Far-Infrared (HIFI)", *Astronomy and Astrophysics*, vol.518, July-August 2010.

NOTES:

Highly-Balanced Quadrature Hybrid with 55% Bandwidth

Chen Chien, Yuh-Jing Hwang*, Chin-Ting Ho

Academia Sinica Institute of astronomy and Astrophysics, Taipei, Taiwan

***Abstract*—A highly amplitude and phase balanced E-plane branch-line quadrature hybrid coupler is realized by using reduced-height, broaden width rectangular metallic waveguides and double-ridge waveguides structure. A design fit to the extended W-band frequency has been designed, fabricated and measured. The design criterion is set to amplitude imbalance < 1.0 dB and phase imbalance < 2 degrees. The simulated and measured results show frequency range over 67.31 – 117.5 GHz and 65.75 – 117.75 GHz fit to the design criterion, respectively. The matching and isolation are typically below -20 dB. A higher frequency version fit to the 125 – 220 GHz frequency range is also designed and now under fabricating and will be measured soon.**

***Keywords*—quadrature hybrid, E-plane branch-line coupler, rectangular waveguide, double-ridge waveguide.**

I. Introduction

THE requirement of the next generation instrumentation for radio astronomy push the development of the millimeter-wave components toward broader bandwidth. For frequency higher than 65GHz, precisely-machined waveguide components are still essential for very low-noise and very low-loss systems, especially for which operated in critical environment like cryogenically cooled receivers for millimeter-astronomy [1]. In this work, we try to push the fractional bandwidth of the 3-dB waveguide quadrature hybrid couplers as wide as possible. The possibility of using single- and double-ridge waveguide to enhance the bandwidth of quadrature hybrid with even better impedance matching, isolation, amplitude/phase balance is explored.

II. Design Consideration

The demand of the ultrabroad bandwidth leads to the development of the ridge waveguide components, and a broadband 45 – 116 GHz ridge waveguide system has been demonstrated [2]. The previous work on the waveguide quadrature hybrid coupler to achieved over 50% bandwidth demonstrated that based on the rectangular waveguide, the bandwidth can be achieved but either the amplitude imbalance or the phase imbalance would be degraded, for example, the quadrature hybrid coupler based on the reduced height, broaden width rectangular waveguide can either achieve 1-dB amplitude imbalance but the phase difference within 90 +- 6 degrees, or 1.6-dB amplitude imbalance with 90 +- 1.5 degree phase imbalance [3]. If maintain the amplitude balance within +- 1dB, and keep the phase difference between through path and couple path within 90 +- 1.5 degrees, for 12th order branch hybrid, the bandwidth is reduced to 44% [4].

The goal of this work is to seek for the E-plane quadrature hybrid 3-dB coupler design with ultra-broadband (BW >55%), low amplitude imbalance ($\Delta A < 1$ dB), low-phase imbalance ($\Delta\theta < 2$ degree), the input/output matchings and isolation better than -20 dB, and the design should be easily machined for mass production. Due to the fact that ridge waveguides are capable to provide 88.2% bandwidth [2], we try to utilize the ridge waveguides with as simple tee junction structures to form the branch waveguide coupler to fulfill the above requirement. The ridge waveguide provides the very broad bandwidth by reducing the guide wavelength dispersion. The tee junction in [5] is a good reference for broadband waveguide circuit design, but its structure is difficult unless using metallic 3-d printing techniques.

To enhance the fractional bandwidth wider than 50% and maintain both amplitude and phase balance in good performance, three different configurations of the E-plane 12th-order branch-line quadrature hybrid coupler are explored: (i) single-ridge waveguides for main arms with rectangular waveguides for branches, (ii) single-ridge waveguides for main-arms with double-ridge waveguides for branches, and (iii) reduced height and broaden width rectangular waveguides for main arms with double-ridge waveguide for branches.

Among these three different configurations, case (i) is not possible to expand the bandwidth, its performance is very similar to the previous work shown in [3]. Both (ii) and (iii) is possible to achieve the fractional bandwidth up to 54%. However, for the 12-branch design, the branch waveguide height will be smaller for the configuration of case (ii) which is very difficult for fabrication and scale toward higher frequency. The design of case (iii) the bandwidth is slightly wider and the cross section of the branch waveguides is possible to be fabricated. The main-arm waveguide section between branches are fine-tuned to quarter wavelength of frequency cover the bandwidth as [4], and the length of the branch arms are all identical to quarter wavelength of around highest frequency, the aperture and cross section of the branch arms are also identical.

III. Simulation, Fabrication, and Measured Results

The simulation was realized by using the 3-dimensional EM structure simulator with time-domain finite-difference (FDTD) method and finite-element analysis method.

To ensure the fabricated hybrid coupler can be measured, additional waveguide section with transitions are extended from the intrinsic quadrature hybrid. The extended waveguide sections are in split-block design; thus, the full quadrature hybrid coupler is in 7-piece design, including 2 sets of split-block extended waveguide sections and the three-piece intrinsic quadrature hybrid (top main-arm waveguide, bottom main-arm waveguide, branch waveguide layer).

Even at extended W-band frequencies, the branch waveguide layer is still quite technical challenging on mechanical fabrication, several approaches are proposed for the fabrication, including the laser cutting, the wire cutting, and the dry-etching on the silicon substrate. The very high aspect ratio of the branch waveguide leads to the difficulty to be realized by dry etching, and the laser cutting could not produce the good ridge waveguide cross sections, thus the wire cutting is the only solution to produce the branch waveguide layer. Fig. 1 shows the mechanical fabrication of the branch waveguide layer.

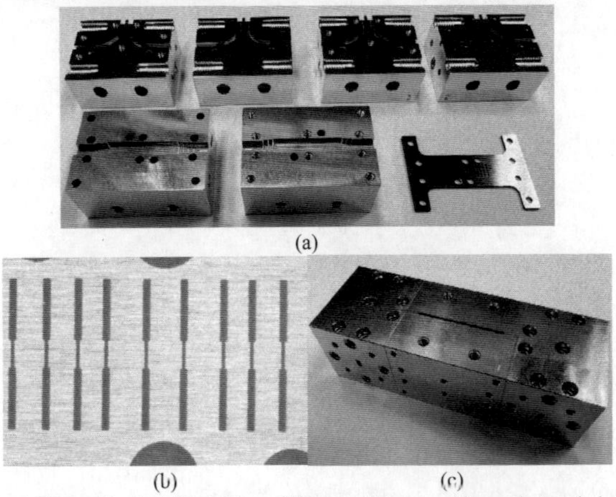

Fig. 1 The mechanical fabrication of the branch waveguide quadrature hybrid, (a) individual components before assembling, (b) wire cutting result of the double-ridge waveguide branches, (c) assembled quadrature hybrid with the transition and taper waveguide sections to WR-10 flange.

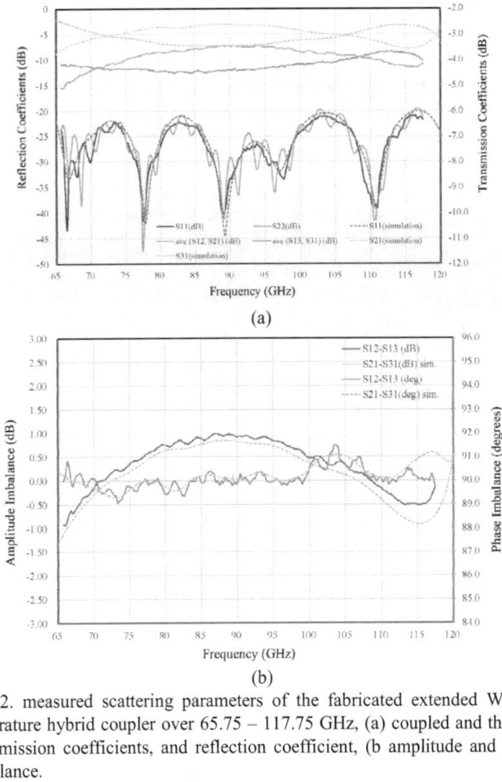

Fig. 2. measured scattering parameters of the fabricated extended W-band quadrature hybrid coupler over 65.75 – 117.75 GHz, (a) coupled and through transmission coefficients, and reflection coefficient, (b amplitude and phase imbalance.

The measurement is realized by using the Keysight N5227B phasor network analyzer with N5292A millimeter-wave controller and a pair of WR-10 frequency extenders. By carefully setting to lower power level, the frequency range can be extended to 65.75 – 117.75 GHz, the upper frequency range is mainly limited by the TE_{01} mode of the WR-10 waveguide, which is 118GHz. The measured scattering parameters are presented in Fig. 2. Please noted that the gold-plating layer would leads to thickness change of the order of 1μm, which could help to maintain the amplitude and phase imbalance in the full waveguide band of 65.75 – 117.75 GHz within + - 1.00 dB and +1.0 / -1.5 degrees, respectively. The fractional bandwidth is 56.67%.

The design of the extended W-band quadrature hybrid is then scaled to shift the frequency range up to 125 – 220 GHz frequency by factor of 0.53 on mechanical size. The simulated performance and is as shown in Fig. 3

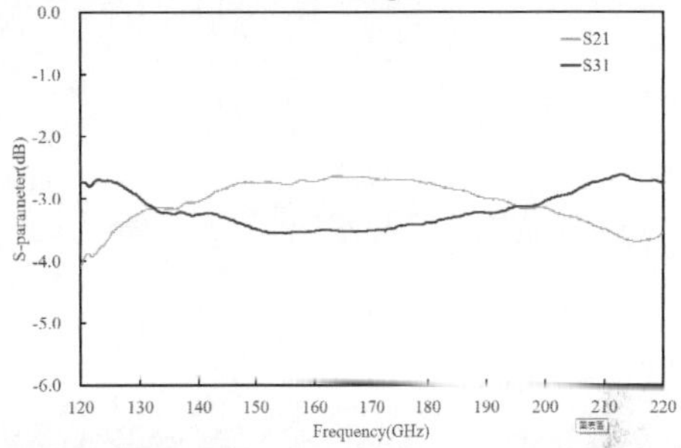

Fig. 3. simulated response of the 125 – 220 GHz branch waveguide quadrature hybrid.

TABLE I. PERFORMANCE SUMMARY†

Item	Simulation	Measured Result
Frequency (GHz)	66.5 – 119.5	65.75- 117.75
S_{21}, S_{31} (dB)	-3.7 to -2.5	-5.0 to -3.5
S_{11}, S_{22} (dB)	< -20.0	< -19.5
S_{41} (dB)	< -20.0	< -19.5
S_{21}/S_{31} (dB)	-1.0 to 0.85	-0.9 to 1.0
$\angle S_{21} - \angle S_{31}$ (deg.)	89.3 – 91.2	89 – 91.5

†Input: Port-1, Transmission: Port-2, Couple: Port-3, Isolation: Port-4 for future application to sideband separation down-converter.

IV. SUMMARY

An E-plane branch-line quadrature hybrid coupler with 55% bandwidth, <1.0-dB amplitude imbalance and < 2-degree phase error from 90 degrees is achieved. The higher frequency version for 125 – 220 GHz is now under fabrication for ALMA 2030 wideband sensitivity upgrade Band-4+5 prototype receivers.

REFERENCES

[1] Y.-J. Hwang, et. Al, "Initial concept and roadmap of the Band-4+5 receiver upgrade of the Atacama Large Millimeter / Submillimeter Array (ALMA)" *24th International Conference on Electromagnetics in Advanced Applications (ICEAA 2023)* Venice, Italy, October 9-13, 2023.

[2] S. Manafi, M. Al-Tarifi, and D. S. Filipovic, "Millimeter-wave double-ridge waveguide and components," *IEEE Trans. Microw. Theory and Techn.*, vol. 66, no. 11, pp. 4726-4736, Nov. 2018.

[3] Y.-J. Hwang, C. Chien, C.-T. Ho, T. Kojima, A. Gonzalez, and T.-W Huang, "E-Plane branch waveguide quadrature hybrids for low amplitude and phase imbalance in extended W-Band frequencies" submitted to *IEEE Trans. Microw. Theory and Techn.*, May 2024.

[4] Y.-J. Hwang, C. Chien, C.-T. Ho, T. Kojima, and A. Gonzalez, "12th-order quadrature 3-dB hybrid coupler for low amplitude and phase imbalance in extended W-Band frequencies," *2023 Asia Pacific Microwave Conference*, pp. 555 – 557, Taipei, Taiwan, Dec. 05 – 08, 2023.

[5] Z. Dang, H.-F. Zhu, J. Huang, H.-D. He, "An ultra-wideband power combining in ridge waveguide for millimeter wave", *IEEE Trans. Microw. Theory Techn.*, vol. 68, no. 4, pp.1376-1389, 2020.